CONCEPTUAL EQUATIONS
FREQUENTLY REFERRED TO IN TEXT

Equation 1, Table 1.1

$$Ve = f(So + Or + To + Cl + Ti)$$

Equation 2, Table 2.1

$$P = f(Bp - Er)$$

Equation 3, Table 2.2

$$Pr_{(SR)} = f(Ha + Di + Pr + Ge + Re + Hu)$$

Equation 4, Table 4.6

$$H_Q = f(Fo + Co + Wa + Sp)$$

WILDLIFE
HABITAT RELATIONSHIPS
IN
FORESTED ECOSYSTEMS

WILDLIFE HABITAT RELATIONSHIPS IN FORESTED ECOSYSTEMS

David R. Patton

Professor, Forest Wildlife Ecology
Northern Arizona University
School of Forestry

TIMBER PRESS
Portland, Oregon

This Book is Dedicated
to the
Hundreds of Forest Wildlife Biologists
Making Thousands of Decisions
Affecting Millions of Acres of Land
That Will be Used by Countless
Future Generations of
Humans and Wildlife

Second printing 1992

ISBN 0-88192-202-1
Printed in the United States of America

TIMBER PRESS, INC.
9999 S.W. Wilshire, Suite 124
Portland, Oregon 97225

Library of Congress Cataloging-in-Publication Data

Patton, David R.
    Wildlife habitat relationships in forested ecosystems / David R.
  Patton.
      p.   cm.
    Includes bibliographical references and index.
    ISBN 0-88192-202-1
    1. Forest ecology. 2. Forest fauna--Habitat. 3. Wildlife
  management. 4. Wildlife conservation.   I. Title.
  QH541.5.F6P37 1991
    639.9--dc20   91-14076
                                                          CIP

CONTENTS

LIST OF FIGURES

LIST OF TABLES

11

FOREWORD

David Patton has training in forestry and wildlife biology, experience in both wildlife management and research, and has served in state agencies, the U.S. Forest Service, the U.N. Food and Agriculture Organization, and as a university professor. This book reflects that breadth of education and experience.

Dr. Patton is one of the fathers of what has become known as "wildlife habitat relationships" programs. These programs are now widely used across North America in state and provincial as well as federal resource management agencies. And, that shows here as well.

Wildlife Habitat Relationships in Forested Ecosystems is a book about basics. These basics are presented in a simple, straightforward manner. Yet there is sufficient detail included to guide the student or practicing natural resource management professional as to the scientific or technical whys and wherefores of the practices and procedures described. I have said elsewhere that "perhaps the greatest challenge that faces professionals engaged in forest research and management is the organization of knowledge and insights into forms that can be readily applied." I still believe that.

David Patton has devoted much of his professional life to meeting that challenge. This latest contribution is another step forward in that process and brings the dream of forest management conducted in the full light of its influence on wildlife one step closer to reality.

Jack Ward Thomas
Chief Research Wildlife Biologist
U.S. Forest Service

READ ME FIRST

The question must be asked: Is there a need for another wildlife textbook and, if so, how will it differ from the others? Most wildlife texts treat forests and forest management as a single chapter or as a subsection under land use. While this book is not entirely about forest wildlife species or forest conditions, its primary purpose is to focus on species that use, are associated with, or depend on trees and forests for part or all of their habitat requirements.

A major difference between this textbook and the others is that a conceptual equation is presented as a hypothesis to focus attention on the factors directly affecting habitats and populations. The equation provides a basis for thinking about wildlife, and its components are repeatedly returned to throughout the book. Second, over 100 direct and indirect practices are described to provide information on how a wildlife biologist goes about the tasks of improving or maintaining habitat. Third, other differences include information on how to develop a local habitat model, relational databases for storing and manipulating data in a microcomputer, and integrating the two into a habitat relationships system.

In writing this book I have drawn from the past using ideas that developed out of the *American Game Policy,* expanded by the *North American Wildlife Policy* (Allen 1973), and including recent thoughts presented in *New Principles for the Conservation of Wild Living Resources* (Holt and Talbot 1978) and *The Ecological Web* (Andrewartha and Birch 1984). Readers should not expect a detailed treatment of each topic since my intention is to provide an introduction to forest wildlife habitat relationships. In some sections I have chosen to present information in summary rather than detailed explanations because in many cases the detailed information has been published elsewhere. For classroom use, material from the *Journal of Wildlife Management, Wildlife Society Bulletin, Transactions of the North American Wildlife and Natural Resources Conference,* and government publications should be consulted for specific examples.

A number of new concepts and ideas about wildlife habitat relationships are included and will be obvious by the paucity of cited literature. During most of my career as a wildlife researcher and manager I have been guided by the principle of Occam's razor (law of parsimony) which states that the hypothesis most likely to be correct is the one that accounts for the maximum number of observations with the minimum number of assumptions—in other words, select the simplest solution to the problem. It has served me well in management and research for over 30 years and has been used to guide the presentations in this book. In the figures and examples I have tried to use an intuitive approach to explanation. Concepts are important to understanding a subject and often a conceptual model with letters or words instead of a mathematical formula better serves to explain a complex system or idea.

The theme pursued is that of relating animals to their physical habitat. The intent is to present new approaches to managing forest wildlife by using plant-animal relationships in a form that is useful for decision making. Because this book has been conceived to be used both in forestry schools by students not majoring in wildlife management and as a reference and refresher for the practicing forest wildlife biologist, I have included limited information on wildlife techniques not usually included in a basic wildlife textbook.

Although I have worked for the Fish and Wildlife Service, two state wildlife agencies, and The Food and Agriculture Organization in Zambia, it is obvious that I have drawn on my background and experience in the USDA Forest Service for most of my material and so the content should not be construed to imply that other agencies such as the Bureau of Land Management and Fish and Wildlife Service have not developed comparable systems or techniques.

Some of the ideas and techniques discussed are original but many are derived from the work of others which I have referenced as appropriate. The topics I have included are those that as a forest manager and researcher I have found most useful. Thus, in some cases theories (for example, population models) may not have a practical use in the field but they do illuminate what is possible. Although my thinking has been influenced and fostered by many practicing field biologists and research scientists in North America the bulk of my experience has been in Arizona and New Mexico. Consequently I have selected examples from this area. However, despite the regional data the theory as well as the techniques for developing forest habitat relationships are completely general and so can be transported to countries where forests are part of the landscape.

At the end of each chapter I have listed books and articles as recommended reading. Some of the publications suggested were from the 1930s to the 1950s and are listed because they present examples, contain theory yet valid in the 1980s, have historical value, or present the topic in a clear manner. A complete listing of all scientific and common names used in this book is presented in Appendix E.

I wish to extend appreciation to the following people who have contributed to this book by influencing my thinking about wildlife populations and habitat throughout my career; in addition, several reviewed selected chapters or the entire book: Jack Adams (deceased), Forest Service; Richard Behan, Northern Arizona University; Erwin Boeker, Fish and Wildlife Service; Ward Brady, Arizona State University; Maurice Brooks, West Virginia University; Dave Brown, Arizona Game and Fish Department; Bill Burbridge, Forest Service; Wally Covington, Northern Arizona University; Donald G. Dodds, Acadia University; Peter Ffolliott, University of Arizona; Bruce Fox, Northern Arizona University; Greg Goodwin, Forest Service; John Hall (deceased), Forest Service; Larry Harris, University of Florida; Howard Hudak, Forest Service; C. Roger Hungerford (deceased), The University of Arizona; Dale A. Jones, Forest Service; Jim Klemmedson, The University of Arizona; Bob Latimore, Forest Service; Burd S. McGinnes, Fish and Wildlife Service; Keith Menasco, Forest Service; Henry S. Mosby (deceased), Virginia Polytechnic Institute; John Mumma, Forest Service; Paul A. Patton (deceased), West Virginia Department of Natural Resources; Neil Payne, The University of Wisconsin; D. I. Rasmussen (deceased), Forest Service; Tom Ratcliff, Forest Service; Hudson G. Reynolds (deceased), Forest Service; Merton Richards, Northern Arizona University; Robert L. Smith, West Virginia University; Jack Ward Thomas, Forest Service; Bob Vahle, Forest Service; Rick Wadleigh, Forest Service; O. C. Wallmo (deceased), Forest Service; and Bill Zeedyk, Forest Service.

Special recognition must be given to Lloyd Swift, former director of wildlife management, Forest Service, for taking the time to write a young forester on the Cleveland National Forest in 1960 giving him guidance about career development as a forest wildlife biologist. His advice started me on the right path to a rewarding career.

Lastly, for a career in the forestry and wildlife profession which I otherwise would not have enjoyed, I thank my wife Doris who sacrificed her own professional advancement to provide the means for my education.

Chapter 1

THE WORLD OF WILDLIFE

The days are ended when the forest may be viewed only as trees and trees viewed only as timber.　　　　　　　Sen. Hubert S. Humphrey

In recent years satellites have returned striking photographs of the earth from thousands of kilometers in space (Fig. 1.1). No other event in modern times has so forced us to think about the conditions on this planet, what we are now doing to it, and how it will be left for future generations. Some of the basic ecological questions that can be asked when looking at earth from outer space are: Where does energy come from? Who or what competes for this energy? Who or what uses the space on the earth's surface? These questions are important to understanding not only human habitat relationships but wildlife habitat relationships as well.

A long-distance view of earth is not only striking but firmly impresses upon us the sense that somewhere on earth we each exist in a place we call home. All wildlife species share the earth with us and also have a home. This book is about wildlife, a subject of many connotations and interpretations, depending on one's particular educational background, experience, and philosophy about living things. For the sake of clarity and continuity, it is necessary to define wildlife so that no doubt exists in the reader's mind as to what is meant.

Wild, in the dictionary, is defined as unrestrained or free-roaming. Life refers to a quality which distinguishes a functional integrated being from ordinary chemical matter. In this context life suggests either plants or animals and in some instances wildlife refers to both. Natural resource agencies define wildlife in a variety of ways depending on the specific laws and regulations under which they operate.

In many agency definitions fish are viewed as separate from wildlife while in a few cases, such as endangered species, invertebrates are often included. Therefore, a discussion of wildlife in any particular

17

Figure 1.1. Apollo 10 view of the earth from 58,000 km (36,000 mi) during the translunar journey toward the moon. (NASA photograph JSCL-605).

instance requires some terms of reference. In North America the working definition of *wildlife* generally refers to all species of amphibians, birds, fishes, mammals, and reptiles occurring in the wild. Occurring in the wild is interpreted as being undomesticated and free-roaming in a natural environment.

The definition of wildlife given must be applied to the species as a whole. For example, while one deer in a zoo may not be able to roam free, it is still considered wildlife because deer as a species occurs unrestrained in the wild. Deer in an enclosure would also be considered wildlife even though their movements might be restricted to a large area. But what about mice and rats which occur in buildings in a city? These city dwellers are not considered wildlife because they are defined as pests which occur in a man-made environment.

No matter what definition is given there will always be classes of animals or situations which do not conform. For this reason it is desirable to state a definition for the particular circumstances whether it be for management plans or environmental impact statements.

1.1 HISTORICAL PERSPECTIVE

Since the beginning of time animals have been a significant part of the human environment. This is evident not only in Egyptian hieroglyphics but also in paintings made by early humans in cave dwellings in various parts of the world. Humans were hunters long before they developed agricultural practices and considerable effort was expended to kill animals for food and hides. Early mankind learned to use animal products for clothing, weapons, and other articles of convenience. Furthermore, the exercise of stalking and killing animals provided experience for men who had to defend their living space, mates, and offspring.

The Bible provides accounts of the use of animals in the Middle East (Farb 1967). In early times wolves, leopards, and lions roamed freely so that herdsmen were forced to keep close watch on their sheep and goats. Predatory animals were subsequently reduced in number, therefore flocks could feed freely and out of danger. As a consequence, domesticated animals increased, resulting in over-browsing, thereby reducing the vegetation to briers and thorns (Isaiah 7:24).

The story of Adam, Eve, and the serpent is well known, and of course Noah had his ark full of animals. Solomon (2:12) speaks of the voice of the turtle (dove), the wild ass as a beast of burden, and pigeons as carriers of messages. Animals clearly were important not only for food but for labor as well, so laws were early mandated to provide for their protection. One of the earliest known laws is the Mosaic law (Deuteronomy 22:6) which reads:

> If a bird's nest chance to be before thee in the way, in any tree or on the ground, with young ones or eggs, and the dam sitting upon the young, or upon the eggs, thou shalt not take the dam with the young; thou shalt in any wise let the dam go, but the young thou mayest take unto thyself; that it may be well with thee, and that thou mayest prolong thy days.

Early Greek and Roman authorities used hunting as a means of keeping armies well trained and physically fit. Xenophon, a Greek leader, said:

> Men who love sport will reap therefrom no small advantage. It is an excellent training for war. Such men, if required to make a trying march will not break down, they will be able to sleep on a hard bed and keep good watch over the post entrusted to them. In advance against the enemy they will obey their orders, for it is thus wild animals are taken. They will have learned steadfastness, they will be able to save

themselves in marshy, precipitous, or otherwise dangerous ground, for from experience they will be quite at home in it. Men like these have rallied and fought against the victorious enemy and have beaten them by their courage and endurance (Leopold 1933).

Many of the victories of the Colonial forces in the American revolutionary war were attributable to the ability of the frontier soldier to hide and ambush the British, a skill learned by surviving as hunters and woodsmen.

Clearly humans and animals have been linked through history but only in the past 100 years or so have we again come to appreciate the importance of wildlife, largely because we have learned that their existence is inextricably tied to our own. This change of view was noted by Leopold, who wrote the first professional book on game management in 1933 and published "A Biotic View of Land" (1939) in which he explained man's relationship to the environment. Basically Leopold proposed that land is not just soil but energy flowing through a circuit of plants and animals and that an ecosystem is much more than the sum of its parts. He also emphasized that wildlife, which was once restricted to game animals, now included predators, songbirds, and all wild things. In Leopold's view the biotic pyramid had an "up-circuit" for energy that started with the soil and progressed upward through food chains; the "down-circuit" returned nutrients to the soil from death and decay of living organisms.

Management of game animals alone was emphasized from the early 1930s to the 60s largely due to the powerful influence of hunters who were paying the cost through purchase of hunting and fishing licenses and special taxes on equipment. Since 1960 a number of laws that set a new direction for management of federal lands culminated in 1973 with an event marking the change from game management to wildlife management. In that year the Wildlife Management Institute reissued *The American Game Policy of 1930* as the *North American Wildlife Policy* (Allen 1973).

Additional emphasis was given to conserving all forms of wildlife when the World Wildlife Fund sponsored two workshops resulting in publication of *New Principles for the Conservation of Wild Living Resources* (Holt and Talbott 1978). The period from 1960 to 1980 was devoted to interpreting the laws anddeveloping a changed awareness of the environment and what we are doing to it.

Now as we move towards the 21st century the emphasis is shifting slowly to a better understanding of natural resource management as one part of ecosystem functioning. This new emphasis encompasses the interactions between all plant and animal species, including humans. The period from 1980, when we were building on past knowledge and developing new theories and techniques, and continuing into the future might be called the *period of ecosystem management*. The ideas of biodiversity (Wilson 1986) and conservation biology (Soule 1986) are also part of this new emphasis on ecosystem management.

1.2 THE CONCEPT OF ECOSYSTEMS

Scientists, in trying to describe and understand the complicated world in which we live, are constantly developing new ways to classify the enormous number of diverse plant and animal species into recognizable categories of like things. The method which has evolved to comprehend living things is a hierarchical system which progresses in ascending order from individual organisms, through species, populations, and communities, to ecosystems (Fig. 1.2). Classification schemes are a product of humankind's thinking in an organized manner which tends to be hierarchical.

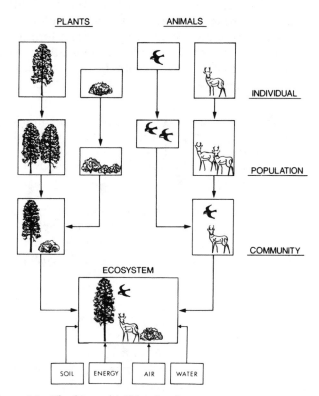

Figure 1.2. The hierarchical levels of an ecosystem.

Hierarchical Organization

Hierarchies are a convenient method of abstraction allowing us to reduce complex systems to simpler units which we can understand and study. But information at each level of the hierarchy can be aggregated upward. This same procedure can be used regardless of the type of data contained in each distinct level. Hierarchical descriptions also provide a way for managers and researchers to communicate. We do not know if in fact nature is actually organized in the way it is perceived, but thinking in hierarchical terms allows us to discover how systems interact.

One way to visualize the differences in the organizational levels of this hierarchy is to progressively add to the complexity starting with a single organism. As more organisms of the same species are added we soon have a *population*. A population of one species when added to a population of another becomes a *community*. Therefore we speak of several populations of different plant species as a plant community or of animals as an animal community.

Animal communities cannot exist without plant communities, and plant communities cannot exist without nutrients and energy. The addition of these two factors to the combination of existing plant and animal communities is an *ecosystem*. The term ecosystem was early defined as the whole complex of physical factors forming the environment (Tansley 1935). More recent definitions clarify the concept:

> Any unit that includes all of the organisms in a given area interacting with the physical environment so that a flow of energy leads to a clearly defined trophic structure, biotic diversity, and material cycles is an ecological system or ecosystem (Odum 1971).

Basically an ecosystem is an energy-nutrient processing system. Ecosystems have physical *structure* as for example soil, plants, and animals as well as *functions* of energy flow and nutrient cycling involving biotic and abiotic components. In addition, ecosystems have time and space attributes.

Time

The time attribute encompasses changes in plant and animal composition, interactions, and density. Four measures of time are useful in understanding ecosystems and forest-wildlife community changes: instantaneous, present, ecological, and geological.

Instantaneous time: Is used in studying population dynamics, particularly for those species characterized by a high reproduction rate. This measure of time is concerned with rates of change of populations.

Present time: Refers to time affecting animals in the short run, from one to five years. It is the "here and now" time frame used by managers in most management plans.

Ecological time: Is associated with the changes in plant and animal communities and is the time most commonly used in describing plant community replacement (succession).

Geological time: Is the tens and hundreds of thousands of years required to effect regional and global changes in topography and climate. Geological time provides a framework for understanding the past and ways of forecasting what might happen in the future.

Space

The space attribute of natural ecosystems is not set because ecosystems are open systems and their boundaries have to be established on a *pro forma* basis. Ecosystems can be as small as a fish pond, as large as a vegetation type, but the whole earth is considered the tertiary ecosystem. The concept, however, focuses on the biological, chemical, and physical processes and not specifically on space or geographical area. Space within ecosystems exists for plants and animals as they have individual boundaries that can be defined. Locations and boundaries of ecosystems are important because they become the basis for further disaggregation for classification, inventory, and management purposes.

Ecosystem Functioning

In the hierarchical system of organization an ecosystem describes an area and the animals and plants (biotic) within it as well as their abiotic environment. In all cases the ecosystem functions as a unit through nutrient cycling and energy flow. Nutrient cycling differs from energy flow as nutrients eventually return to the system through the death and decay of plants and animals in which they are temporarily bound.

All elements go through a cycling process from their reservoirs of living organisms, air, water, and soil. While about half of the basic elements are essential for life, six are the most important: carbon, phosphorus, nitrogen, oxygen, sulfur, and hydrogen. Carbon is a basic building block of the molecules necessary for life so that the cycling of carbon through the use of CO_2 by plants and respiration by animals is a critical life-sustaining process. Living animals take in O_2 contained in the air and give off CO_2. At death the decay process releases carbon bound in animals to the soil, air, and water reservoirs (Fig. 1.3).

The primal source of energy for all ecosystems is the sun. About 67% of the sun's total energy reaches the earth's surfaceat noon on a clear day (Gates 1965). The air we breathe is composed of 21% oxygen (O_2) and about 0.03% carbon dioxide (CO_2). The chlorophyll in the leaves of green plants absorbs a small portion of the sun's energy which is then used to combine carbon dioxide (CO_2) from the air and water (H_2O) from the soil into starches and sugars (carbohydrates, $C_nH_{2n}O_n$) which are contained in the plants eaten by animals. Thus plants depend on the sun's energy, chlorophyll, and photosynthesis for their existence but animals do not produce their own food. It is important to remember this point when reviewing theories about habitat, wildlife populations, and factors regulating populations.

Energy flow is a one-way transfer from plants to animals in a herbivore relationship or from animals to animals in a carnivore relationship. Energy has been changed in the transfer, but none has been created or destroyed; it has merely changed in form (first law of thermodynamics). During the transfer process a loss of energy as heat through respiration always occurs, so to keep the ecosystem functioning there must be a constant flow of energy (second law of thermodynamics) from the sun through the system.

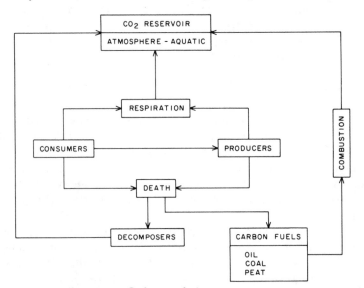

Figure 1.3. Carbon cycle in an ecosystem.

Wildlife in Ecosystem Functioning

Energy flow and nutrient cycling in an ecosystem require pathways. These pathways are to be found in animal *food chains* and *food webs*. In simple ecosystems, as in the arctic, the pathways are short and few in number, but in many ecosystems the pathways are long, numerous, and hence complex. Whatever the length and complexity, every food chain must start with plants (Fig. 1.4). Energy from plants is transferred through a series of animal species each of which eats one or more species preceding it in the chain. The shorter the food chain the greater the energy converted into living matter and metabolic activity.

Figure 1.4. Energy flow in a simple food chain.

In any food chain "waste matter" applies only to an individual organism, since waste such as fecal material from deer provides food and nutrition for other organisms (dung beetles, soil organisms, plants, etc.). Actually there is no waste matter, only artifacts. In the case of humans, some of those artifacts are plastics that may take hundreds of years to decompose into elements usable by other organisms.

When many food chains exist in an ecosystem an interlocking of the pathways develops into a food web (Fig. 1.5). As more animal components are added to a food web a degree of complexity is introduced. This complexity is thought to induce ecosystem and population stability (Hutchinson 1959) so that the loss of one food species results in adjustments by animals to another food source. The idea of stability is often linked to complexity but mathematical analysis has shown this may not be the case (Goodman 1975; May 1973). The hypothesis is: The more complex a food web the more stable the populations in the web. One of theproblems has been the way stability is defined and this is still a matter of dispute.

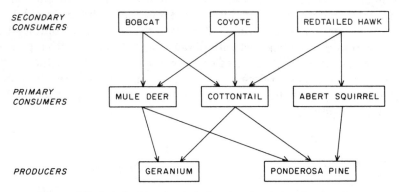

Figure 1.5. Interlocking food chains develop into a food web.

One recent definition of *stability* refers to an ecosystem where the chance of a species going to extinction is low (Colinvaux 1986). For an ecosystem (or population) to be stable it must demonstrate *resilience*—that is, the power, when under stress, to return to its original state when the stress is removed (Holling 1973). Theoretical ecologists discuss stability in terms of the number of species and if this number remains the same over time it is said to be *persistent* (Lewin 1986). On the other hand, a fluctuating density of animals within a species refers to *constancy*. However, the exact relationship of stability and community structure remains an issue of debate.

The Abert Squirrel, Nitrogen, and Fungi

In the southwestern United States there exists a good example of nutrient cycling in a food chain of the Abert squirrel, which returns nitrogen from feeding debris back to the soil for use by ponderosa pine trees. This tree squirrel feeds on pine cones, twigs, and hypogeous fungi growing beneath the duff layer of pine stands.

Research has substantiated that approximately 6 kg/ha/yr (5.32 lb/ac/yr) more nitrogen is returned to the forest floor from squirrels feeding in pine trees in comparison to trees where there is no feeding activity (Skinner and Klemmedson 1978). Other research has documented that squirrels feed on mycorrhizalfungi that are obligate symbionts of the pine root system (States 1979).

By feeding on the fungi the squirrels, through their fecal material, spread spores of the fungi that are root inoculum for bacteria. The bacteria produce mycelium which facilitates the cycling of nutrients from the soil to the root system and trees. This same association of fungi and rodents has been documented for other areas such as the Pacific Northwest (Trappe and Maser 1978).

Wildlife and Human Welfare

In the context of ecosystem complexity, wildlife species and their contribution to human well-being take on a new meaning. Animals facilitate energy flow and nutrient cycling through food chains and food webs and thus are critical components in the functioning of ecosystems. Humans lack the ability to digest and assimilate many of the chemical compounds found in plants, but since other animal species do, the elements contained in these compounds may eventually become part of our own nutrient supply. In this regard, short food chains provide progressively higher energy levels to humans. For example, in Asia where populations are high, the food chain is short—rice to humans. In turn, if larger amounts of energy are available, populations usually will increase to use them.

There is no doubt that human populations can survive without wild animal species but one only has to observe conditions inlarge cities, for example, Hong Kong and Mexico City, to ask: What quality of life do its inhabitants enjoy? While quality of life is clearly a value perceived differently from one individual and one generation to the next it is equally clear that the long-term welfare of our earth and any chance for a life of quality depends directly on a vigorous, diverse environment. It is time that we take seriously the role of wildlife in the functioning of ecosystems and that enlightened management of wildlife and other renewable resources is critical to the conduct of a good life and the future existence of the human species.

1.3 THE FOREST ECOSYSTEM

The forest ecosystem is a complex community of trees, shrubs, fungi, animals, and so forth. but they are not static components because one of the attributes of ecosystems is time which allows for change. Changes in ecosystems occur as a result of biological, chemical, and physical processes acting singly or in combination. These processes facilitate nutrient cycling and energy flow which are primarily responsible for the successive replacement of one plant community (seral stage) with another (*succession*). A *sere* refers to all the stages in the successional process starting with bare soil and ending in a final, climax community. While the definition of succession was formulated for plants it is equally applicable to animal communities and is one of if not the most important principle of forest wildlife management.

Forest Succession

Succession occurs because as one plant community becomes established it slowly creates conditions which allow new species from other communities to invade, grow, and reproduce (Clements 1916). In this manner the preceding community is gradually eliminated by a new community where the plant species composition, abundance, and structure are different from those of the past. The last seral stage in succession is the *climax*. In addition to replacement of one plant community by another there may be instances where one plant community inhibits the establishment of the next and so in this case succession does not progress.

Succession can be caused either by autogenic or allogenic forces. *Autogenic* forces are directed from within the ecosystem and are the agents of change caused by plants and animals, while *allogenic* succession is brought about by outside influences, for example, wind or fire.

In addition to the agents of change, the starting point of succession can be either primary or secondary. *Primary* (starting from bare soil) successional processes have a beginning but the end may be difficult to determine. However, there comes a time in an undisturbed plant community when a balance between all the factors causing vegetational change is achieved. Under equilibrium conditions the wastes from living matter are recycled so that the ecosystem is self-perpetuating until a catastrophic event such as fire, wind, or disease occurs; then the process begins again (secondary succession) progresing toward climax (Knapp 1974).

There is no disagreement by ecologists that succession occurs but the idea of a final self-perpetuating stage, the *climax*, has generated much discussion in ecological literature. In general three theories have been advanced to explain succession. The view just discussed is the *monoclimax* theory of Clements (1916). It holds that regional climate controls the composition of the vegetation in the region, hence the same array of plants repeatedly goes to climax via the successional process until a major climate shift occurs in the region.

A second theory (*polyclimax*) proposes that any factor (fire, soil, animals, etc.) can influence the direction of succession and array of plants in a community, not just climate (Nichols 1923; Gleason 1926; Tansley 1935). As a result, the regional pattern is one of a vegetation mosaic with different seral stages leading to a different climax community. The factors influencing succession can be expressed in a conceptual equation (Table 1.1).

The monoclimax and polyclimax theories assume that vegetation forms a constant and repeating identifiable and distinct unit on the ground. However, a third theory suggests that vegetation forms a continuum along environmental gradients so individual species are replaced and combined in different ways to form communities of *climax patterns* (Gleason 1939; Whittaker 1953; Curtis 1959). This theory emphasizes that species incorporated in a climax community may change gradually or abruptly over an environmental gradient under the influence of the factors in Conceptual Equation 1 (Table 1.1).

The concept of a stable climax based on the idea that succession is unidirectional has been challenged particularly as it relates to secondary succession (Noble and Slatyer 1977). Stable in this context means that the climax stage persists longer than any of the other seral stages. *Secondary* succession starts when the climax stage has been disturbed. Studies have shown that in some ecosystems (e.g., English moors and grasslands, balsam fir forests) patches of vegetation seral stages not uncommonly replace themselves in a cyclic manner (Watt 1947, 1955; Sprugel 1975) or randomly (stochastic) in hardwood forests (Horn 1976).

Of particular interest in forest wildlife management is the gap phase replacement process identified in forest stands (Watt 1947). In this situation individual trees die, thereby initiating a micro-succession which proceeds as a primary succession from bare ground. Such gaps can be small, as in the case of a single tree space, or larger, as in the case of a group of trees removed in a "blowdown." The idea of gaps leads to the notion of continuing diversity even in a relatively stable climax stage. In reality changes attributable to external factors, fire or human intervention for example, universally disturb climax or seral stages. Therefore, ecosystems are never stable in the true sense of the word, but are *resilient*.

Table 1.1. Factors influencing succession (modified from Jenny 1958).

Conceptual Equation 1
$Ve = f(So + Or + To + Cl + Ti)$
where
f = a function of
Ve = Vegetation
So = Soil
Or = Organisms
To = Topography
Cl = Climate
Ti = Time

Forest Seral Stages

In forested ecosystems, vegetation advances through four general seral stages (Fig. 1.6): bare soil, grass-forb, shrub, and tree. The tree seral stage in turn progresses through five distinct structural phases: seedling, sapling, pole, mature, and old growth (old forest conditions). Succession is a powerful force which shapes vegetation types, and a thorough understanding of the processes involved is critical to successful wildlife management.

The ability of humans to modify succession using the axe, plow, and cow is well documented, but we now have other and more precise means, gained by the accumulated knowledge of not only the development of plant communities but also the invention and use of tools, to redirect succession over large areas in a short time. Ecological knowledge together with the tools of the chain saw, bulldozer, and fire have truly made humankind an *ecological dominant"*, an organism which by its presence can radically alter the environment of other species.

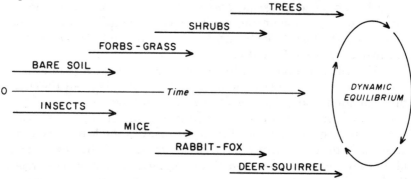

Figure 1.6. A general scheme of plant and animal succession in a forest.

Changes in Animal Species

Coupled to plant community successional stages are the associated changes in animal species. These changes occur because different animal species, like different plant species, have different life requirements and therefore different tolerances (*ecological amplitude*) to the existing environmental conditions. As plant composition changes through succession, there is in turn a change in animal species (Table 1.2). The change in species is not abrupt but rather a gradual replacement process. Because of the differences in ecological amplitudes of species, many will have overlapping habitat require-

Table 1.2. Changes in bird species as a result of succession on abandoned farmland in Georgia (after Johnston and Odum 1956).

Species	Grassland	Shrubs	Pine Forest	Hardwood Forest
Grasshopper sparrow	**************			
Meadowlark	***************			
Yellow-breasted chat		*************		
Cardinal		**********************************		
Pine warbler			*************	
Carolina chickadee			**************************	
Crested flycatcher				**************
Acadian flycatcher				*********

ments during the changes in successional stages and even after climax is attained. The degree of overlap is a function of ecological amplitude and niche separation.

In many instances, animals which man most desires to perpetuate, such as deer, are members of early seral rather than climax stages. Thus, some wildlife species and many of the tree species (e.g., aspen) which are valuable to man thrive best in temporary communities.

Because an organism cannot survive without at least the essential parts of its community, wildlife managers must learn how to manipulate plant succession and so maintain the desired local communities in a more-or-less permanent state. One alternative is to allow succession to proceed normally but to manage so as to have a sufficient number of areas continually coming into the desired seral stages. This idea is the basic premise of silviculture and forest management.

An example of succession and how it affects wildlife can be found in the history of deer populations in the western United States. Woody shrubs, which are the staple diet of mule deer, are characteristic of early stages in forest succession. Prior to settlement of the West, deer were most abundant in the transition between forest and grassland, or in recent burns within the forest but scarce in dense timber (Leopold 1950). As forests were logged or burned, brushy plant species increased and provided the palatable shrubs necessary to maintain large deer herds.

The change in plant and animal species composition over time on the same piece of land results in a natural cycle of both. Populations in an area may not change in total numbers of individuals but will change their location within that area. This concept is important to wildlife management because it introduces the element of ecological time which in most cases is beyond the span of the "here and now" typically incorporated into management plans.

Vegetation Classification

Classification is the grouping together of things with similar characteristics. The delineation of forest vegetation into specific types provides the wildlife manager with a valuable tool for management purposes because of the obligatory relationship that exists between animals and vegetation. Classification reduces heterogeneity depending on the scale or level of abstraction of the groupings. In recent years there has been a major effort by the United Nations Education, Scientific, and Cultural Organization (UNESCO) to develop a worldwide vegetation scheme (United Nations 1973). The system is open-ended and hierarchical using plant physiognomic characteristics (appearance, life-form).

Under the UNESCO system there are five mutually exclusive vegetation classes in the first level of the hierarchy:

I. Forest
II. Woodland
III. Shrubland
IV. Dwarf-shrubland
V. Herbaceous vegetation.

In dealing with forest wildlife relationships we are interested primarily in Class I (Forest) and II (Woodland) because they contain trees as the dominant life-form. The distinction between the two classes is the degree of canopy closure. The woodland class has a more open canopy in the range of 25–60%.

The UNESCO classification is further subdivided into subclass, group, formation, series, and association (Table 1.3). At the associa-

Table 1.3. World vegetation classification scheme (United Nations 1973).

Symbol	Description	Example	
I.	Class	a.	Forest
		b.	Woodland
A.	Subclass	a.	Evergreen forest
		b.	Evergreen woodland
1.	Group	a.	Temperate and subpolar needle-leaved
		b.	Needle-leaved
a.	Formation	a.	Conical crowns
		b.	Rounded crowns
(1).	Series	a.	Engelmann spruce-subalpine fir
		b.	Ponderosa pine
(a).	Association	a.	Engelmann spruce-subalpine fir/grouse whortleberry
		b.	Ponderosa pine/antelope bitterbrush

tion level the first name is the apparent dominant or codominant plant species in the tallest life-form. The second name may be a dominant species in a different layer with a third name added if there are three layers with a conspicuous plant component.

Because the focus of this book is forest-wildlife habitat relationships, a common understanding is needed of what constitutes a forest and what are its major components. Intuitively, most people recognize a forest without resorting to a formal definition but for clarity a *forest* is characterized as a vegetation community made up of a more-or-less dense and extensive tree cover (Ford-Robertson 1971).

The definition used in the Resources Evaluation Techniques Program of the USDA Forest Service (Driscoll et al. 1984) is: A forest is a community formed by interlocking tree crowns with a canopy cover of 60% or more at maturity. In reality a firm dividing line between a forest and land with some trees growing on it is what one defines it to be.

1.4 VALUES OF WILDLIFE

Historically, many species of wildlife were hunted for food and clothing (fur). More recently forests have been cleared to make room for crops, domesticated species have replaced wild animals as sources of meat, and cotton and wool have replaced the need for fur. In developed countries hunting is pursued today largely as a form of recreation. For example, in 1985 in the United States alone more than 16.7 million hunters spent over $10.1 billion pursuing their sport; 46.4 million people spent $28.1 billion on fishing; while 134.7 million people were in the woods not to hunt or fish but simply to observe wildlife at a cost of $14.3 billion (U.S. Dept. of Interior 1985). Clearly wildlife possesses considerable value for a significant portion of the U.S. public.

However, there is a sense of a different kind of value of wild animals that goes beyond their utilitarian value, a feeling which is difficult to explain to those who have never experienced an elk bugling, a coyote barking, an eagle screaming, or air whistling through a hummingbird's wings. This form of aesthetic or metaphysical value is virtually priceless to some people. Yet such subjective values are difficult to include in a management plan explained in economic terms.

Since 1960 a number of environmental laws have been enacted around the world, notably in the developed countries, all of which changed the processes by which wild lands and wildlife populations are managed. These various legislative initiatives and the resulting policies had in part the effect of establishing the view that all wild animals have value in the functioning of ecosystems and must be managed accordingly. Understanding of this role of animals is not new, as any historic review of ecology proves, but it has now advanced to a higher level of public acceptance and understanding.

While resource agencies lack a uniform and generally accepted system for placing a value on wildlife, they are still under constant pressure to assign a monetary measure of value to compare trade-offs with other forest uses in the planning process. Monetary values for some wildlife species have been arrived at by calculating the cost of hunting the species. For example, in Colorado the value of a mule deer has been set at $709 and blue grouse at $20.79 (Norman et al. 1976).

Hunting elk on the Fort Apache Indian Reservation in Arizona requires a permit costing over $10,000 and when other costs, such as travel, lodging, and equipment are added, the typical hunter probably spends over $15,000 to kill a single animal. The problem of assigning a monetary value to nongame species is even more difficult but some techniques have been developed such as the amount people are willing to pay to see a particular species in the wild (Martin and Gum 1978).

I cannot offer the reader contending with the problem of defending the inclusion of a wildlife species or a group of species and their economic value in a management plan an easy way to place a value on these forest inhabitants. They clearly have a substantial worth to the public in the short term, witness the numbers cited above, and one of critical value in the long-term maintenance of ecosystems. Somehow these short- and long-term values must be recognized and incorporated in management plans despite the difficulties in assigning an economic value in the decision-making process as to whether an animal species deserves perpetuation.

Techniques for arriving at a value for a given wildlife species in a given ecosystem have proven elusive but the effort continues to attract the attention of economists. The most promising approach currently is to compare the value of resources (habitat) used for wildlife production against the value of those resources for other purposes (Peterson and Randall 1984).

Justification for including wildlife in management plans already exists in current laws and regulations mandating the perpetuation of viable populations of all native and desirable nonnative species. So fortunately monetary value is not a factor when dealing with species which are dependent upon a very narrow ecological base such as those requiring "old-growth" (Thomas et al. 1988).

An issue related to the value of wildlife is the controversial topic of animal rights versus animal welfare. Basically people belonging to an animal rights organization believe that animals, including those classified as wildlife, have rights similar to humans (Schmidt 1990) and must be given equal consideration along with humans (Singer 1980). Members of these organizations oppose the use of animals for any type of research, such as in agricultural or biomedical fields, and they definitely are against trapping animals for furs and killing for sport hunting.

Animal welfare organizations, on the other hand, concentrate their efforts on reducing the suffering and pain caused by humans for utilitarian purposes including those just mentioned. Animal welfare advocates do not place the value of a species above that of human health and welfare. Wildlife professionals must be convincing welfare advocates because the general public needs to be aware of the differences between the two philosophies and how these differences may affect the quality and way of life for all people. This topic is a hot issue in many sections of the United States and undoubtly the debate will increase in intensity in the coming years.

1.5 WILDLIFE IN THE UNITED STATES

The number of wildlife species within the continental limits of the United States and territories is approximately 2,900 consisting of about 200 amphibian, 900 bird, 1,050 fish, 400 mammal, and 350 reptile species (U.S. Dept. of Agric. 1981). Only a few, such as the white-tailed deer are widely distributed in forest habitats across the continent; for the most part the distribution of species is not so cosmopolitan. The total number of species for North America is well over 3,500 and there is overlap of many species' range between Canada, the United States, and Mexico because of bird migrations and the similarity of habitats.

In the United States the largest number of species of amphibians, birds, and reptiles in all habitats is found in the southern states. The largest number of fish species is found in the Southeast, while the largest number of mammal species is found in the Pacific Coast region. The size, life-form, and distribution of wildlife species inhabiting forested ecosystems creates a diverse and complex environment which is constantly being challenged by habitat changes or destruction by humans.

In the late 1800s wildlife was a resource under the "gun," "axe," and "plow." Seton (1929) estimated that 5 million bison once roamed the eastern forest lands, white-tailed deer numbers were near 40 million, and passenger pigeons were so dense that their flight blocked out the sun. After market hunting became a vocation the situation soon changed; by 1900 the bison was nearly eliminated, white-tailed deer numbers had declined to their lowest known populations, and the passenger pigeon was well on its way to extinction.

Since 1760, 46 species of birds, mammals, and fish have become extinct in the United States. This includes forest species such as Merriam's and Eastern elk, and the Eastern cougar. The current rate of extinction of birds and mammals in the United States, 20 per 100 years, is seven times that estimated for the late Pleistocene period—a time of great ecological change (Council on Environmental Quality 1981).

A new perspective on conserving all species of wildlife in the United States is included in the Endangered Species Act (U.S. Congress 1973). This act provides for the conservation and recovery of species designated as endangered. A species is endangered if it is about to become extinct throughout all or a significant part of its range. A threatened species is one which is likely to become endangered in the foreseeable future. The designation of endangered or threatened species is determined by the Secretary of the Interior. The list of endangered and threatened species is shown in Table 1.4 for March 1990. This listing is published each month. For a current listing the reader can write to the U.S. Fish and Wildlife Service, Washington, D.C.

Table 1.4. Endangered and threatened species in the United States and other countries (U.S. Dept. of Interior 1990).

Species Category	U.S. Only	Foreign Only	U.S. Only	Foreign Only	Listed Total
	Endangered		Threatened		
Mammals	53	248	8	22	331
Birds	74	153	11	0	238
Reptiles	16	58	17	14	105
Amphibians	6	8	5	0	19
Fishes	53	11	33	0	97
Snails	3	1	6	0	10
Clams	37	2	2	0	41
Crustaceans	8	0	2	0	10
Insects	11	1	9	0	21
Arachnids	3	0	0	0	0
Plants	179	1	60	2	242
Total	443	483	153	38	1117

Some ecologists have estimated that one animal species in the world goes to extinction each year while others of a more pessimistic view hold that if plants and invertebrates are included in this category the number will increase to one per day in the near future (Eckholm 1978). Choices about which species to try to save and perpetuate are becoming more difficult. It is far easier to generate concern about animals that people can see in the wild such as the Sonoran pronghorn than it is to generate concern for an animal as small and "non-furry" as the Texas blind salamander.

Humans also seem to be more interested in taxonomically advanced species or those providing some apparent benefit to mankind. The salamander may be just as important in the functioning of an ecosystem as the pronghorn, but convincing a tax-paying public to save both may seem inhuman when some members of society do not have money to put sufficient food on the table. However, if we are concerned about human survival, saving species now going to extinction due to mankind's activities may be the only way to ensure survival of *Homo sapiens* over geological time.

The Wildlife Management Institute (1990) reports that there are more than 3600 domestic species waiting to be considered by the U.S. Fish and Wildlife Service for listing as endangered. The cost to place one plant or animal on the federal endangered list is about $60,000, and the cost of preparing recovery plans for the proposed species would be about 5 billion dollars.

In the near future a system must be devised to determine which species to allow to survive or go to extinction, and this will mean placing a survival value on the candidate species. George Orwell (1946) wrote that all animals are created equal, but some are created more equal than others. This statement is more appropriate today than it was 50 years ago when human populations and activities were much lower. Unfortunately, the endangered and threatened list does not reflect the alternatives to extinction which can be selected, given the direction and funding to do the job. As is witnessed by the emphasis on conservation of game animals in the early 1930s, some major accomplishments in wildlife management are possible if there is the will to do so.

The beaver exists today in all states of the United States, except Hawaii, whereas in 1900 it was found only in limited numbers in a few eastern states (Wildlife Management Institute 1975). The white-tailed deer once extirpated in more than half the states is now common in 48 with an estimated total population of over 14 million. The wild turkey has been restored to huntable populations in 43 states with their numbers totalling over 2 million. Elk common only in and around Yellowstone National Park in 1900 is now doing well in 16 states with populations approaching 500 thousand.

1.6 SUMMARY

Animals have been integral to human history and our existence continues to depend upon their role in recycling nutrients and energy through food chains and webs in plant and animal communities.

Nutrient cycling and energy flow combined with natural forces lead to changes in the environment. Because plants and animals have different tolerances to environmental factors there is an orderly progression of both as one plant community replaces another from bare soil to a mature forest in a process called succession.

In addition to their role in the world ecosystem the value of wildlife species is also found in their educational, scientific, aesthetic, and recreational significance. Their recreational value can be approximated in terms of the money spent on hunting and fishing licenses and equipment, but other values are more difficult to assess.

There are approximately 2,900 wildlife species in the United States. A few of these species are widely distributed but the majority are restricted to regional or local areas. The numbers of several of the game species now in existence, for example wild turkey, beaver, and bison, were once so low that they might well have gone to extinction like the passenger pigeon. But, due to changing attitudes in the early 1930s and new laws in the 60s and 70s, most wildlife species now exist in adequate (viable) populations. Those with populations not in an acceptable range have protection in the Endangered Species Act which requires that an effort be made to restore their numbers.

Wildlife management has progressed from a period of emphasizing only species with value as game, through a time when wildlife was redefined to include all life-forms, to the present-day view in which wildlife is beginning to be appreciated for its ecological role in the functioning of ecosystems. In the future all resources will be managed as essential parts of ecosystems.

1.7 RECOMMENDED READING

Allen, D. L. 1974. *Our Wildlife Legacy*. Funk and Wagnalls, New York.

Drury, W. H. and I. C. T. Nisbet. 1973. Succession. *J. Arnold Arboretum* 54:331–367.

Elton, C. S. 1956. *Animal Ecology*. Sidgwick and Jackson, London.

Elton, C. S. 1958. *The Ecology of Invasions by Animals and Plants*. Methuen and Co., New York.

Kimmins, J. P. 1987. *Forest Ecology*. Macmillan, New York.

Leopold, A. 1939. A biotic view of land. *J. For.* 37: 727–730.

Leopold, A. 1943. Wildlife in American culture *J. Wildl. Manage.* 7:1–6.

Maser, C., and J. M. Trappe. 1984. *The Seen and Unseen World of the Fallen Tree*. Gen. Tech. Rep. PNW-164, USDA For. Serv., Pacific Northwest For. and Range Exp. Sta., in cooperation with USDI Bureau of Land Manage., Portland, OR.

Odum, E. P. 1969. The strategy of ecosystem development. *Science* 164:262–270.

Peterson, G. W., and A. Randall. 1984. *Valuation of Wildland Resource*

Benefits. Westview Press in cooperation with USDA For. Serv., Boulder, CO.

Smith, R. L. 1986. *Elements of Ecology*. Harper and Row, New York.

U.S. Dept. of Interior. 1987. *Restoring America's Wildlife, 1937–1987*. Fish and Wildl. Serv., Wash., DC.

Witter, D. J. 1980. Wildlife values: Applications and information needs in state wildlife management agencies. Pages 83–98 in W. W. Shaw and E. H. Zube, eds. *Wildlife Values*. Inst. Rep. No. 1, Center for Assessment of Noncommodity Nat. Resour. Values, Univ. of Arizona, Tucson.

Chapter 2

THE ENVIRONMENT OF WILDLIFE

The earth is *the* environment, a forest has an environment, and humans have an environment in homes, workplaces, and surrounding their immediate existence. Since the 1960s some people have come to call themselves "environmentalists." So just what is this abstract but pervasive factor which so affects our daily lives, and more specifically for this book, those animals defined as wildlife? We will examine environment starting with a conceptual equation that presents populations as a function of biotic potential and environmental resistance (Table 2.1).

Biotic potential is the theoretical number of young an individual can produce with unlimited resources (food, cover, etc.). It is an innate characteristic of a species and has an upper limit, but in nature, the upper limit is never attained. The biotic potential of a species is a useful term for comparing maximum theoretical production against an actual population. It also gives some indication of what can happen when animals are introduced into a new habitat. Conceptual Equation 2 (Table 2.1) while trivial in itself, started biologists and mathematicians thinking about the factors which control population size.

Table 2.1. Populations are a function of biotic potential and environmental resistance (Chapman 1928).

Conceptual Equation 2
$P = f(Bp - Er)$
where
f = a function of
P = Population
Bp = Biotic potential
Er = Environmental resistance

Environmental resistance is the factor or factors limiting biotic potential, thus reducing wild animal populations. Leopold (1933) separated environmental resistance into decimating factors, welfare factors, and environmental influences. *Decimating factors* (predators, disease, starvation, drought, and accidents) are those that kill directly while *welfare factors* (food, cover, water, and special requirements unique to a species) reduce productivity indirectly by limiting inputs necessary to survival.

Environmental influences are processes or factors that alter the decimating and welfare factors. Environmental factors include succession, weather, and mankind's activities such as timber harvesting and grazing. The use of the terms "biotic potential," "welfare" and "decimating" factors, and "environmental resistance" are still common in wildlife and general ecology literature because they offer an intuitive means of understanding the factors affecting an animal. However, we now possess terminology that is more descriptive in explaining and quantifying the environment of an animal.

Insight into the environment and its effects on animal populations continued to develop in the 1930s and 1940s culminating in 1954 with the publication by Andrewartha and Birch of *The Distribution and Abundance of Animals* which explained environment in detail. These authors later refined and expanded their ideas into a comprehensive theory of environment contained in *The Ecological Web* (1984).

In this book I synthesize and integrate Andrewartha and Birch's (1954, 1984) theory of environment with the ideas of Chapman (1928) and Leopold (1933) along with my own observations into a general model for comprehending the environment of wildlife.

2.1 DIRECT AND INDIRECT FACTORS

The *environment* of an animal consists of all the factors affecting its chance to *survive* and *reproduce*. In discovering what these factors are, focus is directed to an animal's *Umwelt* which is the totality of the surroundings the animal experiences (Klopfer 1969). Umwelt as perceived by humans may be entirely different from that which an animal experiences. The challenge to wildlife biologists is to understand and describe the actual environment affecting wild animals and not that perceived by humans. However, it is a useful exercise to be anthropomorphic for a few moments and consider what factors affect our own chances of surviving and reproducing, and then to transport these factors to different wildlife species and habitat situations. In doing so items will be identified that can be listed under separate categories of factors that directly affect an animal's health and welfare in what Andrewartha and Birch (1984) have termed the *Centrum* of an animal.

The Centrum

The factors affecting an animal's chance to survive and reproduce are contained in a *centrum* of directly acting factors and a *web* of
indirectly acting factors (Fig. 2.1). The directly acting factors are
hazards, diseases, predators, genetics (species characteristics),
resources, and humans. The directly acting factors are, in turn,
affected by a web of indirectly acting factors, both biotic and abiotic in
origin, and through a variety of links and chains create complex
interactions. Five of the directly acting factors are external to the
environment of an animal. However, genetics is an internal factor and
affects the way an animal responds to the five external factors.

The direct and indirect factors affecting an animal, by implication, also pertain to a population. The indirectly acting factors include
living organisms, inorganic matter, and energy. For example, in
Figure 1.4 the coyote is directly acting on the rabbit by consuming it
for food. The rabbit previously ate plants which were directly affected
by soil nutrients and water. Therefore, soil nutrients and water are
part of a web of factors indirectly affecting the reproduction and
survival of the coyote.

An understanding of the centrum and web specific to a particular
animal species emerges from the accumulated and accumulating

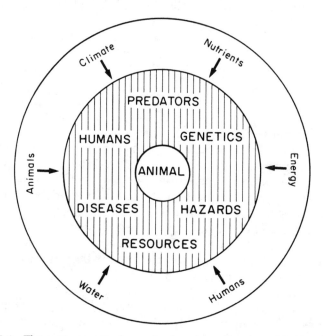

Figure 2.1. The environment of an animal contains directly and indirectly acting
factors (Modified from Andrewartha and Birch 1984).

knowledge of ecological relationships and the growing comprehension of food chains, food webs, and nutrient cycling responsible for ecosystem functioning. The six general factors in the centrum completely describe and account for all the direct factors influencing an animal's chance to survive and reproduce. However, there can be overlap between factors. For example, humans affect the survival and reproduction chances of an animal through hunting and habitat destruction so humans can be classed as either or both a hazard or predator. The intent, however, is to eliminate such duplications and to categorize various factors in a single class so that biologists share acommon understanding of the environment of an animal and what factor or factors they can select for managing a particular species.

With these ideas we can formulate a conceptual equation that replaces Chapman's equation and Leopold's factors as a working hypothesis for the directly acting factors of the centrum (Conceptual Equation 3, Table 2.2). The environment of an animal also includes a web of indirectly acting factors, but in most cases they are obscure and often difficult to identify and quantify. Managers spend 70–80% of their time working with only one or two of the directly acting factors and little time on indirectly acting factors because the former account for a majority of the management problems. For this reason I will not devote any significant attention to indirect factors.

Table 2.2. The directly acting factors of the centrum.

Conceptual Equation 3
$Pr_{(SR)}$ = f(Ha + Di + Pr + Ge + Re + Hu)
where
$Pr_{(SR)}$ = Probability that an individual or population will Survive and Reproduce is
f = A function of
Ha = Hazards
Di = Diseases
Pr = Predators
Ge = Genetics (species characteristics)
Re = Resources
Hu = Humans

Before examining each of the directly acting factors another term requires a definition, namely, habitat. Environment and habitat are often used interchangeably but they are not synonymous. Wildlife habitat includes both abiotic (soil, climate, and water) and biotic (plant and animal) factors which combine in various ways to create a setting in which organisms survive in some kind of community relation.

Habitat is the environment of and the specific place where an organism lives. We speak of the habitat of mule deer as forest or brushland; for trout, cold running water; and for the mallard and muskrat, marshland, but all these habitats have a specific location. Environment is part of habitat. So, the definition of habitat includes *all the factors affecting an animal's chance to survive and reproduce in a specific place*. These specific places are often described by a vegetation type (Oak-hickory, Spruce-fir) or topographic feature (mountains, deserts).

As a result of an accumulation of information from research and experience we now know that habitat is somewhat more complicated than the definition just given. This definition, however, is sufficient until Chapter 7 where it will be expanded upon to better understand habitat relationships.

2.2 HAZARDS

A *hazard* is anything (excluding disease) capable of causing injury or death to an animal. Wildlife must cope with both natural and man-made hazards. Natural hazards include such things as water, snow, wind, and fire. Man-made hazards include roads and fences. In a broader context hazards also include mankind's hunting activity and habitat destruction. But these two activities involve an active agent while the concept of hazards we are attempting to clarify is the encounter of wildlife with the forces of nature, and the artifacts of man (the intention of which is not the injury or killing of animals)—not man himself. Probably the most common and greatest threat to wildlife from natural and man-made hazards is from fire and man-made artifacts, roads.

Fire

Fire has both direct and indirect effects. The direct effect is death by burning as well as suffocation, particularly in the case of small mammals (Lawrence 1966). The air temperature at which animals are killed is 63°C (145°F) (Howard, Fenner, and Childs 1959). Direct mortality has been documented in animals as varied as birds and voles (Bock and Lynch 1970; Hakala et al. 1971), deer, wood rats, and rabbits (Chew, Butterworth, and Grechman 1958; Buech et al. 1977). In most cases, even in large fires, death by burning is rare (Vogl 1977). In general the loss of vertebrates directly to the effects of fire is negligible. While there may be some loss of individuals, the effect on the population is insignificant.

The impact of fire upon an animal is a function of the animal's size and mobility, and the extent, intensity, and speed of the fire. Adult deer and elk commonly avoid injury because they can easily move away from advancing fire but very young animals may not be so fortunate. Chipmunks, shrews, woodrats, and mice are generally reluctant to leave burning woodpiles and grass stands therefore they are probably killed directly (Bendell 1974). However, those species living underground can survive intense fire due to the great decrease in temperature below the ground surface (Howard, Fenner, and Childs 1959).

The indirect effects of fire on wildlife, though difficult to quantify, are substantially greater due to the loss of habitat which leads to starvation, predation, or decimation by weather. These indirect effects usually persist for 1–3 years (Cook 1959; Lawrence 1966). Starvation following fires which burn large areas results from the loss of vegetation used for food, particularly in the case of species of limited mobility (home range). Even when starvation does not result in death an animal's ability to reproduce is frequently severely impacted because, when food is limited, nutrients are used for body maintenance rather than reproduction.

Predation on small mammals following fire is generally increased since a survivor has to search for food at greater distances from cover. Fire may reduce vegetation to an extent that animals are frequently exposed in open areas thereby increasing their vulnerability to predators. It is not uncommon to see hawks and owls in recently burned areas presumably because small mammals are easier to hunt in the open spaces where there is little or no vegetation.

Decimation of animals from weather after fire results from a loss of cover that otherwise provided protection from wind, water, and heat. Although general cover may still exist in the burned area the special types of vegetative cover such as for nesting, feeding, resting, and loafing may be too thin to protect both adults and young from the elements.

Little information is available on the effects of fire on amphibians and the information with respect to reptiles is just as limited. Amphibians have been observed to remain in mud and water until a fire has passed (Komarek 1969). Reptiles may be able to escape due to their ability to detect fire with their heat-sensing pits used in locating prey (Komarek 1969). Birds are most vulnerable during nesting and fledging periods. Fire devastates ground-nesting birds not only because it destroys nests, but also because it removes protective cover, and eliminates insects used for food (Daubenmire 1968).

The impacts of fire on fish are indirectly the result of physical and chemical changes caused by vegetation removal in the surrounding watershed. Removal of stream bank vegetation commonly increases water temperatures, thereby converting a cold-water stream into one of warm water, leading to corresponding changes in species composi-

tion (Lyon et al. 1978; Swanston 1980). A change in water temperature may also affect fish distribution patterns (Koya 1977). Fire
generally results in an increase in erosion leading to increased sedi-
mentation, killing fry or destroying eggs (Swanston 1980).

Roads

Each year the number of miles of paved highways and improved
dirt roads in forested areas increases. For example, in Florida alone
6.4 km (4 mi) of new roads have been built every day for the last 50
years (Harris 1988a). About 18% of the nation's 6.1 million km (3.8
million mi) of roads are located within cities and towns while the
remaining 82% are in rural areas where the chances for encountering
wildlife is increased (Council on Environmental Quality 1981).

The killing of animals along a highway is a direct hazard. Indirect
effects arise from the noise and disturbance caused by traffic that may
prevent animals such as deer and elk from using an area. There are
indications that many species have a tolerance level to roads so that
flight does not lead to displacement (Lyon 1984). Data from a study in
Wyoming on the effects of logging and recreational road traffic on elk
indicate little effect at a distance of about 1 km (0.6 mi) from timber
harvest operations and after activity stops animals move back into the
harvested area (Ward 1976). The same study concluded elk were little
affected by logging and recreation roads during the day within 0.40
km (0.25 mi) once the elk became used to the traffic. Elk seem to avoid
areas of heavy vehicle traffic but the level of avoidance depends upon
the extent of cover (Perry and Overly 1976; Lyon 1984).

Low-mobility species are affected by roads at the time they are
constructed. A direct loss of habitat for small mammals such as mice
and rabbits, as well as a loss of animals in burrows, occurs where a
road is constructed. This negative impact is somewhat offset by the
increase in food and cover plants in the disturbed soil. The practice of
seeding forbs and grasses along paved highways to stabilize berms
increases the probability of small mammals being killed.

The best figures available on death of animals from vehicle accidents are for white-tailed deer. In Michigan deaths exceed 20,000 a
year (Hansen 1983); and in Pennsylvania, 30,000 (Pennsylvania
Game Commission 1985). But wildlife deaths from vehicle collision
are only one part of the story, the other being a loss of human life. In
Michigan over a 5-year period 3,289 motorists were injured and 17
were killed (Arnold 1979). There is no estimate of the total number of
wild animals killed each year by cars but the number must be high
because a book *Flattened Fauna* (Knutson 1987) has recently been published which provides keys to species of dead animals found on highways.

2.3 DISEASES

A *disease* is an impairment of the normal state of a living animal or plant that affects the performance of the vital functions (*Webster's Seventh New Collegiate Dictionary* 1966). Diseases affect individual animals by either killing them directly or so debilitating their physiology or behavior that they die from starvation or become easy prey for predators. Diseases are caused by bacteria (tularemia, botulism, Lyme), fungi (aspergillosis), viruses (foot and mouth, rabies), and toxic chemical substances including pesticides. However, diseases are rarely a simple one-cause–effect relationship but are generally the sum of changes in the environment (Karstad 1979). Diseases transmitted from one animal to another either by direct or indirect contact are called *contagious*. Diseases capable of invading body tissue are termed *infectious* and disease-producing agents are called *pathogens*.

Managers are concerned about diseases in wild animals because there are interactions between humans and wildlife, livestock and wildlife, and between different species of wildlife. An example of a disease transmitted from wildlife to humans and also to domestic animals from bites by skunks, foxes, and small mammals is rabies (*Rhabdovirus*). Rabies quickly overcomes an animal's immunological system because the virus multiplies rapidly; such a disease is described as *virulent*.

Another particularly feared disease is bubonic plague (*Yersinia pestis*) or "Black Death." The fear arises from our knowledge of the large number of deaths that occurred in Europe in the 14th century (Tuchman 1978) and the fact that death from plague still occurs in the southwestern United States where there are large rodent populations. For the period 1950–1989, 46 cases of plague have been reported in humans in Arizona. The disease is not caused by rodents but by the bites of fleas and ticks (called *disease vectors*) inhabiting rodent burrows and dens.

Another disease transmitted by ticks is Lyme disease. Ticks as intermediate hosts get bacterium (*Borrelia burgdorferi*) from the blood of infected animals, particulary rodents, and then transmit the bacterium to humans or other animals when they suck blood (Davidson and Nettles 1988). The disease was formerly called Lyme arthritis and was named for the town of Old Lyme in Connecticut where it was first recognized. Lyme disease has become more prevalent in recent years and is often difficult to detect and treat.

The second form of disease transmission, from livestock to wildlife and vice versa, led at least in one case to the destruction of a deer herd on the Stanislaus National Forest in the 1920s (Allen 1974). Foot-and-mouth disease (a virus) was devastating livestock in California and could only be controlled by killing infected deer. After eliminating over 34,000 deer the disease was brought under control.

Many diseases are transmitted from one wildlife species to another. An example is tularemia caused by the bacterium *Francisella tularensis*. The disease is transmitted by ticks and fleas and occurs in over 45 species of vertebrates but is present most often in rodents and rabbits (Davidson and Nettles 1988).

Another form of disease arises out of naturally produced toxins. A good example is botulism, well known in the wildlife profession because it occurs with such frequency. The disease is caused by the bacterium *Clostridium botulinum* which produces a toxin so powerful that when ingested by waterfowl it results in death. This bacterium typically develops in warm, stagnant water low in oxygen.

Noninfectious diseases include such ailments as tumors (virus) and warts (virus). Also included in the disease category are parasites of roundworms, tapeworms, and flukes occurring in a host-parasite relationship. The immature parasite requires an intermediate host, in this case animals, for a "habitat."

In recent years mankind has added to the disease problem by introducing lead into the environment in hunting waterfowl. Waterfowl feed on lake and pond bottoms where they pick up shot along with food items. The shot is retained in the gizzard, lead is absorbed into the blood stream, and the waterfowl develop lead poisoning which causes them to become emaciated and their ability to walk or fly is impaired (Davidson and Nettles 1988). Raptors feeding on waterfowl with lead shot in their gizzard can also be affected with lead poisoning (Pattee and Hennes 1983).

Pesticides are chemical compounds used to control plants and animals such as weeds, insects, rodents, and predatory animals. Those designed to kill insects are designated as *insecticides* and those designed to control vegetation are called *herbicides*. For an account of the use of herbicides in the United States and how their once widespread application directly and indirectly affected the ability of plants and animals to survive and reproduce, the book *Silent Spring* by Rachel Carson (1962) is an accurate presentation.

One of the best known pesticides is DDT, a synthetic chlorinated hydrocarbon that was made available for public use in 1945 and was banned for most uses by the Environmental Protection Agency in 1972. It is the most persistent chlorinated hydrocarbon followed by dieldrin, toxaphene, lindane, chlordane, heptachlor, and aldrin. Aldrin and dieldrin were banned in 1974 and heptachlor and chlordane were banned in 1975. Other pesticides include the organic phosphorus compounds of parathion and malathion used to kill parasites on plants and animals.

Certain animals are more sensitive to pesticides than others. As a general rule, crustaceans (lobsters, shrimp, crabs) shellfish (clams and oysters), and fish are most sensitive, followed by amphibians (frogs, salamanders), reptiles, birds, and mammals (U.S. Dept. of Interior 1966). Pesticides may kill animals directly but most exert indirect effects through the food chain either by depletion of a food supply or reduction of reproductive capacity after an animal consumes contaminated food species.

Fish and birds are good indicators of pesticide contamination. Birds feed on insects likely to have been sprayed and fish come into contact with run-off water containing the contaminant. Pesticides not only affect wildlife but also humans through the food chain or by direct contact. Most people in North America have measurable amounts of pesticides in their bodies (Council on Environmental Quality 1981).

2.4 PREDATORS

A *predator* is an animal that kills and feeds on other animals. Hawks eat mice, mountain lions eat deer, and foxes eat rabbits. Because of this "kill and eat" trait probably no other topic in wildlife management elicits as much emotion and debate as the subject of predators, particularly as they affect animals which humans want to hunt or view such as deer, quail, squirrels, and rabbits.

A background statement, prepared for a report by an advisory board on wildlife management to the Secretary of Interior, summarizes the attitudes about predators that once existed in the United Sates and how these attitudes are changing (Leopold et al. 1964):

In a frontier community, animal life is cheap and held in low esteem. Thus it was that a frontiersman would shoot a bison for its tongue or an eagle for amusement. In America we inherited a particularly prejudiced and unsympathetic view of animals that may at times be dangerous or troublesome. From the days of the mountain men through the period of conquest and settlement of the West, incessant war was waged against the wolf, grizzly, cougar, and the lowly coyote, and even today in the remaining backwoods the maxim persists that the only good varmint is a dead one. But times and social values change. As our culture became more sophisticated and more urbanized, wild animals began to assume recreational significance at which the pioneer would have scoffed. Americans by the millions swarm out of the cities on vacations seeking a refreshing taste of the wilderness, of which animal life is the living manifestation. Some come to hunt, others to look or to photograph. Recognition of this reappraisal of animal value is manifest in the myriad of restrictive laws and regulations that now protect nearly all kinds of animals from capricious destruction. Only some of the predators and troublesome rodents and birds remain unprotected by law or public conscience.

Since the report was written additional legislation protecting animals classed as predators has been enacted. These laws emphasize the role of animals in maintaining ecological diversity, social and scientific value, and ecosystem functioning.

Over millions of years predators have evolved together with their prey. If predators destroy their prey completely, then the former must necessarily die. So a delicate balance between prey and predator has evolved. In ecological systems, predators are believed to perform the function of removing excess numbers and culling the prey population by removing weak, diseased, or abnormal individuals; however, this function of predators has not been well authenticated in the literature.

The first theoretical work done on predator-prey relations was that of Lotka (1925) and Volterra (1931) who developed the basic equations in the 1920s (Chapman 1931). These equations assumed an increase in prey populations when predators are not present, and a decrease in predator populations when prey is not present. The final result of these equations is a cycle in which as a prey species population increases the predator population increases but with a lag. These increases result in a reduction in prey population lagged by a reduction in predator population. And so the cycle repeats.

The Lotka-Volterra equations do not consider all the interactions taking place in a natural ecosystem, such as populations of other prey species (buffer species) which are available when the primary prey species is reduced by the predator. So as a practical tool for field biologists the equations are of little value. The equations are, however, of historical value because they generated considerable interest among scientists studying animal cycles resulting in a better understanding of the predator-prey relationship.

One of the first studies on predator-prey relationships was in Iowa on mink predation of muskrats (Errington 1967). The results generally confirmed the findings that when a muskrat population was high, predation by mink was also high, but predation was not even across all age classes. Errington also discovered that social interactions within muskrat populations contributed to limiting their populations. In Errington's study the muskrats selected were those suffering from disease and starvation so the predation was *compensatory* not additive. This simply means that the predator replaces disease and starvation as the decimating factor. If predation were additive then the loss of animals would be added to that caused by disease and starvation. However, all predation is not compensatory, for example, bears killing elk calves or their own young, or raccoons eating waterfowl eggs. The idea that predators take only the sick and weak is intuitive but oversimplifies the predator-prey relationship.

Some general propositions of predator-prey relationships that can be formulated from ecological studies are

1. The density of a predator population is always much less than that of its prey.

2. Predators feed on a variety of prey, not a single species.

3. Reproduction rates of prey species is far greater than that of predator species.

4. The prey species is almost always smaller in size than the predator, but when small predators prey on larger species it is usually on the young.

One study of predator control often cited as the reason for an increase in deer populations centered on the Kaibab Plateau in Arizona. Before predator control was begun in 1906 the North Kaibab had an estimated 4,000 deer (Rasmussen 1941). From 1906 to 1923 mountain lions, wolves, coyotes, and bobcats were removed by hunters and trappers. Twenty years later, just before a die-off, the deer population was estimated to be 100,000 on the Plateau, however, the accuracy of this estimate is questionable. Also the results are confounded by the fact that 200,000 sheep and 20,000 cattle were removed from the study area during the same period (Caughley 1970).

An interesting idea that developed from predator-prey studies is that predators can maintain species diversity (Spight 1967). Diversity is enhanced because predators keep abundant species in check so that they cannot out-compete less abundant species for food, shelter, and other necessities. The thought to keep in mind is that, as in the Lotka-Volterra equations, a strict predator-prey relationship seldom provides all the answers to the question. Predators are an important component of an animal's environment, but the effects of predation on a population must be considered in connection with the other factors contained in Conceptual Equation 3.

When considering predator-prey interactions information is needed on five factors before the relationships can be understood (Leopold 1933). These factors are

1. Density of the prey population.

2. Density of the predator population.

3. Prey characteristics, such as reactions to predators, and nutritional condition.

4. Density and quality of alternate foods available to the predator.

5. Predator characteristics, such as its means of attack and food preferences.

As a last point about predators:

All living things are destined to die and be recycled as a part of the flow of energy through the life community or stated another way, a creature must feed, and sooner or later it will be fed upon (Allen 1979).

2.5 GENETICS (SPECIES CHARACTERISTICS)

The innate biological factors leading to the development of a population include spatial, reproductive, and behavioral traits specific to a species. These species characteristics are contained in the general category of genetics as one component of the centrum. At the present time managers may not be able to change the innate characteristics of a species but genetic engineering provides possibilities for the future. Genetic factors determine how a species responds to its environment and in this respect they are directly acting.

Spatial

All wildlife species inhabit an area. Each area contains varying densities of the species but in some areas, patches are unoccupied. Occupied and unoccupied areas make up the *geographical range*. A geographical range may be large as in the case for white-tailed deer or small as that of the Jemez Mountain salamander (Fig. 2.2). The geographical range provides an indication of the adaptability of a species to various climatic, topographic, and vegetative factors.

Within a geographic range an individual moves about in a *home range*. Home ranges can be large as for the mountain lion or small as for the cottontail. One part of the home range is the daily *cruising radius* or the distance an animal travels in one day in search of food and cover. This daily distance over a period of time defines a *center of activity*. The center of activity is generally an area where the habitat is higher in quality. For example, an Abert squirrel with a radio collar had a home range of about 45 ha (110 ac) but spent most time in a 7 ha (17 ac) area (Fig. 2.3). Home ranges can be any shape extending from approximately circular to elliptical. A home range and cruising radius roughly define an animal's *mobility*.

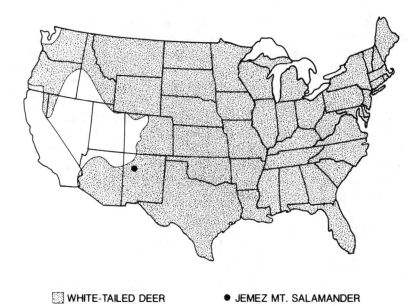

🔳 WHITE-TAILED DEER   ● JEMEZ MT. SALAMANDER

Figure 2.2. Geographical range of white-tailed deer (From U.S. Dept. of Inter. 1987) and Jemez Mt. salamander.

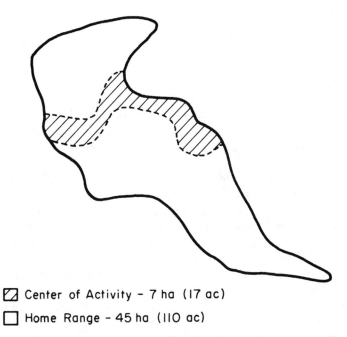

▨ Center of Activity − 7 ha (17 ac)

☐ Home Range − 45 ha (110 ac)

Figure 2.3. Home range and center of activity for an Abert squirrel (Author's original data).

Highly mobile animals are those who travel in distances of kilometers (miles) or those who use large areas of land. An animal of low mobility travels in distances generally measured in meters (yards) or hectares (acres). The area an animal uses in a home range correlates with body weight (Emlen 1973) and the size of the animal is crudely related to reproduction characteristics (Fig. 2.4). Large animals such as deer and elk have large home ranges (highly mobile) and low fecundity rates while small rodents have small home ranges (low mobility) but high fecundity rates. Species of low mobility are at a higher risk from habitat destruction than highly mobile species because of the scale of their home range in comparison to management activities such as timber harvesting.

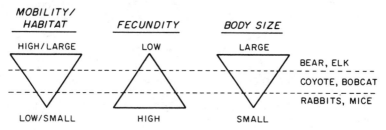

Figure 2.4. Reproductive characteristics as related to mobility and size of animal.

Reproduction

Breeding season, age of sexual maturity, gestation (incubation), longevity, and fecundity are all elements of a species' reproductive characteristics. These characteristics are innate and have a wide range of values.

Breeding season refers to the time of year mating can take place and is variable among species. Breeding season within species occurs at the same time so that there is maximum opportunity for the species to perpetuate itself. Time of breeding is controlled by day length (photoperiod), but breeding is also influenced by temperature and the general physical condition of an individual.

Breeding season within the geographical range of a species may differ; in West Virginia, the white-tailed deer breeds in November, but in Arizona breeding occurs in January. Animals producing many young (mice, rats, etc.) generally breed in the spring and the young are born in the spring or summer when the environmental conditions are most favorable for their survival. Animals, such as deer, producing only one or two young, breed in the fall with the young being born the following spring or summer.

Two general types of breeding behavior are evident within a breeding season. A *monogamous* species is one in which a single male and female pair for the breeding season and maintain strong social bonds. This contrasts to *polygamy* where one male or female breeds with more than one female or male. In polygamous species a social bond forms but is not as strong as in monogamous species. One type of polygamy is *promiscuity* in which one male breeds with many females or vice versa but no social bond is formed. For example, the bald eagle is a monogamous species, elk are polygamous, and cottontails are promiscuous.

Age at sexual maturity determines whether an animal can breed within a breeding season, although poor nutrition may delay its onset. Small mammals generally breed as juveniles. Deer, in contrast, born in the spring generally breed the fall of the next year (1.5 years of age) if they are in good physical condition.

The length of time a female is pregnant defines the *gestation* period. Species with short gestation periods are capable of producing the greatest number of young and some species can breed more than once in a breeding season. Gestation period in mammals is analogous to the incubation period in birds.

The longevity (age) of an animal species determines how many times and how many young a female produces in a lifetime. Longevity is of two types: physiological and ecological. *Physiological longevity* is a species' maximum life span under optimum conditions. Some examples of animal ages acquired from zoo records which should approach physiological longevity are presented in Table 2.3.

Table 2.3. Physiological longevity from zoo records (McWhirter 1981).

Species	Years
Deer mouse	4
Cottontail	10
Mountain lion	20
Cottonmouth	21
Elk	22
Bullfrog	30
Grizzly bear	32
Great horned owl	68

Ecological longevity is the actual or realized longevity of a species in the wild. Long-lived animals generally produce the fewest number of young per breeding season. Contrast the number of young per litter, number of litters per breeding season, minimum breeding age, and longevity of the cottontail rabbit with that of white-tailed deer (Table 2.4).

Table 2.4. Contrasting characteristics between cottontail rabbit and white-tailed deer.

Factor	Cottontail rabbit	White-tailed deer
Young/litter	4–6	1–2
Litters/year	2–4	1
Breeding age	3 months	1.5 years
Ecological longevity	3 years	8–10 years

The number of offspring produced per female of a species is *fecundity* (fertility). Some relative comparisons can be made about fecundity of animals (Table 2.5). The two categories of low and high fecundity are known as *"r-selection"* and *"K-selection"* (MacArthur and Wilson 1967). The former refers to a high intrinsic rate of increase (r) and the latter refers to the ability to maintain a population that fluctuates around a carrying capacity (K).

Table 2.5. General factors relating to low and high fecundity.

Low fecundity	High fecundity
High mobility	Low mobility
Long-lived	Short-lived
Large body size	Small body size
Population stable	Population fluctuates
Late succession	Early succession

Behavior

The behavior of animals affects their interactions with members of the same species particularly during the breeding season. Studies on birds led to the idea of a *territory* as a defended area which is part of an animal's home range (Howard 1920; Nice 1941). More recent definitions of territory explain that it is an exclusive area (Schoener 1968) or any spacing of animals that is more even than random (Davies 1978).

Different definitions have arisen because territory is better defined for some life-forms, such as birds, than it is for others, such as mammals. However, mammals do have territories but they are more difficult to detect and are not always associated with breeding as has been demonstrated for mountain lions in Idaho (Hornocker 1969). In this study, mountain lions had winter territories where all lions shied away from each other but they often crossed territories.

Because territory is an expression of space there has to be some way for the territorial animal to mark and identify the space. In mammals this is done with fecal material or urine but in birds territories are typically defended by displays (an act of aggression) or calls to mark a boundary; however, physical violence is rare because one individual always breaks off the encounter. Territories can be horizontal as is the case of some mammals or vertical for birds. Some large carnivores such as the wolf maintain a group territory and defend it against intruders (Mech et al. 1971).

There is evidence that the nature of territories changes as animal density changes (Brown 1969). At low animal densities territories are located in preferred habitat while at high densities some individuals are forced into less suitable or marginal habitat. Animals at low density defend larger territories but when density is high the territory is smaller (Kendeigh 1941).

Closely related to territory is *hierarchical dominance* which develops in social animals that form groups. These types of hierarchies are found in amphibians, birds, fishes, invertebrates, and mammals (Wilson 1975). Hierarchical dominance establishes a "pecking order" within the social group. This order assures the dominant males with mating opportunities as in wolves (Fox 1980), and with food as in white-tailed deer (Severinghaus 1972). In the case of white-tailed deer, the adult bucks dominate does while does dominate fawns in the order of feeding in winter deer yards.

Territories and dominance hierarchies are elements in the social behavior of some animal species and both appear to provide survival value though this is hard to document. A question concerning population biologists at present is whether these two factors limit animal populations as has been suggested by Wynne-Edwards (1962). In spite of the difficulties of sorting out behavioral problems associated with territories and hierarchies, there are some advantages for animal species possessing this trait:

1. Limited resources are preserved for the small number of individuals in the defended area.

2. Males are assured of reproducing with one or more females.

3. The defending males are the dominant males of the species; therefore, genes of the strongest are passed on to the young.

Defining the role of animals in the functioning of ecosystems sometimes leads to complex theories which are difficult to explain and understand. These difficulties often result in confusing usages among practitioners. One such theory is *ecological niche* (Grinnell 1943). The theory of a niche derives from the way an animal uses its habitat and the special adaptations either in physical structure or behavioral characteristics that facilitate this use. For example, squirrels climb trees, deer browse woody plants, and some rodents create burrows.

The niche of each species is developed through evolution. Each niche therefore is unique and distinct from all others. This in turn leads to an ecological principle which holds that no two species can occupy precisely the same niche and is commonly stated as the principle of *competitive exclusion*. Criteria defining niches are hierarchical and often overlap at higher levels.

One consequence of the principle of competitive exclusion is that if two species compete at the niche level then one species ultimately displaces the other or evolves to develop a new niche. Probably the easiest way to understand a niche is to examine how a particular species uses a resource also used by another species in the same geographical area. For example, the Abert squirrel eats the inner bark of ponderosa pine clipped from twigs. The porcupine also eats inner bark but from the upper trunk and limbs. Although the two species use the same food resource they differ in where they eat and in the size of stems used. The squirrel and porcupine both feed in trees. At the next lower level they feed on bark, so these two species have at least two levels of overlap in their requirements but as the levels are more narrowly defined the two species part company.

Within the geographical range of a certain species, local and regional movements occur coinciding with the onset of fall and spring. This two-way movement is called *"migration"*. Migrations are in some cases vertical as in elk migrating up mountains in summer to avoid heat or to obtain high quality food and thereafter descending to lower elevations in winter to snow-free areas where food is again available. Mule deer and moose also follow vertical movement patterns in many parts of North America.

Probably the best known migrations are horizontal as for example the yearly north/south flights of waterfowl (Linduska 1964). These yearly movements of waterfowl between breeding grounds in northern Canada and their winter grounds largely in the U.S. and Mexico have resulted in the delineation of regional management areas based on seven migratory flyways: Atlantic, Mississippi, Central, Pacific, Alaska, Canada, and Mexico. The Canadian and U.S. Fish and Wildlife Service have established a system of waterfowl refuges to provide food and cover for migrating birds. Migrations occur not only in terrestrial species but also in fish. The cyclic migration of salmon from freshwater to the sea and again to freshwater streams to spawn is well known.

Habitat Size

In recent years attention has been given to the size of habitat area an animal uses and this has developed into a relative distinction between two animal groups. Species with low mobility, high fecundity, are small in size, and that have all their requirements met in identifiable vegetation communities are habitat specialists or *alpha* species (Harris 1988b) (Fig. 2.4). Species large in size, with a low fecundity rate, high mobility, and requiring large landscapes are referred to as *gamma* species or habitat generalists. These two groups of animals will be discussed in more detail in Chapter 7.

2.6 RESOURCES

Resources include the food, cover, and water necessary to maintain basic physiological functions of an animal to survive and reproduce and subsequently for the offspring to become members of a population. Because of this important fact managers spend most of their time dealing with the resource (Re) factor of Equation 3 to maintain stable wildlife populations. The need for food, cover, water, and space for a particular species is as variable as the size, mobility, and niche of the species. Successful managers understand this variability and so direct their management activities to provide a diversity of resources to meet the needs of a variety of species. For management purposes resources can be separated into terrestrial and aquatic.

Terrestrial Resources

In 1798 Malthus proposed in his book *An Essay on The Principle of Population* that all animal species tend to reproduce to a point in excess of available resources so some resources limit population density. Food is an important factor needed by an animal to survive and reproduce. While it is possible to control populations by reducing cover, food is more likely to be a limiting factor because it maintains body heat.

Food

The food resource involves not only the availability of food necessary to obtain energy, but the conservation of this energy once it has been obtained. The consumption of food is determined by *availability* and *palatability*. *Availability* refers to what is obtainable during the various seasons of the year (Fig. 2.5). Availability also relates to the physical location of the food, such as browse out of reach to deer and elk.

Food *palatability* includes qualities of preference, nutritive content, and digestibility. Preference describes the hierarchical ordering of food (Fig. 2.6); the highest being the item selected more frequently than any other. This hierarchical priority can generally be ordered as:

1. High Value (Preferred)—eaten most frequently and in large amounts when available from an array of food plants.

2. Moderate Value (Staple)—second choice but still provides nutrition. Eaten in moderate amounts when an array of food plants is available.

3. Low Value (Stuffing-emergency)—fulfills short-term needs but generally low in nutritional value. Generally eaten in small amounts unless it is the only food plant available.

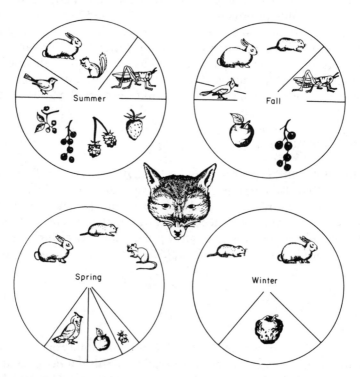

Figure 2.5. Food availability for the red fox differs by season (From Cook and Hamilton 1944).

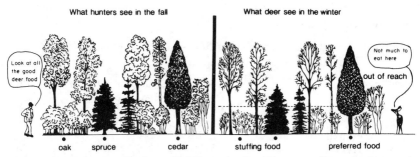

Figure 2.6. Deer see food items differently than humans. (From Jenkins and Bartlett 1959)

Food habits of various animals indicate that a variety of plants are consumed, and even when high value plants are available and being eaten, small amounts of low-value plants remain in the diet.

Nutrients are needed by all living organisms for body maintenance and reproduction but requirements vary by species. More information is available on the food requirements of wildlife species than any other topic in the wildlife profession. One has only to read the *Journal of Wildlife Management* from the early 1940s through the 60s to obtain an appreciation of the effort expended by biologists in determining food requirements.

Comprehensive publications are now available describing food habits at the broad level (Martin et al. 1951), regional level (Patton and Ertle 1982), and local level (Stephenson 1974). The nutritional requirements of game animals have been most extensively studied because the wildlife profession generally began by managing for hunted species such as deer, elk, turkey, rabbits, squirrels, and quail.

Due to the great variation in the nutritive content of plants, wild-life biologists are still seeking a single index of plant nutritional value. Animals need vitamins, fats, proteins, minerals, and carbohydrates. But protein seems to be the most likely indicator of plant nutritional quality. Some of the factors affecting the crude protein content of plants are soil moisture, canopy closure, soil nutrients, grazing intensity, and burning. While protein is a good indicator of plant quality, the complete nutritive value of a plant must be compared with the requirements of a species. Table 2.6 presents examples of the daily protein requirements for a few animals, including domestic livestock, for a comparison.

The study of plant nutrition by the chemist Justus von Liebig (1840) lead him to formulate the principle of limiting factors or the *law of the minimum*. Essentially the law states that the nutrient in least supply is the *limiting factor* for plant growth. Liebig's law of the minimum was extended to wildlife when research began to show that

any one of the resource factors—food, cover, or water—limits the
survival of wildlife species.

Table 2.6. Protein requirements for domestic and wild animals.

Species	Protein (%)	Reference	
Beef cattle	8.3–9.3	Nat. Acad. Sc.	1963
Sheep	7.6–8.2	Nat. Acad. Sc.	1964
White-tailed deer	6.7–16	French et al.	1956
Wild turkey	15–20	Halls	1970
Bobwhite quail	12–28	Nestler	1949
Fox squirrel	15–20	Nat. Acad. Sc.	1962

Furthermore, each species possesses a range of tolerance to its
environment, meaning there can be too much as well as too little of a
specific factor such as temperature or succession (Fig. 2.7). The
inability to cope with any factor which exceeds the limits of such a
range is often referred to as the *law of tolerance*. Any resource factor
exceeding the limits of tolerance for a particular organism becomes a
limiting factor for that organism. In short, all organisms require a
complex of resources and environmental conditions and have a range
of tolerance to any one of them (Odum 1971).

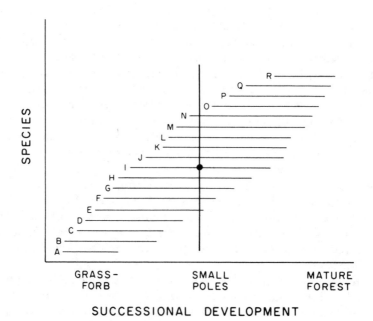

Figure 2.7. Range of tolerance to environment.

Water

Water is a resource necessary for animals to survive and reproduce and like nutritional requirements, water requirements vary by species (Table 2.7). Water is needed by different species for metabolism (drinking) in terrestrial mammals, as a medium for reproduction and rearing of young (waterfowl), or both (fish, salamanders, frogs).

Table 2.7. Water requirements of animal species.

Species	Liters/day	(Gallons/day)	Spacing Km (Mi)	Reference
Range cattle	28–38	(7.4–10)	—	Stoddart and Smith 1955
Sheep	0.95–5.7	(0.2–1.5)	—	Stoddart and Smith 1955
Mule deer	2.3	(0.6)	1.6–4.8 (1–3)	Bissell et al. 1955
Pronghorn	3.8–7.6	(1–2)	3.2–4.8 (2–3)	Payne and Copes 1988
Elk	18.9–30	(5–8)	1.6–4.8 (1–3)	Payne and Copes 1988
Moose	18.9–30	(5–8)	—	Payne and Copes 1988
Quail	2,840	(750/covey/yr)	0.8–1.6 (0.5–1)	Payne and Copes 1988
Chukar	2,840	(750/covey/yr)	0.8–1.6 (0.5–1)	Payne and Copes 1988
Turkey	1,890	(500/flock/sum.)	1.6–3.2 (1–2)	Payne and Copes 1988
Mourning dove	7.6–18.9	(2–5/flock/day)	4.8–8.0 (3–5)	Payne and Copes 1988
Pheasant	7.6–18.9	(2–5/flock/day)	0.8–1.6 (0.5–1)	Payne and Copes 1988
Songbirds	3.8–7.6	(1–2/group/day)	0.4–0.8 (0.25–0.5)	Payne and Copes 1988

Species needing free water for metabolism must find it in their home range. Examples of research to determine water requirements of wild species are provided by Wesley et al. (1970) for antelope and Turner (1973) for bighorn sheep. Species can adapt to regions with low precipitation by conserving water; by going underground (rodents) or seeking shade to avoid heat (deer and elk); by morphological adaptations (body size and shape as in the jackrabbit); or by using metabolic water (kangaroo rat).

Research has shown that precipitation in the form of rainfall has an effect on the distribution and reproduction of some species; for example, in Arizona the number of young quail is related to precipitation (Fig. 2.8) (Gallizioli 1965); in South Carolina drought reduced the number of bobwhite young per female (Rosene 1969); in California when rainfall is less than 16 cm (6.3 in) the number of young quail are reduced (McMillan 1964); and deer densities in Texas are related to the precipitation of the previous year (Teer, Thomas, and Walker 1965).

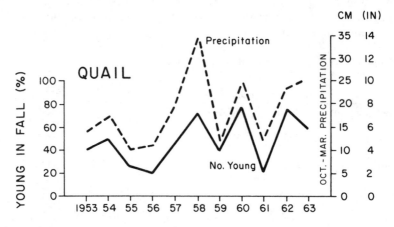

Figure 2.8. Number of young quail is related to precipitation (Gallizioli 1965).

Many factors affect the distribution, availability, and potability of water for use by wildlife and, excluding rainfall patterns, most result from mankind's activities such as dams and reservoirs, livestock use, wetland drainage, pollution in many forms (sewage, acid rain, chemicals, oil spills, etc.). All these are part of the web of indirect factors.

In studies to determine the food and water requirements of a species investigators must consider the following questions:

1. What is available by season?

2. What is accessible (height of browse plants, plants in rocky areas, waterholes, lakes, etc.)?

3. What is palatable and potable?

4. What nutritional levels of plants and amounts of water are needed for physiological processes at each season?

5. What is the individual adapted to eating (physical adaptations)?

Cover

Once energy has been obtained through a proper diet, an animal must then conserve this energy through some form of cover. *Cover* is the physical habitat component or landscape feature providing protection from hazards (Ha) and predators (Pr). The cover requirements of individual species are as variable as species' food requirements. Cover is furnished by vegetative structure (trees, shrubs, grasses, and forbs) (Fig. 2.9), as well as by stages within life-form (seedling, sapling, etc.), topographic features (aspect, hills, valleys, soil, etc.), and water.

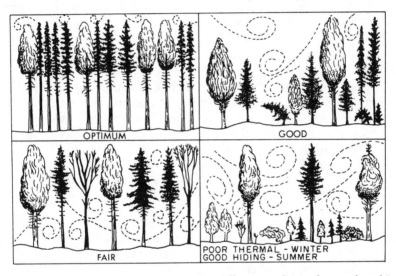

Figure 2.9. Vegetation structures provides different qualities of cover for white-tailed deer (From Verme 1965).

Although cover provides protection there are special types of cover associated with specific functions for each wildlife species. Some of the cover types that have been identified in the literature are thermal (heat and cold) cover for deer and elk, roosting cover for turkeys, nesting cover for grouse, resting cover for quail, and escape cover for squirrels. For example, thermal cover for elk in Oregon and Washington is defined as any stand of coniferous trees 12 m (40 ft) or more in height with an average canopy closure exceeding 70%; and for deer, coniferous trees either pole or sapling size, at least 1.5 m (5 ft) tall with 75% canopy closure (Thomas, Black, et al. 1979). These various forms of cover are not mutually exclusive; for example, hiding cover for deer and elk often serves as thermal cover.

Like food, cover is subject to annual and long-term changes (ecological time), as in the change in location of forest stands. A *stand* is defined as trees sufficiently uniform in composition, age, structure, and spatial arrangement to be distinguishable from adjacent communities (Ford-Robertson 1971). Care must be taken in describing cover in general terms. For example, thermal cover for elk in the summer may not exist in the same stand in the winter if the stand is composed of deciduous trees.

In addition to general types of cover, for some species there is a subset that can be identified as *shelter*. Shelter is associated with the rearing of young, such as in a nest. For example, the Abert squirrel builds a nest in a single ponderosa pine where young are born. This nest occurs in a tree surrounded by a group of trees. The tree group and individual nest tree provide cover from weather and predators, but the nest (shelter) provides additional protection for the young. Not all species have specific shelter requirements but some types of shelter used are rock crevices, tree cavities, hollow logs, ground dens and burrows.

When discussing cover and shelter requirements of specific species, *CAUTION* is urged in not transferring the requirements across geographical boundaries or vegetation types unless studies or experience have shown that the requirements are similar in the hierarchical levels being compared.

Spatial Attributes

The minimum arrangement of food and cover in close proximity to one another for a species is termed *juxtaposition* (King 1938). Food available in one area (opening, meadow, etc.) and cover (tree stand, shrubs, etc.) nearby constitute a *habitat unit* and clearly reduce the energy required for and the natural hazards associated with the search for food away from cover. A habitat unit is an abstract entity and is not always easily defined. Equally clear is that the expenditure of energy to search for food and cover cannot be greater than that obtained from the food source or the animal loses weight. Another term related to juxtaposition is *interspersion* which is the distribution of habitat units for a species over the landscape. The arrangement of stands to provide food and cover in a checkerboard pattern provides landscape diversity to meet the habitat requirements for a large number of species.

Soil

While resources (Re) directly affect an animal's chance to survive and reproduce, soil nutrients directly affect the ability of plants to survive and reproduce. Therefore, soil is one of the indirectly acting factors in the Web of an animal. It is well documented that agriculture production is directly related to the level of nutrients in the soil. Nutrients in the soil are transferred to nutrients in plants which are in turn transferred to animals when consumed. If the nutrients in plants are in sufficient quantity and proper relationships the consuming animal remains healthy. This is as true for wildlife as it is for domesticated animals. Studies in Missouri have shown that the highest density of game animals (rabbits, raccoons, etc.) is in areas of the highest soil fertility (Crawford 1950).

When mule deer are removed from low-fertility California chaparral and fed optimum diets in confinement they reach a larger size after the first year of feeding than some adults in the wild attain in a lifetime (Dasmann 1981). Results of research in Pennsylvania on white-tailed deer followed the same trend as in California (French et al. 1956). Other research has shown that the number of young produced by cottontails (Hill 1972) and white-tailed deer (Cheatum and Severinghaus 1950) is correlated to increasing soil fertility. In other studies on white-tailed deer body weight has been shown to be related to soil fertility (Smith, Michael, and Wiant 1975).

Allen (1974) emphasized that in terms of fertility, soil and water are equivalent since water is needed to make the soil nutrients available. The use of measured amounts of fertilizers to enrich water for fish production is a common practice in farm ponds. Nitrogen (N), phosphorus (P), and potassium (K) fertilizers are used to increase phytoplankton which are eaten by zooplankton or minnows that are in turn eaten by bass, bluegills, or trout. However, care must be used in fertilizing ponds so that the fertilizer does not become a polluter.

Managers must remember that soil is a major factor influencing vegetation and that vegetation is the driving force for wildlife habitat which translates into food and cover. Perhaps the most important soil-wildlife relationship is that wildlife production, like any other crop of the land, is directly proportional to fertility of the soil and that land which is not sufficiently fertile to produce vegetation for food and cover is marginal for wildlife (Denney 1944). Dr. Hugh Bennett (1950) summarized the soil, water, wildlife relationship this way:

All land—with the water which nourishes it—is wildlife land. All soil and water conservation—when properly planned and carried out on the land—is wildlife conservation.

Aquatic Resources

The definition of wildlife includes species of amphibians, fishes, mammals, and reptiles that require water either as a cover medium, food source, or both, in streams flowing through or lakes and ponds situated within forested areas. Excluding energy from the sun, the primary force controlling the forested ecosystem is trees and similarly the force controlling the aquatic environment is water; therefore, an understanding of the properties of water is a necessary requisite for management.

In their classic text on freshwater biology Needham and Lloyd (1937) summarize the importance of water this way:

> Water, the one abundant liquid on earth, is, when pure, taste-less, odorless and transparent. Water is a solvent of a great variety of substances, both solid and gaseous. Not only does it dissolve more substances than any other liquid, but, what is more important, it dissolves those substances which are most needed in solution for the maintenance of life. Water is the greatest medium of exchange in the world. It brings down the gases from the atmosphere; it transfers ammonia from the air into the soil for plant food; it leaches out the soluble consti-tuents of the soil; and it acts of itself as a chemical agent in nutrition, and also in those changes of putrefaction and decay that keep the world's available food supply in circulation.

Water Properties

The properties of water important in ecological processes are *density*, *viscosity*, and *transparency*. Water has its highest density at 4°C (40°F); therefore, it expands and becomes lighter above and below this point, so ice floats allowing life to continue in the deeper water (Hickman 1966). This unique property prevents some lakes from freezing solid and also thermally stratifies others. Related to density is specific heat. Water is capable of storing large quantities of heat with a relatively small rise in temperature. This property prevents wide seasonal temperature fluctuations and tends to moderate local environments. Thus, water in ponds and lakes warms slowly in the spring and cools slowly in the fall.

Viscosity is the source of resistance to objects moving through water. The faster an aquatic organism moves through water the greater is the stress placed on the organism's surface and the volume of water that must be displaced. Replacement of water in the space left behind by the moving animal adds drag on the body similar to the drag on an airplane moving in air. Fish, such as trout, with a short, rounded front and a rapidly tapering body, meet the least resistance in water.

Below the water's surface water molecules are strongly attracted. Above the surface, air molecules are less strongly attracted to the water, therefore surface water molecules are drawn down into the liquid, creating a surface tension and forming a water skin. This skin can support small objects and animals such as water striders and spiders. Some mayflies and caddisflies, on the other hand, cannot escape from the skin and so become food for fish.

Transparency is the property that allows light to penetrate water. Aquatic plants need energy from the sun for photosynthesis the same as terrestrial plants. Any impediment to clearness of water reduces photosynthesis and therefore the amount of plant food available for aquatic organisms. The degree of clearness is termed *turbidity* which is caused by the increase of soil particles of silt and clay or chemicals (Bennett 1986). The inorganic soil particles are, however, the leading cause of turbidity derived from the surrounding watershed.

Water contains oxygen, the amount of which is directly related to temperature. As water temperature decreases its capacity to hold oxygen increases. At 0°C (32°F) water can hold about 15 ppm (parts per million) of oxygen but drops to 7 ppm at 40°C (104°F) (Welch 1948). Oxygen can be absorbed from the air or derived from photosynthesis by aquatic plants. Warm-water fish need at least 5 ppm of oxygen while cold-water species such as trout require amounts greater than 5 ppm (Bennett 1986). Acidity of the water affects the ability of fish to extract oxygen. A pH range between 6 and 8 seems to be about right for the best fish production.

Freshwater Habitats

Freshwater habitats are of two types: *lentic*, meaning calm water (lakes, ponds), and *lotic*, meaning running water (springs, streams, rivers). Both have characteristics that influence the aquatic life they contain. These waters can be classified into three categories based on temperature (Forbs 1961):

1. Cold water, <18°C (65°F) in summer; habitat for salmon, trout, whitefish, and grayling.

2. Cool water, 18–24°C (65–75°F); habitat for smallmouth bass, northern pike, walleye, muskellunge, sturgeon, and shad.

3. Warm water, >24°C (75°F); habitat for largemouth bass, sunfish, suckers, and carp.

Fish are sensitive to temperature because it affects their oxygen supply and rate of growth. The ability to tolerate any given temperature is species dependent. For example, rainbow and brook trout can thrive in temperatures of 21°C (70°F) and may be able to tolerate higher temperatures for short periods of time (James, Mechean, and Douglas 1944), but they survive better in colder water (Everhart and Youngs 1981). This is probably because there is more oxygen in cold water. The ability to tolerate temperature changes seems to depend on whether the increase or decrease is rapid or slow; slower changes allow time for acclamation.

Lakes: In lakes as the surface temperature of the water rises in the summer a warm layer, the *epilimnion*, develops (Fig. 2.10). Oxygen and food are plentiful in the epilimnion as a result of the exchange between the water surface and the air and because sunlight is available for food production. This layer circulates but it does not mix with the lower layers.

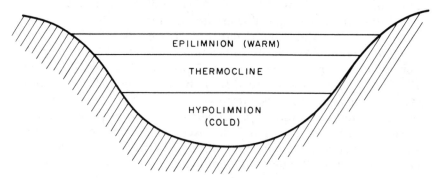

Figure 2.10. Lake stratification in the summer.

Beneath the epilimnion is the *thermocline* which is a thin layer, the temperature of which drops rather rapidly with depth, about 1°C (1.8°F) per meter (Black 1968). The thermocline is also defined as the plane of maximum rate of decrease in temperature (Reid 1961).

The *hypolimnion* is the lowest and coldest layer in a lake. If the thermocline is below the photosynthetic zone then light is reduced and the oxygen supply is depleted in the noncirculating hypolimnion. This condition most commonly occurs in summer leading to lake stagnation. In the fall the epilimnion cools until it is the same temperature as the hypolimnion so circulation of oxygen and nutrients in the whole lake is restored by what is called the *fall overturn*. The overturn is caused by the mixing of the water layers by wave action.

When air temperature is sufficiently cold the temperature of the surface water drops to below 4°C (40°F), causing the water to expand, become lighter, and then freeze. If snow accumulates on the surface of the ice, light is cut off stopping any photosynthetic processes. As a result, oxygen is depleted leading to winter kill of fish.

In the spring the water warms and along with wind action creates a *spring overturn* before summer stagnation sets in again. Thermal stratification does not occur in running water so the plant and animal species are quite different between lentic and lotic systems. Oxygen and temperature are significant qualities in both systems, but water current is a factor only in streams except in those lakes and ponds which are small relative to the volume flowing between inlets and outlets. But in general, lakes and ponds are closed systems.

Because lakes have little if any water flow (closed system), *eutrophy* sets in—that is, lakes move from an unproductive state with low levels of plant nutrients (*oligotrophic*) to a state with high levels of nutrients (*eutrophic*). All lakes experience some natural eutrophication as a result of runoff containing nitrates and phosphates from their drainages but accelerated eutrophication leads to pollution when these nutrients are at high concentrations. Highly eutrophic waters are unsuitable for many aquatic species.

A new lake is oligotrophic, containing low concentrations of nutrients and plankton in the limnetic zone leading to low fish populations (Fig. 2.11). The *limnetic* zone is the zone of light penetration and consists of two areas, the *littoral* zone containing rooted plants, and the *open water* zone containing floating plants. Beneath the limnetic zone is the *profundal* zone where light does not penetrate. As the lake ages, nutrients become more plentiful, the profundal zone accumulates sediments so becomes shallower, and fish density increases due to increased vegetal growth resulting from increased nutrients and light.

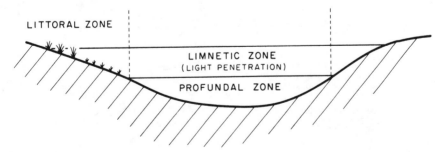

Figure 2.11. Biotic zones of a lake.

Streams: The biological characteristics and development of lakes are completely different than those of running water. The current of a stream determines the stream's physical characteristics. The velocity of a stream is controlled by gradient, width, bottom roughness, and shape (straight or meandering). Nutrients entering a stream directly from its headwaters and from adjacent areas pass through the system to the mouth where they are deposited (open system).

A multitude of micro-environments are formed within a stream system arising out of the combination of pools and riffles produced by rocks, boulders, and bottom formation. These combinations of physical characteristics affect the amount of oxygen and carbon dioxide in the system. In addition, the chemistry of streams is strongly influenced by the soil, vegetation, and land uses along their borders. Water percolating through the soil picks up nutrients which in turn affects the acidity and nutrient qualities of the water. Streamside or riparian vegetation is a major determinant of water temperature by

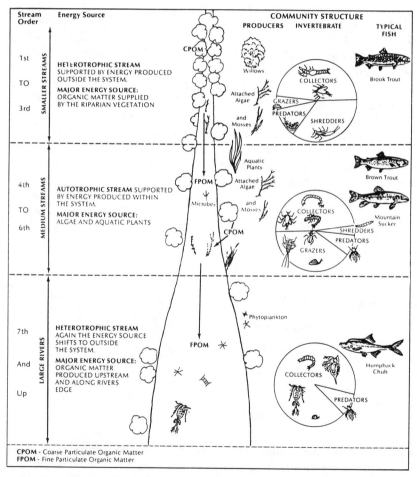

Figure 2.12. The river continuum concept (Hoover et al. 1984 adapted from Cummins 1973).

exposing the water to sun or shade. In addition, riparian vegetation provides micro-habitat for aquatic insects completing part of their life cycle on land.

Vannote et al. (1980) have proposed the river continuum concept whereby the biological and dynamic processes of the stream are determined by the physical gradient and drainage network (Fig. 2.12). Basically the concept means that streams are small at the headwaters and get progressively wider, but in the process there is a change in the kinetic energy which in turn determines the type of organisms that are present in any given stream reach. The concept is not unlike the continuum concept proposed for vegetation succession (Gleason 1939).

In the headwaters of a stream riparian vegetation is important because it provides shading and produces detritus. In the middle reaches algae and rooted plants provide the nutrients. As the stream widens the effect of riparian vegetation is reduced, but there are large quantities of upstream organic matter entering the system. All the interactions between a stream's physical characteristics and its biotic components produce a state of dynamic equilibrium similar to that produced in the successional processes on land (Fig. 1.6).

Riparian

Vegetation immediately adjacent to streams or along the edges of lakes and ponds is characterized by plant species and life-forms differing from those of the surrounding forest and is termed *riparian* (Fig. 2.13). Riparian tree composition depends on elevation but typically consists of deciduous trees from the genera Populus, Acer, Salix, and Alnus. The marked contrasts between riparian and upland vegetation, produces structural diversity and edge characteristics enhancing its utility for wildlife.

Because of their long narrow shape, stream riparian areas contribute few acres to the total available habitat; however, they are highly productive so their value to wildlife is well out of proportion to their small area (Thomas, Maser, and Rodick 1979a). The highest density of birds ever recorded in North America was found in riparian vegetation in Arizona (Johnson 1970). The density of 1324 pairs per 40 ha (100 ac) of noncolonial birds is also among the highest densities in the world including tropical forests.

Not only are riparian areas important to terrestrial animals, but they also control the associated lotic habitat for amphibians and fish. Canopies provide shade, root systems stabilize banks, and plant detritus and insects provide nutrients for stream organisms (Meehan, Swanson, and Sedell 1977). Riparian areas create an oasis effect in dry

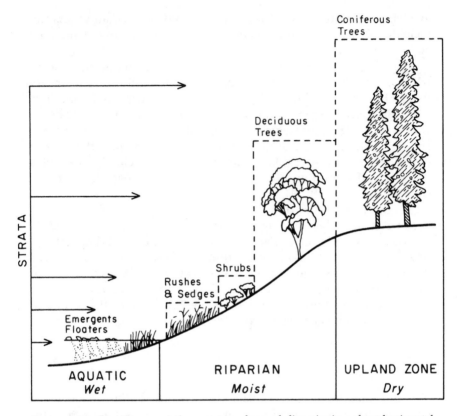

Figure 2.13. Riparian vegetation creates edge and diversity in a short horizontal distance.

lands and because of their cooler microclimate and free water they are major resting places for many north-south migrating birds. In addition, tall trees along a stream or watercourse at times create a fluelike condition causing an updraft which brings in air underneath the vegetation and over the water.

Riparian vegetation provides a natural travel corridor for local species such as deer, raccoons, turkey, and bear (Fig. 2.14). In eastern United States the riparian zone is not as distinct as it is in the West where higher rainfall east of the 100th meridian results in more vegetation perpendicular to the aquatic region of the riparian zones. The increased moisture and deeper soils integrate the vegetation types lessening the distinction between where riparian starts and ends.

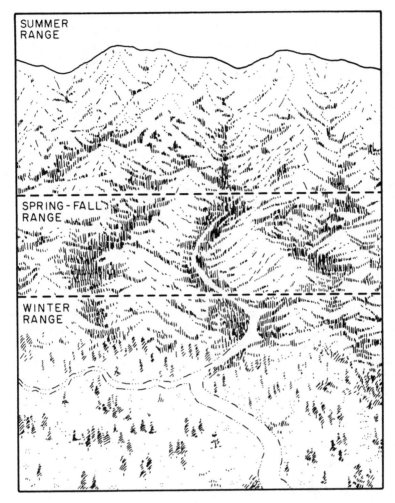

Figure 2.14. Riparian vegetation is used as a travel corridor by many wildlife species (Thomas, Maser, and Rodiek 1979a).

2.7 HUMANS

It is obvious that resources are being used more extensively and intensively by expanding human populations. In many parts of the world humans are now in conflict with and will increasingly be in conflict with other animal species for the use of the earth's limited resources. An understanding of the conflicts between humans and wildlife is one of the keys to the well-being of both.

There is no doubt that humans directly (legal and illegal hunting) and indirectly (habitat destruction) affect the reproduction and survival of a large number of animals. Also there is something about the presence of humans which causes undomesticated animals to flee or to maintain their distance. Animals can become adapted to human presence after several generations of being in a local population where the interaction occurs, but these interactions often lead to problems as animals such as deer first become tame and then a nuisance.

Man as a hunter belongs in the centrum of an animal, but man as a destroyer of habitat properly belongs in the web. While hunting, both legal and illegal, affects animals directly, we can assume that legal hunting, where it is part of a continuing system, is under control, but illegal hunting is a different situation. The direct effect from habitat destruction is a loss of resources but man remains indirectly the cause of the effect on wildlife.

Habitat Destruction

Habitat destruction by humans undoubtedly poses a severe indirect impact upon wildlife. But care must be taken in describing destruction because the loss of habitat for one animal is often a gain for another. So destruction must be considered in terms of the species affected. However, there are three forms of habitat destruction that adversely affect all wild species:

1. Urbanization.

2. An increase in clean agriculture (removal of all plant material).

3. Soil loss.

Conversion of forest land to cropland and urban uses such as roads and reservoirs, is projected to result in another 5% of the U.S. land sections occupied by the year 2010 with the most significant being conversion of bottomland hardwoods to cropland in the lower Mississippi River area (National Academy of Science 1982). In 1937 there were an estimated 4.8 million ha (11.8 million ac) of bottomland hardwoods in that region; about 2.1 million ha (5.2 million ac) remained in forest in 1978, with a projection that another 0.5 million ha (1.3 million ac) will be lost by 1995. In Florida alone, 60,700 ha (150,000 ac) of forest land are converted each year to urban use (Harris 1988c).

Large losses in habitat for big game species have been documented in the West. For example, in Wyoming a state of wide open spaces, it is estimated that from 1967 to 1977 approximately 388 km^2 (150 mi^2) of forest land were urbanized and lost as deer and elk habitat (Rippe and Rayburn 1981). In the Southwest, riparian vegetation consisting of large deciduous trees is dwindling due to the activities of woodcutters, recreationists, grazing, and an increase in home construction. Only in the past 15 years, riparian areas have been recognized as important and critical habitat for many species so loss of riparian habitat has not been documented adequately due to lack of long-term inventory data.

As human populations increase there must be a corresponding increase in living, work, and commercial space as well as in food growing areas, railroads, and highways for transport. Since the land base is fixed, such increase must necessarily be appropriated from land used by wildlife species. The loss of habitat to urbanization will inevitably proceed as long as human populations continue to increase.

Illegal Hunting

The impact of illegal hunting (poaching) on populations of wildlife species is difficult to determine. Market hunting as it existed in the 1800s to the 1930s is no longer a threat. All U.S. and Mexican states and Canadian provinces to varying degrees now have conservation or other law officers in place to enforce hunting and fishing regulations. In some countries special antipoaching units exist and the public, particularly in the United States and Canada, is encouraged to report violations.

Estimates from some midwestern states in the United States indicate poaching may exceed the legal harvest (Bennitt and Nagel 1937). Approximately 10,000 white-tailed deer are illegally killed each year in Missouri (Witter 1980). Studies in Idaho have provided estimates that in 1967 out-of-season deer kills were 8,000 at a time when the legal harvest was 66,000 mule and white-tailed deer (Vilkitis 1968). On the Kaibab Plateau in Arizona a mortality study using radio-collared mule deer indicated that poaching could account for as much as 10% of the loss of instrumented deer (McCulloch and Brown 1986). The evidence while not well quantified does confirm that poaching continues to be a problem in some areas. The problem is aggravated by evidence that local residents often tolerate illegal hunting (Decker, Brown, and Sarbello 1981).

2.8 ENVIRONMENTAL DENDROGRAM

Populations are constantly changing from year to year and the direct factors affecting them generally are known (Table 2.2). A systematic analysis of these factors by using the theory of environment provides an approach to analyzing population problems. Andrewartha and Birch (1984) have provided a way to analyze population problems through the use of a *species environmental dendrogram.*

The directly acting factors affecting the chances of an Abert squirrel to survive and reproduce are listed in Figure 2.15. These factors are the same as those mentiond in Conceptual Equation 3. Under each factor is one element or a set of elements that directly act on the species. The centrum is the first level of influence but there are also second level environmental influences (the web) affecting the first. For example, one of the diseases affecting the squirrel is bubonic plague but the plague is transmitted by fleas which are also carried by other small rodents. These small rodents are in the third level of influence. Other levels can be added as more information becomes available to trace the source of the problem. By listing all the direct

SPECIES : <u>Abert</u> <u>Squirrel</u>

CENTRUM	LEVEL 1	LEVEL 2
1. HAZARDS:		
ROADS		
SNOW AND ICE		
2. PREDATORS:		
GOSHAWK		
REDTAIL		
3. DISEASES:		
BUBONIC PLAGUE	FLEAS	RODENTS
4. GENETICS:		
BEHAVIOR	INTERACTION WITH OTHER TREE SQUIRRELS	
5. HUMANS:		
HUNTING	SEASONS	
6. RESOURCES:		
FOOD AND COVER	TREE SIZE	SILVICULTURE
PONDEROSA PINE,	TREE DENSITY	FIRE
GAMBEL OAK	TREE GROUPING	
	REGENERATION (SEED CROP)	MOISTURE TEMPERATURE
FOOD	SPORE DISPERSAL	RODENTS
FUNGI		(INCLUDING SCIABE)

Figure 2.15. Environmental dendrogram for the Abert squirrel.

and indirect factors affecting an animal's chance to survive and reproduce, a weak link may be found where management might intervene to correct the problem.

The species dendrogram is a valuable tool for identifying the indirect factors affecting an animal's well-being. In developing a dendrogram for any species it is best to list only those factors having a significant influence. Unless restraint is exercised the dendrogram can become so cluttered it will be useless. The same procedure can be used to analyze habitat problems with the exception that most of the factors affecting a specific habitat will be from the web of indirectly acting factors.

2.9 SUMMARY

The modern understanding of animal population dynamics was inaugurated in the late 1920s when Chapman formulated a conceptual equation to explain populations as a function of biotic potential and environmental resistance. Aldo Leopold later separated environmental resistance into welfare and decimating factors. Insight into the factors affecting animal populations culminated in 1984 with Andrewartha and Birch's comprehensive theory on the environment of an animal.

Environment consists of all those elements (the umwelt) affecting an animal's chance to survive and reproduce. The notion of a centrum made up of six factors which directly act upon wildlife—hazards, diseases, predators, genetics (species characteristics), resources, and humans—surrounded by a web of indirectly acting factors is presented as a working hypothesis. The six factors replace Chapman's equation and Leopold's welfare and decimating factors in explaining the environment of an animal. While environment affects an animal's chance to survive and reproduce, habitat is the environment of and the specific place where an organism lives.

Hazards include all those factors that can injure or kill an animal or so challenge an animal's physiological capacity as to prevent it from reproducing successfully. Hazards include the natural factors of fire, rain, and wind as well as man-made structures such as roads and fences. Diseases disrupt the normal physiological functions of an animal. While some diseases in the wild are confined to wildlife species, managers must also be concerned with those that can be transmitted to humans, livestock, between species, and vice versa. A third decimating factor is predation. Predators and prey have evolved together but a one cause–one effect relationship is too simplistic to describe the relationship since predators can turn to buffer species as food items to support predator populations.

Each animal species is uniquely distinguished by a variety of genetic (innate) characteristics that influences the extent and nature of home range, breeding season, longevity, territory, and niche, which are collectively referred to as species characteristics. Another of the centrum factors is resources. This is the factor most amenable to management. Resources include not only food and water to obtain energy to properly carry out physiological processes but the conservation of this energy by adequate cover and the proper juxtaposition and interspersion of both. Habitat destruction is the indirectly acting factor while both legal and illegal hunting are the principal directly acting human factors affecting wildlife populations.

One way to analyze problems involving the direct and indirect factors affecting an animal's chance to survive and reproduce is by a species environmental dendrogram. If properly developed the dendrogram can identify areas for management action. The same approach can be extended to habitat analysis.

2.10 RECOMMENDED READING

Andrewartha, H. G., and L. C. Birch. 1984. *The Ecological Web: More on the Distribution and Abundance of Animals.* Univ. of Chicago Press, Chicago, IL.

Carson, R. 1962. *Silent Spring.* Fawcett Publications, Greenwich, CT.

Dice, L. R. 1952. *Natural Communities.* Univ. of Mich. Press, Ann Arbor.

Everhart, W. H. and W. D. Youngs. 1981. *Principles of Fishery Science.* Cornell Univ. Press, Ithaca, NY.

Klopfer, P. H. 1969. *Habitats and Territories: A Study of the Use of Space by Animals.* Basic Books, New York, NY.

Leopold, A. 1933. *Game Management.* Charles Scribner's Sons, New York.

Pennsylvania State University. 1986. *A Trout Stream in Winter.* Movie. 25 minutes. University Park, PA.

Rudd, R. L., and R. E. Genelly. 1956. *Pesticides: Their Use and Toxicity in Relation to Wildlife.* Game Bul. No. 7, Calif. Dept. of Game and Fish, Sacramento.

Stokes, A. W. 1974. *Territory.* Benchmark papers in animal behavior. Vol. 2. Dowden, Hutchington and Ross, Stroudsburg, PA.

U.S. Dept. of Agriculture. 1979. *Habitat: A Special Place.* Movie. 30 minutes. National Audio Visual Center, Capital Heights, MD.

Chapter 3

POPULATION DYNAMICS

Conceptual Equation 3 (Table 2.2) focuses on factors that influence an individual animal's chance to survive and reproduce. Individual animals, however, do not exist alone, but are always part of a community organization (Fig. 1.2). Therefore, all species function as a population within a habitat. Populations can only increase or decrease as a result of four factors: natality, mortality, immigration, and emigration. Such a simple statement does not describe changes in the population over time, which is the concern of *population dynamics*.

The study of populations is fascinating but can be complicated. A detailed population analysis requires mathematical tools including calculus which deals with rates of change and limits. However, in this section I approach population dynamics largely on an intuitive basis depending only slightly on advanced mathematics.

3.1 POPULATION CHARACTERISTICS

A population (P) is a social, interbreeding group of animals occupying space and characterized by such qualities as natality (birth rate), mortality (death rate), and density, which are not characteristic of individuals. Fecundity of individuals underlies the natality of a population. *Natality* is the number of new individuals added per year in a population and is expressed either as a crude birth rate or a specific birth rate. A *crude birth rate* describes the number of individuals born per unit of population; for example, 10 births per 100 individuals. *Specific birth rate* adds a factor such as age; for example, the number of young produced per year by 3-year-old females.

The number of individuals in a population dying during a given period of time, usually a year, is the *mortality* of that population. Mortality, is expressed as a rate per unit such as number per 100, 1000, etc., and can be given for the total population or specific age classes. *Immigration* and *emigration* obviously refer to animals which join a group from an outside breeding population or leave a breeding population to join another. If natality and immigration in combination are higher than mortality and emigration, a population will increase; otherwise it will decrease. In any case, the number of animals present per unit of area is reflected in *population density* (P_D) as

$$P_D = \text{Immigration} + \text{Natality} - \text{Emigration} - \text{Mortality}$$

Two types of density can be defined and it is important to distinguish between them. *Crude density* is the number of animals per unit area regardless of what resources and other factors the area contains. An example is the number of deer per square kilometer (km²) (247 ac) or square mile (mi²) (2.59 km²) even though part of the area may include a town and roads which are not deer habitat. On the other hand *ecological density* includes only those areas considered suitable habitat for a particular species. An example is the number of Abert squirrels per hectare in ponderosa pine.

The density of a population is given in number of animals per unit of area but there is no simple measure for *distribution*. Three different areas may have the same population density but one may be homogeneously distributed, another randomly distributed, while the third may be distributed in a randomly clumped manner (Fig. 3.1). The pattern of animal distribution is important because pattern influences the manager's choice of inventory technique.

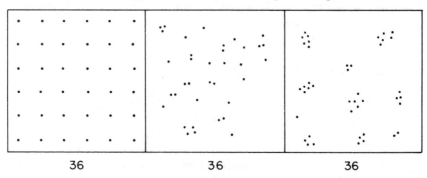

| 36 | 36 | 36 |

Figure 3.1. Populations with the same density but three distributions.

Populations are of two types: natural and local (Andrewartha and Birch 1984). A *local population* is a group of animals restricted to the use of a particular area. Thus a local population may exist on the side of a mountain or on several mountains, but use the same place consistently over time. Local populations probably arise out of the imprinting of young to the area thereby enhancing survival and reproduction.

Occasionally one or more animals in a local population mate with an individual from another local population so there is exchange of genetic material. In West Virginia I used to hunt a local population of black, eastern gray squirrels. This population was confined to an area on one side of a mountain containing an oak-hickory forest. Evidently the continued survival of this melanistic black phase population was possible because there was very little or no breeding with other local populations of normal gray squirrel color.

A *natural population* is the set of all local populations that are not separated by geographical or ecological barriers which prevent breeding. In some cases the natural population may encompass large areas such as the species' entire geographical range but as long as there are no barriers to breeding between local populations the definition holds. The natural population typically displays a roughly steady-state density without large yearly fluctuations despite the fact that local populations may be high in one area and low in another. Thus, it is possible for a local population to go to extinction while the natural population continues to survive and reproduce. This is an important concept because from it the risk of extinction of a natural population can be determined.

Recent legislation has introduced the term "viable population" but no clear definition is given. Biologically, a *viable population* is the number of individuals sharing a common gene pool and able to reproduce and maintain the population from one generation to the next as in a local population. Viable population concepts are only beginning to appear in the literature so there will undoubtedly be a time of testing, evaluating, and changing definitions and procedures of measuring before the concept is fully mature. The question that must be answered is: What is the number of local populations necessary to maintain a natural viable population? The question is important for managing wildlife because of the tradeoffs that may be necessary to maintain viable populations of other species in the same area.

What options do managers presently possess to guide them in the planning for and management of a viable population of a given species? Two seemingly appropriate techniques are currently in use: *estimated protection levels* and *risk analysis*. The former is a scheme to identify hierarchical levels of protection at abstract levels of populations. At present, nine levels of protection have been established (Table 3.1). One of the major problems in assigning a population level to one of the nine categories is making a reasonable estimate of the population density. Clearly, when populations are high and well distributed the probability of maintaining a viable population is high, and equally clearly the situation is reversed in the case of low, dispersed populations, but much uncertainty arises in assigning species populations to some of the intermediate levels.

When uncertainties in making a decision occur a *risk analysis* may help in making management decisions. Risk analysis deals with two types of uncertainty: Scientific and decision making (Marcot 1986).

Table 3.1. A hierarchical scheme to identify levels of protection associated with an abstract level of population (Schonewald-Cox 1983).

	Description
Level 1.	Individual survival—low likelihood of survival for any time beyond a few decades. Several individuals or pairs isolated on the area with no interchange with the species off the area.
Level 2.	Family survival—survival likely for several decades. A family, social group, or small population isolated on the area with a total of 10–20 adults.
Level 3.	Family resilience—survival for a half century or longer likely. Several reproductive or social groups partially to fully isolated in the area with a total of 20–50 adults.
Level 4.	Population resilience—continued existence on the order of many decades is likely. A well-distributed local population, but fully isolated from the rest of the species. Adult numbers range from 50 to several hundred.
Level 5.	Short term adaptability—continued existence beyond a century is likely. A well-distributed local population with an effective biological population (Ne) in the low to mid 100s.
Level 6.	Mid-term adaptability—continued existence beyond a century is likely. A well-distributed population with an Ne in the high 100s.
Level 7.	Long-term adaptability—continued existence for many centuries likely. A well-distributed population with an Ne approaching 1,000.
Level 8.	Evolutionary viability—continued existence for many centuries is likely. Populations could diverge genetically into a new species under gradual environmental change. Well-distributed population with an Ne approaching several 1,000s.
Level 9.	Evolutionary viability—continued existence on the order of millennia is likely. Populations are fully capable of evolutionary change. Well-distributed local populations; parts of a biological population with an Ne that exceeds the low 1,000s.

Scientific uncertainty centers on the lack of biological information about a species and the consequent inability to adequately predict either long-term natural or short-term catastrophic events. Uncertainty in decision making results from the use of inadequate or inaccurate information. In dealing with the uncertainties, both scientific and decision-making, risk analysis involves estimating probabilities of various events, estimating values of outcomes of various possible decisions, and finally applying the resulting probabilities of outcomes to estimate a range of results for a variety of possible decisions (Fig. 3.2).

The use of a risk analysis reduces some of the uncertainty in making a decision. However, given the intractable uncertainty of the future it is clear that science cannot furnish wildlife managers with an absolute deterministic scenario of events. "A precise forecast of the future is excluded if we can have only an inaccurate measurement of present circumstances" (Weaver 1975).

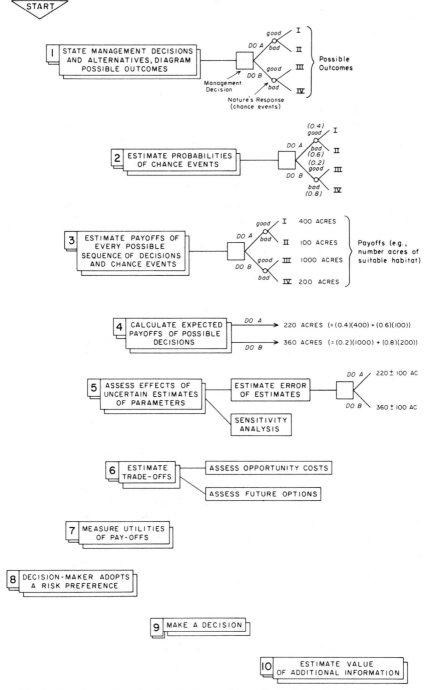

Figure 3.2. Procedure for developing a risk analysis. (Marcot 1986).

3.2 UNINHIBITED GROWTH MODEL

How does the mating of several individuals produce a population? To provide a theoretical answer to this question and to demonstrate how a population might grow, a hypothetical animal population can be developed conceptually by imagining the outcome of placing 1 male and 1 female with a physiological longevity of 10 years in an enclosure of 1 km² (247 ac). The enclosure is stipulated to be predatorproof, free of all hazards and diseases, and possessing the optimum resources for the species. Further, it is stipulated that two young are produced in June each year (female was bred the previous fall) with a 50:50 sex ratio at birth, and one breeding season per year. At the end of the first year the enclosure will contain 4 animals consisting of 2 males and 2 females (Table 3.2). In the second year both females breed and produce 2 young each for a total of 4 new animals.

Table 3.2. Uninhibited growth (exponential) for a hypothetical population with a 50:50 sex ratio, one breeding season per year, and a physiological longevity of 10 years.

End of Year	Adult		Young		Subtotal		Total
	M	F	M	F	M	F	
1	1	1	1	1	2	2	4
2	2	2	2	2	4	4	8
3	4	4	4	4	8	8	16
4	8	8	8	8	16	16	32
5	16	16	16	16	32	32	64
6	32	32	32	32	64	64	128
7	64	64	64	64	128	128	256
8	128	128	128	128	256	256	512
9	256	256	256	256	512	512	1024
10	512	512	512	512	1024	1024	2048

At the end of 10 years there are 2,048 animals in the enclosure. A population increasing in this manner develops a characteristic J shape when plotted over time (Fig. 3.3) and has been termed the *uninhibited growth model*. Uninhibited or *exponential growth* (also called geometric because the exponent is in the series 1, 2, 4, etc.) results when animals are introduced into unoccupied habitat, called *pioneering populations* or under laboratory conditions. Individuals born into a population at the same time are referred to as a *cohort*. In our hypothetical animal population, a cohort turnover occurs at 11 years when the first pair introduced dies.

It is important to notice in this animal population model that although there is a constant growth rate, 2 offspring for each mature female, the number of individuals added each year changes dramatically. Adding 10% to a population when it is low ($0.10 \times 100 =$

10) is truly insignificant while adding 10% when it is high (0.10 × 1000 = 100) may in the same environment be catastrophic. This factor leads directly to a consideration of rates of population change.

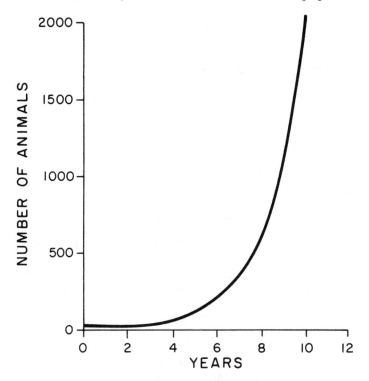

Figure 3.3. Uninhibited growth model.

3.3 RATES OF CHANGE

Net population increases are a function of the number of young born to females of specific ages. The *net reproductive rate* (R) is the average number of females born per female in each age group. The value of R is included in fecundity tables but it is difficult to determine and is almost never available to wildlife managers. However an approximation of R is provided by the finite rate of change. The total yearly change in a population can be expressed as a growth rate in the ratio of population numbers from one year to the next. This rate is the *finite rate* (Greek letter lambda, λ) or the rate of change of population size. Finite rates are observed or actual rates obtained from population estimates. For example, using information from Table 3.2, the finite rate of change in population from year 2 to year 3 is:

$$\lambda = \frac{P_3}{P_2} = \frac{16}{8} = 2.0 \qquad [3.1]$$

where

P_3 = population at year 3
P_2 = population at year 2

The base population of 8 changed by a factor of 2 (100%) in one year. Subsequent years can be calculated by the general equation:

$$P_{t+1} = \lambda \times P_t$$

where

P_t = population for a given year
t+1 = population for the next year

For year 5 the calculation is:

$$P_5 = \lambda \times P_4 = (2.0 \times 32) = 64$$

Notice that the rate of change in our hypothetical population is constant between years.

A population increasing at a constant growth rate follows a sequence of increasing power of λ such that:

$$P_t = P_0(\lambda)^t \qquad [3.2]$$

where

P_0 = initial population
P_t = population in year t

This equation assumes that the species will breed only once during the breeding season and does not have overlapping generations. For example, the animal population in year 5 with a λ of 2 is:

$$P_5 = 2 \times 2^5 = 64$$

A field example of exponential growth occurred when 2 male and 6 female pheasants were introduced on Protection Island in 1937 (Einarsen 1945). The population increased from 8 to 2,000 pheasants in six years. Other examples of uninhibited growth include deer introduced into the George Reserve (McCullough 1979) and reindeer on St. Matthew Island (Klein 1968).

To simplify rates it is convenient to use delta (Δ) notation to express a change in quantity to distinguish finite changes from instantaneous changes (d) used in derivatives in calculus. A change in quantity can occur either in the time or quantity factor. Another way to express rate of change of a population is to determine an average rate per individual (r_0) per unit of time as:

$$r_0 = \frac{\Delta P}{P\,(\Delta t)} \qquad\qquad [3.3]$$

where

r_0 = rate of change per individual per unit of time or specific growth rate

ΔP = change in population

Δt = change in time

The rate of change for our animal population changing at a constant yearly rate is (select any two years, such as year 2 to year 3) (Table 3.2):

$$r_0 = \frac{16 - 8}{8 \times 1} = 1$$

The interpretation is that every animal (male and female) in existence in year 2 will account for 1 animal to be added to the population in year 3. The rate of population change per unit of time can be rewritten as:

$$\frac{\Delta P}{\Delta t} = r_0 P \qquad\qquad [3.4]$$

By knowing the rate of change (r_0) per individual we can grow a population from an initial number over time. For example, the population in year 3 is:

$$P_3 = P_2 + r_0 P_2 = 8 + (1 \times 8) = 16$$

The finite rate of increase (λ) may be expressed in exponential form as follows:

$$\lambda = e^r$$

where

e = 2.71828 (the base of natural logs)

r = natural log λ = intrinsic rate of increase

In this equaion "r" is an exponent to which e is raised to equal λ and is a rate of increase that can be used in population equations. For our hypothetical population λ is 2; therefore, the value of (r) is 0.693147 (the natural log of 2). Finite rates are always positive and have a range from zero to infinity but when λ is greater than 1 the population will increase, and when λ is less than 1 the population will decrease (Table 3.3). In population analysis (r) is used instead of λ (Caughley 1977) because:

1. It is centered at zero and therefore a rate of increase has a positive (+) sign while an equivalent rate of decrease has a minus (−) sign.

    Example: In a rabbit population of 100 animals

    50% die before 1 month old
    90% die from 1–12 months

    50% + 90% does not = 140%
    100 to 50 = 50% loss or $r = -0.693$
      50 to 5 = 90% loss or $r = -2.303$ (45/50)
                             total $\overline{-2.996}$

    $\lambda = 1.0 - e^r$
      $= 1.0 - e^{-2.996}$ ($e^{-2.996} = 0.05$)
      $= 0.95$ or 95%

2. It converts easily from one time unit to another. Thus, to convert a yearly rate to a daily rate divide by 365.

3. It can be used to calculate the doubling time of a population by the formula:

    $$\frac{0.6931}{r} = \text{doubling time where}$$

    where

    0.6931 = natural log of 2

Example: From year 3–4, $\lambda = 2$, $r = 0.6931$

$$\text{Doubling time} = \frac{0.6931}{0.6931} = 1$$

Whenever the percentage of animals in each age class of a population remains constant over time the result is a stable age distribution. From the uninhibited growth data (Table 3.2) of our hypothetical animal population, a table (Table 3.4) can be developed that shows when the population will become stable. This happens at about the 7th year when age class 1 stabilizes at about 50%. When a stable age distribution is attained the population will increase according to the equation:

$$P_t = P_o e^{rt} \qquad\qquad [3.5]$$

where

P_t = population at time t
P_o = initial population
e = 2.71828 (base of natural logs)

r = instantaneous or intrinsic rate of change

t = time in years

Table 3.3. The relationship between % change, lambda (λ), and exponential rate (r).

P_{y+1}	P_y	% change	Lambda	r
50	100	−50	0.50	−0.693
100	100	0	1.00	0.000
150	100	+50	1.50	+0.406
200	100	100	2.00	+0.693
500	100	400	5.00	+1.6094

y = year of interest

Table 3.4. Stable age distribution for a hypothetical animal population (Table 3.2).

Year	Number at Age								Total Pop.	% of age 1 in population
	1	2	3	4	5	6	7	8		
1	2	0	0	0	0	0	0	0	2	100
2	4	2	0	0	0	0	0	0	6	66
3	8	4	2	0	0	0	0	0	14	57
4	16	8	4	2	0	0	0	0	30	53
5	32	16	8	4	2	0	0	0	62	52
6	64	32	16	8	4	2	0	0	126	51
7	128	64	32	16	8	4	2	0	254	50
8	256	128	64	32	16	8	4	2	510	50

Because the value r is the instantaneous rate of change per individual per unit of time, it accounts not only for the difference between births and deaths, but for emigration and immigration over time as well. The instantaneous rate was referred to earlier as instantaneous time because it can be used in very short intervals of weeks and days.

Using the above equation our hypothetical population for year 5 is:

$$P = P_o e^{r(5)}$$
$$P = 2 \times 2.71828^{(0.693147 \times 5)} = 64$$

The same equation can be used to reconstruct a population growth rate if the initial and ending population numbers are known. For example, if a pioneering population of pheasants increased from 8 to 2000 in 6 years (Einarsen 1945) when introduced into unoccupied habitat, then what was λ and r? The calculations are:

Basic Equations:

$$P_t = P_o e^{(rt)} \tag{3.5}$$

$$\frac{P_t}{P_o} = e^{rt} = \lambda$$

$$2000 = 8 \times e^{rt}$$

$$\frac{2000}{8} = e^{rt} = 250$$

$$r = \text{natural log of } \lambda = 5.5215$$

$$r = \frac{5.5215}{6} = 0.9203 \text{ (for 6 years)}$$

substituting figures into 3.5

therefore $P_6 = 8 \times e^{(0.9203 \times 6)} = 2000$

Since the world is not overrun with animals, as the uninhibited model seems to imply, there must be something controlling populations so that numbers do not increase without bounds. But before approaching this question several factors which condition population growth must be noted.

First, the illustration used to derive the J curve assumed females which first breed at one year. If instead the females in the hypothetical animal population first breed at 2 years the increase in numbers is dramatically reduced so the shape of the growth curve changes to reflect a slower growth rate. Further the example assumed a 50:50 sex ratio at birth which seems to be normal for most vertebrate species (Caughley 1977). But what if the number of young born is greater or lesser than two? Table 3.5 shows how annual production changes with a change in litter size (note the assumed 50:50 sex ratio). Now I return to the hypothetical model to identify those factors limiting population growth.

Table 3.5. Effects of a change in litter size on population growth. Assume a 50:50 sex ratio, 1 litter per year, and no mortality.

Average litter size	Year	
	1	5
1	3	15
2	4	64
5	7	1,052
10	12	15,552

3.4. INHIBITED GROWTH MODEL

Obviously populations cannot continue to expand indefinitely. Sometime during the uninhibited growth of a new population, the rate of increase r must start to decline. This change of rate results from the interactions and consequences of the factors contained in Conceptual Equation 3 (Table 2.2). After these factors become fully effective, a point is reached where the population levels off and reaches an equilibrium with its environment (Fig. 3.4). Exponential growth moving to a steady state over time is reflected by an S-shaped curve. The upper part of the curve, however, can take several shapes in addition to the one shown.

The inhibition of the growth of a population can be dealt with mathematically by inserting a variable in the uninhibited growth formula (Equation 3.4) to reduce the value of r as the density increases. This expression is:

$$\frac{K - P}{K}$$

where

K = carrying capacity (resources)

and shows the same S-shaped sigmoid (logistic) curve as in Fig. 3.4.

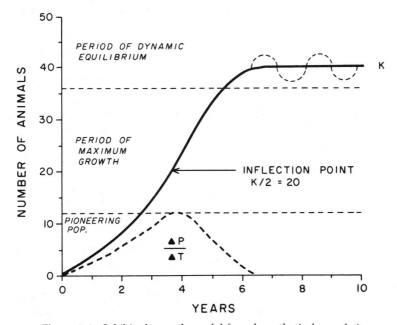

Figure 3.4. Inhibited growth model for a hypothetical population.

If the enclosure containing the hypothetical population possesses resources sufficient to support only 40 animals, the closer the population approaches 40 the closer the value of the expression approaches and eventually equals 0 as a limit. At carrying capacity (K), population growth is zero. To calculate the annual population for column 2 of the logistic growth curve in Table 3.6 the following formula is used:

$$\frac{\Delta P}{\Delta t} = r_o P \; \frac{K - P}{K} \qquad [3.6]$$

Table 3.6. Inhibited growth of the population in Table 3.2 with a K of 40.

End of Year	(1) Exponential Growth	(2) Logistic Growth	(3) Yearly Addition
1	4	4	4
2	8	8	6
3	16	14	9
4	32	23	10
5	64	33	6
6	128	39+	1
7	256	40	0
8	512	40	0
9	1024	40	0
10	2048	40	0

Column 3 is the yearly addition for logistic growth. Columns 2 and 3 are rounded to the nearest whole number.

For example, the logistic growth (yearly addition) from year 6 to year 7 is:

$$r_o = \frac{\Delta P}{P \, (\Delta t)} = \frac{256 - 128}{128 \times 1} = 1 \quad \text{(See Equation 3.3)}$$

substituting figures into Equation 3.6

$$\frac{\Delta P}{\Delta t} = 1 \times 39 \; \frac{40 - 39}{40} = 0.97 = 1 \text{ (rounded)}$$

One of the interesting features of the logistic curve is that maximum production occurs at about K/2, which approximates the *inflection point* where the population is still increasing but at a decreasing rate. For the hypothetical population in the example the inflection point occurs somewhere between year 4 and 5. The sigmoid growth curve is a conceptual model of what could happen but contains many marginally questionable assumptions, such as:

1. Mortality is entirely dependent on population density.

2. Natality and mortality are continuous, not discontinuous functions.

3. The environment never changes.

4. All individuals in the population react in the same way to changes in density.

5. The expression $\dfrac{K - P}{K}$ has a biological basis.

The causes underlying the change from a J-shaped to an S-shaped growth curve have led to several differing schools of thought about the effects of animal density on animal populations. One way to analyze the effects is to categorize the influences into *density-dependent* and *density-independent* factors.

Independent factors are those affecting the population regardless of size—the hazards, diseases, and predators of Conceptual Equation 3 (Table 2.2). Density-dependent factors on the other hand are those factors affecting the ability of individuals to reproduce: stress, behavior, and the shortage of resources as a result of competition. In the wild the rate of increase of a population is reduced not by any single density-dependent or -independent factor but by a combination of both.

Population regulation describes a combination of feedback mechanisms which bring about resilience or equilibrium of a population. Density-independent factors do not regulate populations but can reduce numbers on a yearly basis. The important point is that populations of any species cannot increase indefinitely without some regulating factor or factors intervening to control the density. Chapman (1931) defined the intervening factors as environmental resistance.

With so many assumptions and difficulties in measuring factors it hardly seems worthwhile to view the logistic equation and sigmoid growth curve as a useful tool for wildlife management. Nor will most wildlife managers have occasion to use the formulas or to determine an *r* value because virtually all of the populations they work with are already somewhere between K/2 and the carrying capacity. Despite the theoretical difficulties, *these concepts give the wildlife manager an intellectual framework suitable to roughly determine the limits within which management activities can be conducted and to understand new approaches as they are developed.*

Intuitively, the initial growth in both the uninhibited and inhibited models make sense because animals reproduce and add young to a population but eventually something happens to restrict that growth. The growth models provide the foundation for concluding that if a given population was once high and is now low, then presented the right conditions that population can return to high densities rather rapidly. Wildlife management seeks to discover: What is that mix of right conditions? One of the great practical benefits of understanding growth curves and their characteristics is that they give the wildlife manager a place to start thinking about and discussing population problems.

3.5 CARRYING CAPACITY

Critical to the inhibited growth model is the concept of *carrying capacity* (*K*). The idea that an area can support only so many animals is well established in wildlife literature but an exact meaning of the term has defied attempts at clarification (Edwards and Fowle 1955). Most managers would agree that *carrying capacity* is a balance between vegetation and animals, characterized by an expression of animal density (Caughley 1979).

While the idea of carrying capacity is simple, the actual definition is more complicated because some type of constraint is needed for quantification. Carrying capacity has been described and used in many ways in wildlife management but three ways have been primarily used in the past (Dasmann 1981):

1. The number of animals of a given species actually supported by a habitat, measured over a period of years.

2. The upper limit of population growth in a habitat, above which no further increase can be sustained.

3. The number of animals a habitat can maintain in a healthy, vigorous condition.

Most of the ideas about wildlife carrying capacity were first developed by range managers where the emphasis was on density of animals. In the case of livestock the term *grazing capacity* seems more appropriate (Stoddart, Smith, and Box et al. 1975). Carrying capacity is largely a concept rather than a real, quantifiable entity. Extending the concept to amphibians, birds, small mammals, and reptiles is not common but if carrying capacity is real then an extension to these life-forms should be possible.

As a population produces young each year, animal numbers (as for deer and elk) may exceed the year-long carrying capacity until some environmental factor (Conceptual Equation 3, Table 2.2) checks the population increase. As the habitat quality of a given vegetation stage increases for a particular species, the carrying capacity for that species increases. Carrying capacity is not a fixed factor of the land and will change as food, cover, and water changes with the seasons or over long time periods, so the process is dynamic, not static as is often suggested by the many definitions. For herbivorous animals such as deer, carrying capacity is difficult to determine, and if capacity is exceeded for several years, food and cover can be affected adversely. The problem is that of detecting when animal density is starting to exceed carrying capacity. Most ungulate populations cannot be maintained at a level above carrying capacity for long periods without damage to the resource base.

In addition to reducing the impact of the factors that reduce the value of r in the logistic equation, the ecosystem manager's job is to keep animal numbers at a level that will not damage the basic soil and plant resources. In reality, animal populations fluctuate above and below carrying capacity because the environmental factors of hazards, predators, disease, resources, and humans (Conceptual Equation 3, Table 2.2) are never constant. A continued fluctuation suggests a process leading to a balance or equilibrium and feedback between resources (primarily vegetation) and animal density. This fluctuation of population numbers might be termed the *process of dynamic equilibrium* (the same term is used in plant and animal succession and the river continuum concept) in place of the term carrying capacity (Caughley 1979).

There are three parts to the process of reaching dynamic equilibrium for herbivores (Caughley 1979):

1. Growth response of plants. The rate of increase of plant biomass as a function of plant density.

2. Functional response of herbivores. The rate of intake per animal as a function of plant density.

3. Numerical response of herbivores. The rate of increase per animal as a function of plant density.

When an equilibrium is reached between plant growth and animal density in the absence of hunting then the *ecological carrying capacity* has been reached (Caughley 1979). By interjecting hunting a new state of equilibrium is achieved that is below the ecological capacity; this is termed *economic carrying capacity*. The economic carrying capacity is determined by quantifying the density of animals desired to meet certain objectives.

3.6 LIFE TABLE

So far the discussion of population dynamics has centered on natality, but what about mortality? What tools are available to help understand the changes in a cohort of animals over their lifetime? Fortunately the study of human populations (demography) has provided techniques to deal with mortality by specific age classes. These techniques were developed by life insurance companies which were forced to estimate how long a human in a certain age class will live. Wildlife managers can use these same techniques to deal with questions of mortality among wildlife species.

A life table follows a *cohort* of animals born at the same time through its various outcomes until the last member has died. A life table developed in this manner is an *age-specific* or *dynamic* life table. However, biologists seldom have the luxury of studying a group of animals all born at the same time. The next best option is to reconstruct from either census data or death records what happens to groups of animals. These data lead to a *time specific* life table in which animals in one period of time are captured and their distribution of ages determined. The mortality factor is calculated from the distribution of the age classes, based on the assumption that a stable age distribution exists.

A third kind of life table is a *composite* life table developed from mark-recapture data for two or more periods of time. Techniques for constructing this type of life table can be found in Schemnitz (1980). By way of example, Table 3.7 shows my findings for a cohort of 58 Kaibab squirrels which I tracked for a period of eight years, 2 of which lived to be 6 years old (fx column).

Table 3.7. Composite life table for the Kaibab squirrel(combined sexes) (author's original data).

Age x	Frequency fx[1]	Survival lx[2]	Mortality dx	Mortality rate qx	Survival rate
0–1	58	1000	552	0.552	0.448
1–2	26	448	207	0.462	0.538
2–3	14	241	69	0.286	0.714
3–4	10	172	86	0.050	0.500
4–5	5	86	52	0.605	0.395
5–6	2	34	34	1.000	

[1]Number of animals remaining in each age class from an initial population of 58.
[2]Number of animals converted to a base cohort of 1000.

The Abert squirrel has a litter averaging about 3.5. While fecundity is high, so is the mortality in the 0–1 age class which is 55%, decreases to mid-age (2–3), and then increases in the older (4+ years) age classes. This mortality rate indicates an unhunted population in which some adults are able to approach the limits of physiological longevity. Caughley (1966) suggested that mortality in mammals follows a U-shaped trend with age; mortality in the Kaibab squirrel life table follows this pattern. Life tables, while a useful tool in analyzing a population, are difficult to construct and the data are expensive to collect; therefore, they are almost never used in real management situations.

3.7 FLUCTUATIONS

Fluctuate, in wildlife management, means a change in quantity in a short time. The most common type of fluctuation is the yearly peaks brought about by high populations of small mammals born in the spring to the winter trough when animals die due to various environmental factors (Fig. 3.5). While there is yearly change in most small mammal populations for example, cottontails and quail—there is no great departure from the peaks and troughs between years, therefore, the population tends to oscillate over time giving the curve of annual change a flat appearance. Leopold (1933) has termed this type of annual change a stable population.

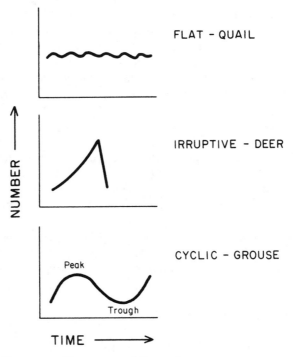

Figure 3.5. Three types of fluctuation.

The populations of some animal species fluctuate through peaks and troughs separated by several years in a predictable pattern; such fluctuations are known as *cycles*. Two distinct multispecies cycles have been described: a short cycle of 3–4 years and a long cycle of 9–10 years. Lemmings and voles in the arctic and subarctic along with their predators, the arctic fox, red fox, and snowy owl, are short-cycle species. Elton (1942) was the first to discover and report on the short cycle. Short cycles are exhibited by animals having a limited breeding season or short life cycle and occur in simple ecosystems whose communities include only a few species.

The long cycle became evident from the Hudson Bay Company fur records which indicate a 9–10 year cycle for the Canada lynx (Elton and Nicholson 1942). A snowshoe hare population increase is accompanied by an increase in the lynx population which lags the snowshoe peaks and valleys. Other long-cycle species include ruffed grouse, muskrat, bobwhite, and pheasants (Errington 1945; Hewett 1954; Kozicky, Hendrickson, and Homer 1955).

A second type of fluctuation is an *irruption* (or eruption) which is marked by a sharp increase in a short time followed by a population crash with a return to stability. The Kaibab deer herd crash of 1940 is considered the classic example of an irruption (Leopold 1943; Leopold, Souls, and Spencer 1947) but irruptions have been reported for jackrabbits (Wagner and Stoddart 1972), quail (Gullion 1960), and thar in New Zealand (Caughley 1970). Irruptions can occur when new habitat is created by disturbance as happened in California in the Devil's Garden deer herd (Salwasser 1979).

Theoretically fluctuations can be caused by any of the external factors of Coneptual Equation 3. In addition to these factors, however, certain other events or conditions seem to correlate with high populations and have been hypothesized, at least in part, to be the cause of cycles. Some of these theories are the ozone theory, sunspot theory, stress theory, predatory-prey theory, and random theory (Christian 1950; Calhoun 1952; Cole 1954; Krebs et al. 1973).

At one time cycles were a subject of major concern and study, but the topic now seems to be of little interest. This is because most biologists accept the validity of crude cycles but also know that cycles are not a factor that can be managed. This is not, however, the case for irruptive species such as deer where increases can be detected and management action taken before a crash occurs.

3.8 DEMOGRAPHIC VIGOR

Conceptual Equation 3 (Table 2.2) identified six factors acting directly on an animal or a population:

$$Pr_{SR} = f(Ha + Di + Pr + Ge + Re + Hu)$$

This expression accounts for all factors which affect an animal's chance to survive and reproduce. It is clear that the value r in the exponent of Equation [3.5] has a tremendous effect on a population because it increases animal numbers exponentially as in compound interest. What is it then that affects r? The answer is to be found in Conceptual Equation 3. The rate of increase, r, integrates the six factors of the equation into a single value characterizing the probabilities of a population to survive and reproduce. It reflects the action of environmental influences and summarizes quantitatively the vigor or "health" of a population in given circumstances. r is a quantification of Chapman's Biotic Potential (BP) in Conceptual Equation 2 (Table 2.1).

Because of the difficulties inherent in establishing a real value for r, biologists must search for a reasonable indicator to use in management decisions. For years wildlife biologists in the United States and elsewhere have been counting cow:calf ratios, doe:fawn ratios, number of eggs hatched, hen:poult ratios, and other measures of productivity. These types of ratios and counts if done consistently in the same environmental conditions can be a good indicator of r as demographic vigor and can form the basis for practical management decisions. However, it is possible that populations with a high r value can still decrease because female young may not live long enough to reproduce and add offspring to the population. If reproduction is not a problem then the other factors of disease, hazards, predators, etc., may be the ones keeping a population in a declining state.

3.9 SUMMARY

Population dynamics deals with changes in populations over time. This change is brought about by such individual and population factors as birth rate, mortality, immigration, and emigration. Populations possess characteristics of natality and density which are not properties of individuals. Individuals produce young that are part of a population and under ideal conditions the population will grow, initially exponentially, until the decimating factors of Conceptual Equation 3 come into play to keep the population at carrying capacity. Understanding population dynamics involves rates of change from one year to the next.

Two theoretical models, inhibited and uninhibited, help to explain how a population grows and then levels off. Several examples for pheasant and deer indicate that populations of wild animals do increase in keeping with theoretical models. These models while not useful for decision-making, do provide the conceptual basis for understanding what is possible when animals are introduced into favorable environment or reduced to low numbers through some decimating factor.

The opposite of natality is mortality and this latter population characteristic can be modeled with the use of life tables. Life tables indicate the loss of animals by age class until the last animal in a cohort has died. Life tables are also useful in identifying age classes that might be subject to differential mortality.

Carrying capacity has been defined and used in many ways and applies more often in the case of ungulates than to other species. Animal populations fluctuate above and below carrying capacity because of the environmental factors identified in Conceptual Equation 3. The most pronounced fluctuations are cycles and occur in simple ecosystems where they have characteristic shapes and predictable intervals. Theories relating to the causes of cycles include sunspots, predators, stress, and random events.

In the inhibited and uninhibited growth models the value r has a strong influence on determining population density. Values of r are difficult if not impossible to determine in the field but the measured values of cow:calf, doe:fawn, hen:poult ratios can be used as fairly accurate indicators of r. These counts and ratios if done consistently are a measure of demographic vigor and can be used in making management decisions.

3.10 RECOMMENDED READING

Caughley, G. 1977. *Analysis of Vertebrate Populations.* John Wiley and Sons, New York.

Caughley, G., and L. C. Birch. 1971. Rate of increase. *J. Wildl. Manage.* 35:568–663.

Krebs, C. J. 1972. *Ecology: The Experimental Analysis of Distribution and Abundance.* Harper and Row, New York, NY.

Schemnitz, S. D. 1980. *Wildlife Management Techniques Manual.* 4th ed. Wildl. Soc., Wash., DC.

Vandermeer, J. 1981. *Elementary Mathematical Ecology.* John Wiley and Sons, New York.

Chapter 4

INVENTORY, EVALUATION, AND MONITORING OF FOREST HABITATS AND WILDLIFE

In the past 20 years the number of techniques available to inventory wildlife species and habitat has increased to where a manager has a variety of tools from which to select. While the availability of a large number of techniques may be desirable it can create a problem in selecting one that is cost effective and biologically sound. As a result, biologists must spend more time planning the job before starting field work for an inventory.

Since passage of resource management legislation in most of the developed countries, a new emphasis on inventory has resulted because several different types of data are needed to fulfill all the decision-making requirements. The emphasis is now on inventorying all resources at the same time, in other words—*multiresource inventory*. Inventories are basic to management and so must be planned carefully. Scientific management is impossible without an inventory, but too often an inventory is made before adequate planning and objective setting have been undertaken.

Planning must start by setting objectives. Experience has shown that without clear, well defined-objectives, one of two things happen—either too much or too little data is collected, or the wrong types of data are collected. Either of these results in the loss of both time and money. The technique must fit the resources available to do the job.

4.1. MULTIRESOURCE INVENTORY

Once a decision has been made to conduct an inventory then one task needing early attention is to determine if data can be collected as a part of another resource inventory, such as for timber or watershed management. If no such inventories are planned it will have to be initiated as a single purpose inventory—as for example wildlife. In short, do not collect data if it is already available. Make use of existing sources such as timber surveys, allotment analysis, recreation plans, and watershed and soil surveys, even if the data needs interpreting or revision.

In any case, the actual data collection must be consistent. Use of the following guidelines will help to meet this goal:

1. Collect data using the same methods.

2. Collect data at the same time of year.

3. Collect data on the same plots if the sample was systematic as in previous inventories.

Use a field form that forces data entry into an allocated space for a given topic in the order it will be entered into a computer. This procedure reduces the possibility of mistakes by a data entry operator. Another alternative is to use an electronic field recorder programmed for data transfer to a microcomputer and into a database management system. *Field data recorders* (FDR) are portable, battery-operated microcomputers. Most of the new FDRs can be programmed with rules allowing only data with certain characteristics to be entered into storage. The advantage of some types of recorders is that the data can be transmitted electronically by telephone at night to a home base computer.

4.2 CATEGORIES OF INVENTORY

Category of inventory has to do with purpose and how the information collected will be used. In general, there are three hierarchical categories of inventory: national, regional, and management unit. These categories are not all-inclusive in that there can be overlap between them, nor are they the only categories that can be identified. However, they are intuitive because of the commonality with geographical and political boundaries.

National inventories usually provide limited information consisting of lists of resources used for formulating legislation, determining policy, setting management and research priorities, and supporting budget presentations. Information collected just for a national inventory generally cannot be disaggregated to a lower category.

Regional inventories typically focus on a geographical area or major vegetation types that cut across political boundaries. This type of inventory is used to develop programs, allocate funds, assure distribution of resources, and provide for regional diversity. The information becomes part of a planning document that can be revised and updated to conform to changing conditions.

The size of the third category of inventory is a management unit which will vary. It can be as large as a national forest or as small as a watershed. Management unit inventories are the basis for management decisions and most often include detailed resource information. Information collected at the management unit level generally contains too much detail for all of it to be aggregated to higher categories. However, in almost all cases some information is suitable for regional use.

The categories of inventory are additive from the local level to a national level such as:

$$\text{Local} + \text{local} \ldots \text{n}$$
$$\text{Region} + \text{region} \ldots \text{n}$$
$$\text{National.}$$

This scheme has many advantages particularly if all the data are collected at the same detail, format, and consistency at each level.

4.3 INTENSITY OF INVENTORY

The intensity of an inventory must be controlled by the purpose and objective of the inventory, but all too often it is controlled by the money available. Ideally the level of detail increases with the level of intensity from low to high.

Low-intensity inventories consist largely of lists of species, grouped by some characteristic of interest such as threatened and endangered, game and nongame, taxonomic group, or economic value. Low-intensity inventory information is frequently used at the national level but it is also the first step in a high-intensity inventory for a management unit. Data for low-intensity inventories is largely derived from literature surveys or expert opinion.

A moderate-intensity inventory builds on the species list and classification by adding information on where a species feeds and breeds. It also can include an estimate of population abundance using general terms of low, moderate, or high, and a population trend of increasing, decreasing, or stable. Data for moderate intensity inventories is derived from literature reviews, expert opinion, and limited field surveys.

High-intensity inventories are the most detailed, most expensive, and take the greatest amount of time to complete. Data are collected by field crews with a knowledge of the animals and their habitat. Data include plant species composition and abundance, animal population characteristics (density, sex and age ratios, etc.), and quantity and quality of habitat. If data are collected properly they can be aggregated with other management units to provide information for regional and national purposes. The intent and scope of high intensity inventories must be carefully planned, priorities defined, precise cost estimates established, and the right types of data clearly identified.

An inventory of resources is the first step in the management process and without an inventory there can be no management except by default. The manager when considering category and intensity of inventory can select from one of nine combinations but some combinations are not as practical as others because the detail is not necessary or the cost is too high. The possible combinations of category and intensity are shown in Table 4.1.

Table 4.1. Combinations of category and intensity of inventory.

	Category		
Intensity	National	Regional	Management Unit
Low	x	x	x
Moderate	—	x	x
High	o	—	x

x = Possible
— = Maybe
o = Not likely

4.4 MAP SCALE

After a selection has been made for category and intensity of inventory the next major job is to select a working map at a scale that provides an acceptable level of resolution. *Scale* is the ratio of actual distance on the ground to the same distance on the map. Scale is a representative fraction where the map and ground distances are shown in the same units such as 1:63,360 which means 1 cm (or 1 in) on the map equals 63,360 of the same units on the ground.

A scale of 1:24,000 is the standard map scale for a 7.5 minute latitude (angular distance north or south from the earth's equator measured through 90 degrees) and 7.5 minute longitude (great circles running through the north and south poles) map. Each quadrangle (square sheet) covers approximately 16,270 ha (40,215 ac). Large-scale maps such as 1:7,920 or 1:24,000 are useful when small areas need to be delineated with detailed information. Intermediate-scale maps range from 1:50,000 to 1:100,000, cover larger areas, and are suitable for land management planning (Steger 1986). Maps of 1:250,000, to 1:1,000,000 are small scale and cover very large areas on a single sheet.

Scales of 1:7920 and 1:15,840 are often used because they provide acceptable levels of resolution for different levels of inventory but they are not the only scales used by modern map makers (Table 4.2). These two scales are also commonly used in aerial photography in the United States because conversion from inches, miles, and acres is readily made, so detail from photographs is easily transferrable to the maps. The outline of the management unit drawn on the map provides the basic reference for sampling and determining area.

Table 4.2. Common map scales with equivalents.

Metric Units			
Scale ratio	Meters per centimeter	Centimeters per kilometer	Hectares per square centimeter
1:1,000	10	100.00	0.01
1:2,000	20	50.00	0.04
1:5,000	50	20.00	0.25
1:7,920	79	12.62	0.62
1:15,840	158	6.32	2.50
1:20,000	200	5.00	4.00
1:24,000	240	4.16	5.76
1:31,688	316	3.16	10.00
1:33,300	333	3.00	11.10
1:50,000	500	2.00	25.00
1:63,360	633	1.57	40.31
1:100,000	1,000	1.00	100.00

English Units			
Scale ratio	Feet per inch	Inches per mile	Acres per square inch
1:500	41.667	126.72	0.0399
1:1,000	83.333	63.36	0.1594
1:2,000	166.667	31.68	0.6377
1:5,000	416.667	12.67	3.9856
1:7,920	660.000	8.00	10.0000
1:15,840	1,320.000	4.00	40.0000
1:20,000	1,666.667	3.168	63.7690
1:24,000	2,000.000	2.640	98.8270
1:31,680	2,640.000	2.000	160.0000
1:63,360	5,280.000	1.000	640.0000

In the United States the National Cartographic Information Center can provide information about maps, charts, aerial and space photographs, satellite images, geodetic control, and digital and other cartographic data. A catalog is available from:

National Cartographic Information Center
U.S. Geological Survey
507 National Center
Reston, Va. 22092
Tel. (703) 860-6045

4.5 GEOGRAPHIC INFORMATION SYSTEMS

A *Geographic Information System* (GIS) is a computerized means of
making base maps with overlays (Fig 4.1). Every forest biologist
appreciates the value of overlays in decision-making because of the
many hours spent in using colored pencils to code habitat characteristics, delineate hunt unit boundaries, and locate water developments on acetate film. Overlays present a simultaneous visual interpretation of the several factors involved that would otherwise be difficult to explain in written or oral form.

GIS maps built from data transferred to a computer
electronically—by tracing on the map with a handheld "digitizer"
creating overlays—require but a fraction of the time expended using
the manual method. A digitizer converts line data to points that can be
connected and later converted back to lines on a map. One major
advantage of a GIS is that revisions can be made without revising the
entire map. Another advantage is that a combination of overlays can
be used to show where characteristics coincide.

When a GIS is coupled with a *Database Management System*
(DBMS) containing detailed information and relates it to items on the
overlays the total system becomes a powerful management tool. For

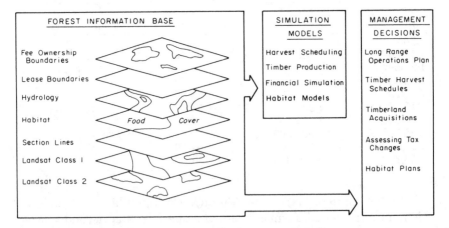

Figure 4.1. Geographic information systems can integrate resource data (Modified from Schwaller and Dealy 1986).

example, a base map of boundaries of a group of timber stands is prepared by GIS. In an associated DBMS the characteristics of the stands are recorded (species, acres, tree density, quadratic mean diameter, etc.). The DBMS also contains another file with quantified data on habitat conditions for a given wildlife species. Knowing that the stands and habitat characteristics have been defined, overlays can then be plotted which identify stands with the defined habitat characteristics. If the DBMS also contains a file describing tree growth and yield, then stands can be manipulated in the computer to model habitat conditions for future dates and alternative management practices.

4.6 EXPERIMENTAL DESIGN

After the map scale has been selected a sampling design can be superimposed over the management unit delineated to show location of plots, or points of a transect. Sampling design is a detailed and complex topic and it cannot be adequately presented in a few paragraphs. There are however several useful sampling and plot guidelines which should be followed. These are:

1. When possible, use nested plots for sampling different resources in the same area.

2. Use circular plots to reduce border area.

3. Use a systematic sample with a random component to reduce the cost of sampling.

4. Make a preliminary sample to determine variance to estimate the number of plots needed in a statistical sample.

Types of Systems

In general there are four types of sample systems that can be used in vegetation inventory:

1. Plotless
2. Point
3. Plot
4. Transect

Plotless sampling is used primarily for trees. The sample is obtained by using a calibrated prism to decide whether a tree is to be counted. This type of sampling is fast and accurate if the sample is sufficiently large to overcome variance.

Points are commonly used for sampling herbaceous plants and

grasses at given distances along a permanently established line. Only the species under the point is recorded.

Plots are probably the most common system in use for all types of sampling of vegetation or animals. Plots have dimension so the density of organisms is easily calculated.

Transects can be of several types—a graduated line for use with points, a long line with a width of centimeters or meters, or a line with plots of any size or shape at intervals along the line. As can be seen, many combinations of sample systems are possible and because of this sufficient time must be taken to determine the correct system for the job to be done.

Nested Plots

Vegetation sampling to measure the attributes of different life-forms of trees, shrubs, and herbs can be done using nested plots. *Nesting* means that a large plot may enclose a smaller plot inside (Fig. 4.2). Nested plots are advantageous because they incorporate different plot sizes at one location for sampling different factors. The center of the large plot may or may not serve as the center for smaller plots. Smaller plots can be systematically located within the larger plot. An alternative to nesting is to cluster the subplots around the outside of the primary (largest) plot.

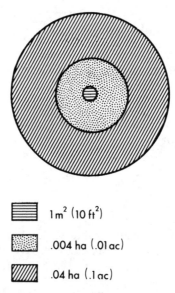

$1 m^2$ ($10 ft^2$)

.004 ha (.01 ac)

.04 ha (.1 ac)

Figure 4.2. Nested plots are useful in habitat inventory.

Shape of Sample Plots

When sampling vegetation in habitat surveys data is often collected from within circular plots on a transect. Circular plots have proved satisfactory in sampling because they are the easiest to use when deciding what is to be counted if an organism is partially in a plot. This is because the length of the circumference of a circle is smaller than the length of the sides of a square of the same size, so fewer border decisions have to be made.

Plot Size

The plot size selected for sampling vegetation must be based on the spatial characteristics of the vegetation. In general, the radius for the sample plot is equal to the average distance between the items to be measured. One way to determine plot size is to do a nested plot study and then construct a density area curve (a modification of the

Table 4.3. Circular plot sizes for sampling.

Size of plot	Radius
Metric Units	
0.10 m^2	0.17 m$^+$
1.00 m^2	0.56 m
10.00 m^2 (0.001 ha)	1.78 m
40.00 m^2 (0.004 ha)	3.56 m
100.00 m^2 (0.01 ha)	5.64 m
400.00 m^2 (0.04 ha)	11.28 m
1000.00 m^2 (0.10 ha)	17.84 m
10,000.00 m^2 (1.00 ha)	56.41 m
English Units	
0.001 ac	3'8.7"
0.01 ac	11'9.3"
0.025 ac	18'7.4"
0.05 ac	26'4.0"
0.10 ac	37'2.8"
0.25 ac	58'10.5"
0.50 ac	83'3.2"
1.00 ac	117'9.0"
Other Conversions	
* 0.96 ft^2	6.63"
** 9.60 ft^2	20.97"
*** 96.00 ft^2	66.33"

* 1 g/ 0.96 ft^2 = 100 lbs/ac
** 1 g/ 9.60 ft^2 = 10 lbs/ac
*** 1 g/96.00 ft^2 = 1   lb/ac

$^+$20 × 50 cm is the standard plot configuration

species area curve) (Mosby 1963). For trees, four or five different plot sizes are nested and trees are counted on each plot. The average of a sample of individual plot sizes is then expanded to a per unit basis and the density compared.

The smallest plot that comes the closest to the density of the largest plot is the most efficient. A similar procedure can be used to develop a species area curve where the number of species are counted on different plot sizes until no more new species are added and this plot size is the one selected.

For habitat inventory work a .04 ha (.10-acre) circular plot is adequate for trees, a .004 ha (.01-ac) plot for shrubs, and a 1 m² (10.76 ft²) for herbaceous vegetation (Table 4.3). A plot of .1 m² (1.07 ft²) is also a common size for sampling herbaceous vegetation because it can be divided into small squares for estimating percentages. In addition, some federal agencies in the U.S. use a 9.6 ft² (.89 m²) plot because dry weight of vegetation in grams converts directly to pounds per acre on this scale.

Sample Size

Once the sizes of plots have been selected the next problem is to determine how many plots will be needed for a statistically valid inventory. The easiest approach is to make a pre-inventory sample and apply a variance formula to estimate numbers (Table 4.4). For example, when counting browse plants on a pre-sample of 10 plots S = 6, t = 2.262 (9 degrees of freedom at 0.95 confidence), and d = 1.5 (mean = 15, accuracy is designated at 0.10) then:

$$N = \frac{36 \times 5.11}{2.25} = 81.76 = 82$$

The number of sample plots needed as computed from the pre-sample data is 82. There is a restriction in using the formula that the data must follow a normal distribution.

Table 4.4. Estimating number of plots needed for an inventory (Snedecor 1956).

Variance Formula
$N = \dfrac{S^2 \times t^2}{d^2}$

where

N = number of samples required,
S = standard deviation,
t = confidence interval at given degrees of freedom, and
d = margin of error (arithmetic mean times designated accuracy).

The number of sample points depends entirely on the variance of the item being sampled, but in many cases the amount of money available to do the job will be the deciding factor. Statistical techniques based on the cost of establishing plots as a factor in determining how many to establish are available. However, as a general rule a sample size from 2 to 5% of the area is a minimum.

Sampling

Random samples are the best of all possible choices because they eliminate bias. However, such sampling is often difficult to do in the field because of problems in transferring a sample drawn from locations on a map to finding this location on the ground. To select a random sample for inventory, a dot grid is placed over the standard map or on aerial photographs and each dot then becomes a sample point. Numbers are then drawn from a table of random numbers to select the points to sample.

A systematic sample is more convenient to execute in the field and most statisticians accept it with constraints, such as adding a random start component. The general procedure for a systematic sample is to randomly select starting points along the border of the area to be sampled and then locate the circular plots systematically at a given distance along parallel transects.

Stratification

One of the easiest ways to reduce variation in any sample scheme is to stratify the area into like units. For example, if two vegetation types are present in an area statistics can be calculated for each type separately rather than including both as one set of sample data. Stratification of vegetation types can be done with the aid of aerial photographs whereby the types are delineated on overlays for selecting the sample. Strata can be identified for many types of physical measurements including vegetation types, physical size of trees in stands, administrative units, etc. For whatever the criteria, the strata to be sampled should (Lund 1978):

1. Logically be related to the information that managers want sampled.

2. Exist in nature or have been artificially determined.

3. Represent a homogenous condition that can be defined and is easily recognizable.

4. Have a definition which conforms to accepted standards.

4.7 INFORMATION TO BE COLLECTED

There have been many attempts to collect wildlife data in multi-resource inventories in the past but the rate of success has been variable. Estimating the density of animals, including amphibians, birds, mammals, and reptiles, cannot in most cases be done on plots established for a multiresource inventory because each animal group possesses unique behavioral traits requiring different sampling procedures. An integrated, multipurpose approach works well when the objective is to collect habitat information applicable to many species, such as the number of dead trees, stumps, and rotten logs used for cover.

The information to be collected in most inventories can generally be grouped into one of four categories. These categories do not contain the entire domain of items to be listed or sampled but are intended as a preliminary list from which to begin selection. Information for specific areas can be added to it.

1. Vegetation
   a. Dominant overstory species, basal area, percent crown cover, dbh, trees per acre, and degree of grouping. Basal area is the cross-section area of a tree measured at diameter breast height (4.5 ft) (1.37 m). Crown cover is the area of crown of trees or shrubs measured as a percent of a plot.
   b. Understory measurements for shrubs, small trees, and herbaceous plants. Browse plant count, including form and age class, or a measure of basal area or crown density.
   c. Forage production estimate. Clip and weigh material from small plots. Separate by grass and forbs.
   d. The number of vegetation layers present and a measure of the amount of live foliage in each layer.

2. Physical Features
   a. Rock areas.
   b. Cliffs.
   c. Seeps and springs.
   d. Caves.
   e. Lakes, ponds, and streams.
   f. Rainfall patterns.
   g. Slope and aspect.
   h. Topographic features (hills, valleys, etc.).
   i. Burned areas by date and acreage.
   j. Hunt or management units.

   k. Cultural features of roads, trails, buildings, camp-
       grounds, and recreation areas.

3. Signs of Wildlife Occupancy
   a. Fecal material or other signs (tracks, feathers, etc.).
   b. Tree nests (squirrels and birds).
   c. Ground nests.
   d. Hollow trees.
   e. Rodent mounds and tunnels.
   f. Roost trees (wild turkey, pigeons).
   g. Squirrel middens (caches).
   h. Ground dens (rabbit, bear, badger, coyote).
   i. Snags (cavity-nesting birds and mammals).
   j. Elk wallows, deer rubs.

4. Wildlife Observations.
   a. All species seen or heard.
   b. Monitor nests, water holes, and special areas using
       video cameras, or time-lapse photography.
   c. Radio telemetry to determine movement.
   d. Aerial counts and photography from aircraft.

4.8 USES OF HABITAT DATA

After habitat data has been collected the next step is to consider
how the information is to be analyzed and displayed. Thanks to the
availability of computers, consideration should be given to using
graphics software to develop histograms and other forms of
frequency diagrams for visual presentation. However, probably the
most useful way to transfer information is by maps. Habitat data
collected on sample plots can be used to develop three types of maps:
(1) food, (2) cover, and (3) water. These maps should show at least the
following items:

1. Food Map
   a. Shrub areas by species and acreage. Form and age
       class of woody plants should be shown.
   b. Isograms of forage production in pounds per acre
       based on data collected from sample plots. An *iso-
       gram* is a line connecting points of equal quantity,
       similar to an elevation contour on a map.
   c. Locations of big trees that are likely mast producers.
       Mast is all fruits, nuts, and berries.
   d. Concentrations of plant species that produce berries.
   e. Natural openings and wet meadow areas by acreage
       and forage production.

2. Cover Map
   a. Plant reproduction thickets by species and acreage.
   b. Location of caves, den trees, snags, and rock outcrops.
   c. Dominant overstory species by acreage. Information can be drawn from timber type maps.
   d. Roost trees and special features such as bear dens.

3. Water Map
   a. All streams both perennial and intermittent.
   b. Location of current or potential pollution sources.
   c. All springs, wells, and permanent water bodies.
   d. Isograms of the zones of influence around all water sources. Water deficient areas will become apparent.
   e. Important amphibian habitats (moist areas, seeps, and springs).

These maps can be made as overlays on aerial photographs or existing topographic maps of suitable scale or, even better, the data can be digitized and incorporated into a GIS and DBMS.

4.9 ESTIMATING ANIMAL NUMBERS

Without doubt the most difficult job of any wildlife manager is to determine the density or total number of animals on a given unit of land. Until the details of an animal population are accurately known the management of a unit lacks any clear direction. Numbers or an indication of abundance are needed to evaluate carrying capacity, to determine the allowable harvest of game animals, to establish viable populations, and to determine the dynamics of the population.

A scale factor controls the estimation of animal populations just as it controls the making of habitat inventory maps. Recall that scale refers to the degree of resolution needed to meet management objectives. For example, a low-intensity habitat inventory generally requires only the same intensity of sampling for animals. The same relationship is true of high-intensity surveys. Clearly there are two general ways of estimating animal numbers: counting animals or their sign.

Counting Animals (Direct)

Direct methods depend upon making a true *census* which is a count of all the animals in an area or an estimate of the number by sampling. A manager seldom has the time and personnel to do a complete census. One direct method that can be used when staff is available is an animal drive, particularly if the area is small. Animals which depend upon a territory and can be easily seen or heard—birds, for example—can be censused by mapping defended area. A census is approximated in a capture-recapture technique by trapping animals over two or three weeks until no more unmarked animals are recorded; the accumulated total is the animal population.

Aerial counts work well for large animals such as deer, elk, and antelope when in open habitats or for species which flock. Counts for flocks should be augmented by aerial photographs from which animal numbers can later be checked against observations.

Another direct way to estimate animal numbers is to use sample plots and mark-recapture methods. With the widespread availability of computers many versions of mark-recapture formulas have evolved but are beyond the scope of this book. Many of the formulas for estimating animal numbers from mark-recapture methods are based on the equation known as the *Peterson or Lincoln index*. The Lincoln Index is a good place to start when planning a survey which requires estimating populations because it is a simple ratio of marked to unmarked animals in a given time (see Appendix A).

All population estimation methods have advantages, disadvantages, and assumptions which must be considered when selecting the one method best suited to the objectives of the survey. Direct methods are only used when there is a need for a high level of resolution, such as for setting hunting seasons or for associating numbers with habitat factors for research purposes. Estimating numbers of wildlife using direct techniques is seldom done except for certain game and threatened and endangered species. The reasons for this are that the techniques for arriving at a population estimate are time consuming, expensive, and the accuracy of the data is often not at a level that provides confidence in decision-making. Care must be observed when discussing census figures or population estimates to not convey an accuracy that is not there. This is easy to do after time has been spent analyzing data with some of the very sophisticated statistical formulas that exist for this purpose.

Counting Signs (Indirect)

Indirect methods rely on indexes developed from animal signs such as tracks, fecal droppings, nests, or calls. One common technique is fecal (pellet) counts for deer in as much as deer commonly defecate a fixed number of times a day. The number of droppings in sample plots is rapidly converted into deer per unit area or deer days use per unit area for a given number of days, ranging from 365 for yearlong use to as few as several weeks if the time of residence can be observed (Appendix A).

In addition to *signs* left by an animal, other indirect techniques depend upon a frequency of observations to yield a population assessment as abundant, common, uncommon, or rare (Fig. 4.3). Frequency classes can also be used to approximate abundance (Table 4.5). In this method, stations are established to observe the presence or absence of animals for a given number of times throughout the year or by season. If an animal is observed 7 times in a sample of 10 observations the probability is 70% of seeing an animal in that habitat and in similar conditions. This type of indirect abundance estimating depends upon an observer who knows the habitat requirements of an animal and its behavior characteristics in using the habitat (time of day, season of year, etc.). Frequency counts to estimate relative abundance are best suited for low-intensity surveys not associated with setting hunting regulations.

LARGE MAMMALS

A = Very likely to be seen in large numbers every time by a person visiting its habitat at the proper season.

C = May be seen most of the time or in smaller numbers in its habitat at the proper season.

U = May be seen quite regularly in small numbers in the appropriate environment and season.

R = Occupies only a small percentage of its preferred habitat or occupies a very specific limited habitat.

SMALL MAMMALS

A = Frequently observed or collected in numbers.

C = Often observed or collected.

U = Infrequently seen or collected, present in low numbers.

R = Seldom seen or collected, present in very low numbers.

A = ABUNDANT
C = COMMON
U = UNCOMMON
R = RARE

Figure 4.3. Examples of statements to determine relative abundance of mammals.

Table 4.5. Abundance of animals based on frequency of occurrence.

Abundance	Frequency Sightings/Efforts	Probability
Very abundant	8/10	> 80%
Abundant	6–8/10	61–80%
Common	4–6/10	41–60%
Uncommon	2–4/10	21–40%
Rare	1–2/10	1–20%
Very rare	< 1/10	< 1%

Signs of animals can be important in establishing what is in an area if it has never before been inventoried. Reconnaissance inventories are often made using plots to record presence or absence information to establish a base of relative abundance which can in itself be used as an index to compare between years or as a tool in designing more detailed inventory techniques.

4.10 EVALUATING HABITAT

The resource (Re) factor of Conceptual Equation 3 contains subfactors of food, cover, and water needed for a species to survive and reproduce. *Habitat evaluation* is an assessment of the "health" of these factors as they affect a single species or a group of species using the same resources. Before an assessment of the health of habitat can be undertaken, a scale factor specifying the detail and levels of resolution must be determined. When the scale is decided any number of the habitat evaluation techniques which have been developed over a long period of time can generally be used to make an assessment against set standards.

Ecosystem Functioning

Ecosystem functioning involves energy flow and nutrient cycling through food chains and food webs. Evaluating an ecosystem is a complex process and ideally should employ systems analysis techniques. This approach to evaluating an area specifically for wildlife is not practical now because of a lack of understanding of ecosystem components nor will it be practical for some time in the near future. However, it should remain the goal of researchers and managers. Wildlife specialists should routinely be adding to the knowledge and database every new insight into the ecological processes as they occur. Data and knowledge of processes are accumulative so every piece of new information moves us along in understanding how ecosystems function.

Habitat Quality and Populations

Estimating the numbers of animals has been discussed previously. But how do numbers relate to habitat? A basic problem in
managing wildlife is a lack of ways to determine the effects of environmental change, both natural and human-induced, on wildlife populations. Once effects are known and quantified, they can be used to
evaluate land treatment alternatives.

One way to approach the problem is through habitat capability
models. These models are based on the premise that wildlife populations are a function of their habitat, and habitat quality in part determines population density (Conceptual Equation 4, Table 4.6). The
need to develop techniques for determining the present or future
wildlife habitat conditions is becoming more important as forest
managers develop their land-use plans. The need is there because
vegetation is constantly being changed to benefit humans, often at the
expense of wildlife species which need it for food and/or cover.

Table 4.6. Wildlife populations are a function of habitat quality (Patton
1982).

Conceptual Equation 4
$H_Q = f(Fo + Co + Wa + Sp)$
Where
f = a function of
Fo = food
Co = cover
Wa = water
Sp = space (or some diversity index)

The ultimate tool for evaluating habitat is a quantified system
relateing animal populations to different combinations of food and
cover for each ecosystem: ponderosa pine, oak-hickory, northern
hardwoods, and so forth. Habitat capability models are not without
problems. One criticism is that by basing population density only on
quality of habitat, other factors influencing survival and reproduction are not represented. These factors are predators, hazards,
disease, competition within and between species for resources, and
inherited characteristics (the genetics part of the centrum) such as
behavior (for example, territory).

This shortcoming can in part be remedied by common sense. In
habitat models vegetation is assumed to be the most important
environmental factor controlling populations. By deduction populations at some level can be related to habitat quality at some level. For
example, the Abert squirrel needs trees, but how many? Certainly
one tree in the middle of a large clearcut is not adequate nor are two,

three, or four. Obviously such conditions are poor squirrel habitat. On the other hand, 4 ha (10 ac) with 80 trees/ha (200 trees/ac) at an average diameter of 4.3 cm (11 in) might be enough to maintain a viable population and so is something better than poor habitat. There is no doubt that all the factors of Conceptual Equation 3, not just resources (food and cover), have an effect on a wild animal population, but measuring and quantifying the effects has not been a very practical approach for managers.

A Habitat Evaluation Procedure (HEP) to evaluate the requirements of selected species has been developed for use by the Fish and Wildlife Service (U.S. Dept. Inter. 1980). Once the species has been selected, then a habitat suitability index, based on research data and expert opinion, is calculated to compare differences between two or more areas at the same time or the relative value of the same area now and in the future. The assumption is that a *Habitat Suitability Index* (HSI), a numeric value summarizing habitat suitability based on habitat qualitywith a decimal value of 0.0 to 1.0, can be developed for the selected species.

One feature of the HSI is a series of equations that provide the quantitative value as:

$$HSI = [V_1 + V_2 + V_3]/3 \quad \text{Compensatory Model}$$
$$HSI = [2V_1 + V_2 + V_3]/4 \quad \text{Weighted Mean}$$
$$HSI = [V_1 \times V_2 \times V_3]^{1/3} \quad \text{Geometric Mean}$$

Each of the factors (V_1, etc.) pertains to some habitat component such as forage, escape cover, or water. with a decimal rating. A large number of habitat suitability indexes have been developed for both terrestrial and aquatic habitats and can be obtained from the U.S. Fish and Wildlife Service upon request.

Quantified habitat models are very useful but also very expensive and time consuming to develop and validate sufficiently for management purposes. For this reason ecosystem managers and biologists presently have to accept something less than this ultimate tool to get the habitat evaluation job done.

Condition and Trend

The terms condition and trend are well established in the range literature and have become part of the wildlife management terminology as well, primarily for game animals and particularly for the food element in resources. *Condition* refers to present forage composition and production in relation to the kind and amount a particular site or habitat can produce under either accepted management practices or climax conditions. Theobjective of a habitat condition survey is to determine whether the trend in plant development is up, down, or stable by predetermined criteria. Because of the complexity of plant communities, no single factor can be used for determining habitat condition. A host of interacting factors affect plant growth. Further, these factors never remain constant. These characteristics make it difficult to classify entire plant communities consisting of numerous species responding in different ways through time to combinations of factors.

Several methods of estimating wildlife habitat condition and trend based on key browse plants in selected areas have been developed (Dasmann 1951; Cole 1958; Patton and Hall 1966). These methods evaluate use of only key browse species (preferred plants) and do not consider the contribution of either herbaceous vegetation or other factors such as litter or soil stability. The assumption is that the condition of the key browse species is an indication of the state of all factors contributing to habitat condition and trend. It is possible, however, that the trend in habitat condition could be downward before the shift is reflected in key browse species alone. Factors determining habitat condition and trend are:

1.   Plant species composition
2.   Soil properties
3.   Herbage production
4.   Degree of plant use
5.   Available moisture

Plant species composition is important in that there is considerable variation in forage value among and preference for different species. Also, nutritional requirements of big game animals are better met by a mixture of species. Therefore, the better habitats are those providing a variety of preferred forage species.

The reasons for including an evaluation of soils, particularly erosion, is obvious—it is the medium from which terrestrial plants obtain nutrients. Herbage production is dependent on plant density, vigor, and moisture. *Herbage* is all species of browse, forbs, and grasses. All vegetation present must be considered, even though certain plant species are not important constituents of the diet of the targeted wildlife species. In contrast to herbage, *forage* constitutes only the species eaten by a given animal.

Methods of evaluating habitat condition based on ecological principles consider plant composition and use as the major criterion for judgment. Thus, a plant inventory list for the area being evaluated is a basic necessity. An essential feature of the inventory is the classification of species grouped by their successional position—that whether they are *increasers, decreasers,* or *invaders* (Table 4.7). These terms refer to the ways plants react to grazing pressure by ungulates (Fig. 4.4) and are species specific. The terminology was developed in range management but has been extended to include wildlife species of deer, elk, and others.

Table 4.7. Some common terms used to indicate plant value and condition of vegetation.

Condition A	Condition B	Condition C
Decreasers	Increasers	Invaders
Preferred	Staple	Low value
Desirable	Intermediate	Least desirable
Choice	Fair	Unimportant

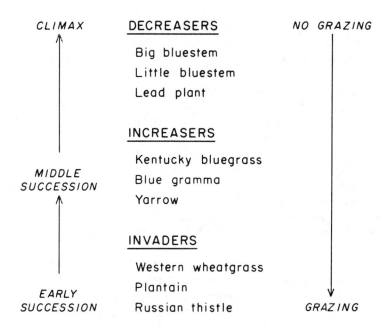

Figure 4.4. Successional changes related to grazing on the North American prairie (Modified from Weaver 1954).

Increasers are those plants that increase in numbers with increasing grazing pressure because other more desirable plantsare consumed first. The opposite is true for decreasers. Invaders are plants that establish as a result of increased grazing but were not part of the original climax vegetation (native). Each condition class is related to successional climax but it should be recognized that climax vegetation is not always the most desirable food or cover on a given site for a given wildlife species. For this reason certain preferred or desirable plant species may indicate equal or better wildlife habitat condition than climax species.

One technique used for evaluating condition and trend for browsing animals is to estimate the degree of use of plant annual growth. Various types of transects or plots can be established and the number of twigs browsed compared to the number not browsed provides a utilization factor (percent) that is compared from year to year (Shafer 1963; Stickney 1966). Another approach is to enclose browse plants in small, protected plots to compare annual growth on similar plots not protected from grazing animals. This same procedure can be used for herbaceous species.

In addition to plant composition and abundance there are other ways to determine habitat condition of which the use of indicators is one. Indicators are only as useful as the accuracy with which they reflect a cause-effect relationship. The terms increaser, decreaser or invader when applied to a specific plant are of the indicator type. Another indicator often used is animal condition.

Since animals are a product of their environment animal condition should also be an indicator of habitat condition (cause-effect situation). Methods used to assess animal condition include indicators of productivity, such as offspring (demographic vigor), weight, and blood chemistry. One of the problems in using animal condition is that by the time the animal's condition begins to decline the decline in the habitat is already well advanced.

Habitat is often evaluated using a *score card* which is another form of a habitat model. Score cards help to determine present condition and relate it to statements of habitat quality. For example, a score card used for gray squirrel habitat in Alabama is based on five criteria with different points (Table 4.8). From this example the following information is available: hardwood composition is 90%; the area contains 1 den tree per acre, in stands that are about 55 years old, with an overstory 75% closed, and a midstory of nut and fruit-bearing trees of 5%. The rating is determined from total point value where 1–5 is poor, 6–10 is fair, and 11–15 is good. The total point value of the 3 categories above is 9 so the rating is fair.

Another evaluation technique frequently used in the West is the classification of shrubby plants (browse) into form and age classes (Table 4.9). Availability takes into account the height of the shrub in relation to how high an animal can reach to obtain leaves and stems. The degree of hedging is a subjective judgement based on experience. *Hedging* is an estimate of the effect of browsing on the regrowth of stems after they have been severed. Several years of continued use cause the stems to clump.

The classification of browse plants into age classes varies according to the species. Ages are often assigned according to main stem diameter size; for example, young might be in the 3–6 mm (⅛–¼ in) class. Plants are considered decadent if more than 25% of their crown

Table 4.8. A score card used to evaluate gray squirrel habitat in Alabama (U.S. Dept. of Agric. 1974).

Criteria

Category I (1 point each)

       <50% hardwood type
x   1 den tree per acre
       Stand age 10–40 years
       Overstory 25–50% closed
x   Midstory 1–10%, nut and fruit-bearing trees

Category II (2 points each)

       51–80% hardwood type
       2 den trees per acre
x   Stand age 41–60 years
x   Overstory 51–75% closed
       Midstory 11–25%, nut and fruit-bearing trees

Category III (3 points each)

x   81–100% hardwood type
       3 den trees per acre
       Stand age >61 years
       Overstory >76% closed
       Midstory >25%, nut and fruit-bearing trees

Table 4.9. Classification of browse plants by form and age class (Dasmann 1951; Cole 1958; Patton and Hall 1966).

Form Classes
1. All available, little or no hedging
2. All available, moderately hedged
3. All available, severely hedged
4. Partly available, little or no hedging
5. Partly available, moderately hedged
6. Partly available, severely hedged
7. Unavailable
8. Dead

Age Classes
S = Seedling
Y = Young
M = Mature
D = Decadent

is dead. Plants classified by form and age class along a permanent transect are compared with previous measurements usually at three-year intervals to arrive at an evaluation for deer, elk, or cattle use of an area.

    The examples given are just two of the many types of score cards (habitat model) that have been and are still being used by wildlife biologists to evaluate habitat. While the criteria for each score card vary, they almost always are based on a system of total points to arrive at a rating using adjective statements (poor to good, etc.).

4.11 MONITORING HABITAT

Monitoring is the process of checking, observing, or measuring the outcome of a process or the changes deviating from some specified quantity, such as in the score card example described previously. Detecting change has always played a significant role in wildlife management whether in a low-resolution population analysis or some higher level of measurement aimed at determining habitat condition and/or trend. In recent years monitoring has taken on a new dimension since it has been incorporated into resource legislation and regulation. As a result, there has been increased attention given to new approaches to monitoring.

The techniques for monitoring both habitat and populations, or indicators of the two, have essentially already been developed, so at this writing no new avenues are evident other than improving the accuracy and efficiency of current techniques, including remote sensing, GIS, and DBMS. Managers needing to monitor an area or the outcome of a management plan are confronted with the same old problems of how much detail is needed to detect change and the cost of obtaining this level of detail.

Monitoring is a feedback mechanism involving five steps (Salwasser et al. 1983):

1. Set objectives.
2. Identify actions and impacts to be evaluated.
3. Collect and analyze data.
4. Evaluate results.
5. Make corrections in the system.

The ultimate goal should be to set in motion a detection network to monitor the resource (Re) factor of Conceptual Equation 3 indirectly through monitoring habitat condition and trend. While the techniques to achieve this have already been developed, though not always refined, the means of constructing an integrated system to monitor all forms of wildlife has not. The effort to correct this shortcoming has led to some interesting and fruitful approaches in systems analysis.

The most direct way to detect changes affecting wildlife is to monitor vegetation quantity (hectares, acres) and quality (succession) over time. This can be done for each management unit on a map documenting acreage and location of vegetation by successional stage. Area size, stand size, and location are the factors monitored over time. Location is as important to monitor as is area size for it depicts the juxtaposition of food and cover.

Verner (1983) has developed a flow chart depicting the relationships among elements of an integrated system for monitoring wildlife populations (Fig. 4.5). While the boxes cannot identify the predictive tools used in action items to be implemented and goals to be evaluated, getting the subsystems within each of the boxes to work will be a long-term process. The flow chart is, nonetheless, a realistic approach to solving the problem of an integrated monitoring system.

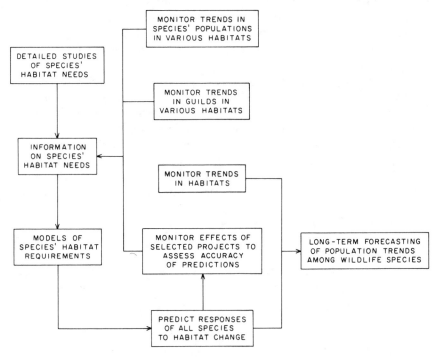

Figure 4.5. A systems approach to monitoring multiple wildlife populations (Verner 1983).

4.12 SUMMARY

The number of techniques to inventory wildlife species and habitat has increased so that a manager now has a variety of tools from which to select. Because of this wealth of information, more time must now be spent in planning. In general, there are three categories of inventory: national, regional, and local, depending on the purpose of the inventory. Any of these three inventories can be conducted at a low, moderate, or high rate of intensity, dependent on the level of resolution needed for decision-making or use. Resolution also pertains to map scale and sampling design which are important components of conducting inventory data collection so must be determined early in the planning process, usually at the time category and intensity are selected.

Information to be collected in the inventory consists of vegetation components and structure, physical features of the landscape, wildlife homes and signs, and wildlife observations. In addition to wildlife observations, the inventory may require an estimation of animal numbers. Such estimations may be made using either indirect or direct techniques. Indirect techniques use indexes developed from animal signs while direct techniques necessitate a census involving mark-recapture methods or aerial surveys for estimating a population.

Habitat data collected in an inventory can be used to develop food, cover, and water maps. All three of these types of maps are critical to the decision-making process. Aerial photography and geographic information systems are useful in compiling management maps. The inventory data together with compiled maps and overlays are used to evaluate habitat. Habitat evaluation is an assessment of the "health" of the habitat components of food, water, and cover based on ecosystem functioning, habitat quality, populations, or condition and trend.

Habitat monitoring is the process of checking, observing, or measuring a process to detect desired changes or unwanted deviation from stated standards. Many of the techniques required to monitor habitat and populations have already been developed but their refinement and integration into a coherent system is needed for practical field application.

4.13 RECOMMENDED READING

Allan, P. F., L. G. Garland, and R. F. Dugan. 1963. Rating northeastern soils for their suitability for wildlife habitat. *Trans. N. Amer. Wildl. and Nat. Resour. Conf.* 28:247–261.

Avery, T. E. 1975. *Natural Resources Measurements*. McGraw-Hill, New York.

Call, M. W. 1982. *Terrestrial Wildlife Inventories*. Tech. Note 349, USDI Bureau of Land Manage., Denver, CO.

Cooperrider, A. Y., R. J. Boyd, and H. R. Stuart. 1986. *Inventory and Monitoring of Wildlife Habitat*. USDI Bureau Land Manage. Serv. Center, Denver, CO.

Davis, D. E. 1986. *Handbook of Census Methods for Terrestrial Vertebrates.* CRC Press, Boca Raton, FL.

Flood, B. M., M. Sangster, R. Sparrowe, and T. Baskett. 1977. *A Handbook for Habitat Evaluation Procedures.* Resource Pub. 132, USDI Fish and Wildl. Serv., Wash., DC.

Gysel, L., and J. Lyon. 1980. Habitat analysis and evaluation. Pages 305–327 in S. D. Schemnitz, ed. *Wildlife Management Techniques Manual.* Wildl. Soc., Wash., DC.

Lund, H. G. 1986. *A Primer on Integrating Resource Inventories.* Gen. Tech. Rep. WO-49, USDA For. Serv., Wash., DC.

Lund, H. G., V. J. LeBau, P. F. Ffolliott, D. W. Robinson. 1978. *Integrated Inventories of Renewable Natural Resources.* Gen. Tech. Rep. RM–55, USDA For. Serv., Rocky Mountain For. and Range Exp. Sta., Fort Collins, CO.

Marmelstein, A. 1977. *Classification, Inventory, and Analysis of Fish and Wildlife Habitat.* FWS/OBS-78/76, USDI Fish and Wildl. Serv., Office of Biol. Serv., Wash., DC.

Myers, W. L., and R. L. Shelton. 1980. *Survey Methods for Ecosystem Management.* John Wiley and Sons, New York, NY.

O'Neil, L. J. 1985. *Habitat Evaluation Methods Notebook.* Instruction Rep. EL 85-3, Dept. of the Army, Corps of Eng. Waterways Exp. Sta., Vicksburg, MS.

U.S. Dept. Interior. 1980. *Habitat as a Basis for Environmental Assessment.* ESM 101, Release 4-80, Fish and Wildl. Serv., Div. of Ecology Serv., Wash., DC.

White, G. C., D. R. Anderson, K. P. Burnham, and D. L. Otis. 1982. *Capture-Recapture and Removal Methods for Sampling Closed Populations.* LA-8787-NERP, Los Alamos Natl. Lab., Los Alamos, NM.

Chapter 5

FOREST REGIONS AND WILDLIFE

The forests of the United States can be broadly separated into the western and eastern regions (Fig. 5.1) (U.S. Dept. of Agric. 1968). The basic difference between the two is rainfall and elevation both of which influence vegetation growth. As a result, forests in the West contain trees that are primarily coniferous while those in the East are mostly deciduous although there are great variations within the Regions. Other countries have similar schemes for grouping forests.

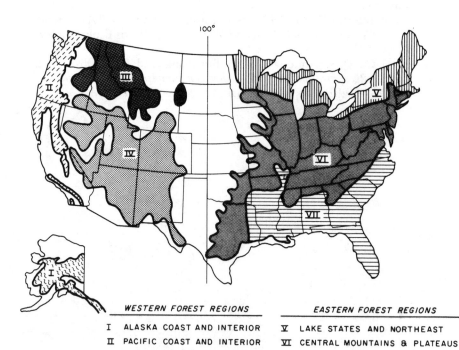

WESTERN FOREST REGIONS	EASTERN FOREST REGIONS
I ALASKA COAST AND INTERIOR	V LAKE STATES AND NORTHEAST
II PACIFIC COAST AND INTERIOR	VI CENTRAL MOUNTAINS & PLATEAUS
III NORTHERN ROCKY MOUNTAINS	VII SOUTHERN STATES
IV SOUTHERN ROCKY MOUNTAINS	

Figure 5.1. Forest Regions of the United States.

5.1 FOREST CLASSIFICATION

In addition to the United Nations (1973) world scheme of vegetation classification (Table 1.3), forests can be classified geographically (western, eastern, southern), economically (commercial or noncommercial), in combination (western coniferous forests), or ecologically by vegetation characteristics. The geographical classification is useful because it integrates a large amount of complex data into a simple picture for quick reference. One common way to ecologically classify forests is by a recognizable cover type derived from the dominant tree species such as oak-hickory, ponderosa pine, or redwood. Within the continental limits of the United States, forests have been defined and delineated into 25 major cover types (U.S. Dept. of Agric. 1958) containing over 850 native tree species.

The terms *cover type* and *forest cover* are synonyms for forest cover type. Tree diversity in the forest types results from the interactions of the range of soil, topography, and climatic conditions (Conceptual Equation 1) found within the geographical areas for each type. The Society of American Foresters (Eyre 1980) has developed a detailed description of 145 forest cover types. The criteria used are as follows:

1. The dominant trees must cover at least 25% of the area.

2. The type must occupy a fairly large area but not necessarily in continuous stands.

3. The type must be based entirely on biological considerations.

Because of the number, size, and scale complexity of the forest types it is difficult to show on small maps the detail of a specific type. For this reason statistics by a geographical area of sections and regions are often used as was done for *An Assessment of the Forest and Rangeland Situation in the United States* (U.S. Dept. of Agric. 1981).

Approximately 320 terrestrial wildlife species in the United States are found on forested lands. Nowhere is the conflict between humans and wildlife more evident than in forested areas where trees used by many wildlife species are also in demand by humans for the manufacture of wood products or to be cleared for agriculture, urbanization, or other purposes. A *tree* is defined as a woody plant with a diameter greater than or equal to 10.2 cm (4 in) measured at breast height (dbh) of 1.37 m (4.5 ft). Trees provide needles, cones, bark, leaves, pollen, acorns, logs, stumps, twigs, mistletoe, and associated insects for wildlife use.

Conversely, over 5,000 products used by humans come from trees. Major uses include wood for construction and the manufacture of furniture and pulp for paper and synthetic fabrics such as rayon

(Fig. 5.2). As the population of humans increases so does the demand for wood products. The trend in the total lumber consumption (Fig. 5.3) in the United States will rise from 323 million cubic meters (approximately 57 billion board feet) in 1990 and peak in the year 2030 at about 368 million cubic meters (approximately 65 billion board feet).

Forest land—at least 10% stocked with trees—in the United States covers about one-third of the total land area or about 298 million ha (736 million ac). Some forest land is found in every state, but the states of North Dakota, South Dakota, Nebraska, and Kansas have less than 5% (Fig. 5.4), while more than 50% of the area of 21 states is forested. Nationally, 195 million ha (482 million ac) are *commercial forests* capable of growing economically valuable trees in excess of 0.56 m^3 ha (20 ft^3).

Of the total commercial land, three-quarters is in the eastern United States, equally divided between northern and southern states. Privately owned land contains more than twice the commercial acreage of federal lands (Fig. 5.5).

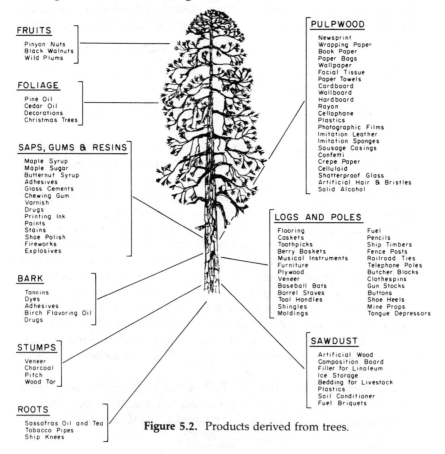

FRUITS

Pinyon Nuts
Black Walnuts
Wild Plums

FOLIAGE

Pine Oil
Cedar Oil
Decorations
Christmas Trees

SAPS, GUMS & RESINS

Maple Syrup
Maple Sugar
Butternut Syrup
Adhesives
Glass Cements
Chewing Gum
Varnish
Drugs
Printing Ink
Paints
Stains
Shoe Polish
Fireworks
Explosives

BARK

Tannins
Dyes
Adhesives
Birch Flavoring Oil
Drugs

STUMPS

Veneer
Charcoal
Pitch
Wood Tar

ROOTS

Sassafras Oil and Tea
Tobacco Pipes
Ship Knees

PULPWOOD

Newsprint
Wrapping Paper
Book Paper
Paper Bags
Wallpaper
Facial Tissue
Paper Towels
Cardboard
Wallboard
Hardboard
Rayon
Cellophane
Plastics
Photographic Films
Imitation Leather
Imitation Sponges
Sausage Casings
Confetti
Crepe Paper
Celluloid
Shatterproof Glass
Artificial Hair & Bristles
Solid Alcohol

LOGS AND POLES

Flooring
Caskets
Toothpicks
Berry Baskets
Musical Instruments
Furniture
Plywood
Veneer
Baseball Bats
Barrel Staves
Tool Handles
Shingles
Moldings

Fuel
Pencils
Ship Timbers
Fence Posts
Railroad Ties
Telephone Poles
Butcher Blocks
Clothespins
Gun Stocks
Buttons
Shoe Heels
Mine Props
Tongue Depressors

SAWDUST

Artificial Wood
Composition Board
Filler for Linoleum
Ice Storage
Bedding for Livestock
Plastics
Soil Conditioner
Fuel Briquets

Figure 5.2. Products derived from trees.

LUMBER CONSUMPTION

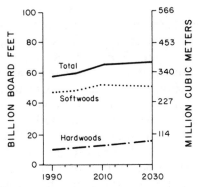

Figure 5.3. Projected lumber consumption for 1990–2030 (U.S. Dept. of Agric. 1982).

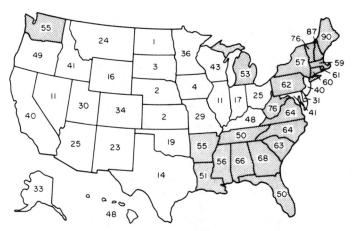

Figure 5.4. Forest land as a percentage of total land area (U.S. Dept. of Agric. 1981).

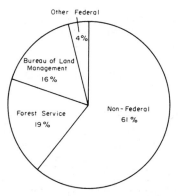

Figure 5.5. Ownership of forest land in the United States (U.S. Dept. of Agric. 1981).

5.2 FOREST REGIONS IN THE UNITED STATES

Coniferous forests occur west of the 100th meridian and consist chiefly of large areas of trees extending over the main Rocky Mountain and Pacific Coast ranges. Stands of trees grow along the coast and inland from Alaska to California. Many comparatively small forested tracts are located on ridges and the higher plateaus, intermingled with treeless stretches. They are widely scattered in great arid districts, especially in parts of the central and southern Rocky Mountains where large areas of the dry foothills support coniferous woodlands.

The western forests can be broadly grouped into four regions (I, II, III, IV) based on the extent of the dominant vegetation (Table 5.1). In the eastern United States there are three regions (V, VI, VII). Florida and Texas have small areas of tropical forests. Large unbroken forest areas still exist in the Maritime provinces, northern New

Table 5.1. Forest regions and cover types.

Region	Description
I.	ALASKA COAST AND INTERIOR (Hemlock/Spruce/Hardwoods) Dominant types:* 23, 24
II.	PACIFIC COAST AND INTERIOR (Douglas-fir/Ponderosa pine/Redwood) Dominant types: 11, 14, 13, 12, 16 Other types: 17, 18, 22
III.	NORTHERN ROCKY MOUNTAINS (Lodgepole pine/Douglas-fir/Larch Dominant types: 16, 11, 17, 14, 18 Other types: 15, 22
IV.	SOUTHERN ROCKY MOUNTAINS (Pinyon-Juniper/Ponderosa pine/Fir-Spruce) Dominant types: 22, 14, 18, 16 Other types: 11
V.	LAKE STATES AND NORTHEAST (Aspen-Birch/Maple-Beech-Birch/Spruce-Fir) Dominant types: 10, 9, 2, 1 Other types: 6, 8
VI.	CENTRAL MOUNTAINS AND PLATEAUS (Oak-Hickory/Oak-Pine) Dominant types: 6, 5, 9 Other types: 7, 8, 9
VII.	SOUTHERN STATES (Loblolly-Shortleaf/Longleaf-Slash pine) Dominant types: 4, 3, 5, 6 Other types: 1, 6, 7, 22

*Numbers listed for dominant and other types refer to descriptions given under forest cover types.

England, northeastern Pennsylvania, and the Appalachian section of the South Atlantic and Gulf states. Elsewhere the forest lands are mainly in small tracts on farms and other small properties.

5.3 FOREST COVER TYPES

Each forest region contains several major and minor forest cover types that indicate on a broad geographic scale the vegetation characteristics of that region (Table 5.1). Each forest region is made up of a distinct plant complex. The forest cover types used to define these regions are derived from several publications (U.S. Dept. of Agric. 1973, 1977; Soc. of Am. For. 1980). Twenty-two of the types are identified by color on a 1:7,500,000 scale map for the continental United States, and two unnumbered types are identified for Alaska (Type 23 and 24) (U.S. Dept. of Agric. 1968). A general description of each type is given in Appendix B.

5.4 FOREST WILDLIFE

The forest region with its associated wildlife species is the first level necessary for understanding forest-wildlife habitat relationships (Table 5.1). The complexity of tree species in a region does not account for the total diversity of wildlife species but rather trees and the understory of grasses, forbs, and shrubs which are food and cover for wildlife and are influenced by overstory trees as the dominant life-form.

Furthermore, there is considerable overlap in the occurrence of animal species from one region to another. This overlap is an indication of the ecological amplitude of the species. For example, 29 species, excluding fish, associated with forests in the United States occur in all seven regions (Table 5.2). On the other hand, 50 species occur in only one region (Table 5.3), but the majority (122 species) occur in more than four regions (Table 5.4).

Summary life history information for wildlife species found in Forest Regions is provided in Appendix C for representative life-forms of amphibian, bird, fish, mammal, and reptile (Conant 1958; Forbs 1961; Chapman and Feldhamer 1982; National Geographic Society 1983; Bull and Farrand 1984; Wenger 1984; Stebbins 1985; Whiting 1985; Sutton and Sutton 1987; Audubon Society 1988; Whitaker 1988). The species listed are those known to be associated with various cover types in the region's geographical area. However, the species presence for each region is potential as it may be absent in a particular local area. Species are listed if they could be present either summer, winter, or yearlong. Species listed may not occupy the entire region but may be in scattered local populations or in some cases the region is at the fringe of their geographical range and nominal habitat.

The wildlife species in Appendix C were chosen because they occur in four or more forest regions, are representative of a large taxonomic group, are important game animals, are examples of species restricted to a given cover type or region, or have aesthetic value. The lists should not be considered definitive but are intended only to show species-vegetation associations over a broad geographical range, to demonstrate the importance of forests for a large number of wildlife species, and to indicate the ecological amplitude of a select group of species. The association of animals with forest regions provides a draft database for developing wildlife habitat relationship systems for management purposes.

Table 5.2. Wildlife species common to seven forest regions.

Common Name
Birds

American redstart
American robin
Bald eagle
Brown creeper
Cedar waxwing
Chipping sparrow
Dark-eyed junco
Downy woodpecker
Fox sparrow
Golden-crowned kinglet
Great horned owl
Hairy woodpecker
Hermit thrush
Northern flicker
Northern saw-whet owl
Pine siskin
Red-breasted nuthatch
Ruby-crowned kinglet
Sharp-shinned hawk
Song sparrow
Tree swallow
White-crowned sparrow
Yellow-rumped warbler

Mammals

Beaver
Black bear
Coyote
Mink
Red fox
River otter

Table 5.3. Wildlife species common to a single forest region.

Common Name	Region
Amphibians	
Appalachian woodland salamander	VI
Arboreal salamander	II
California newt	II
Cascades frog	II
Clouded salamander	II
Foothill yellow-legged frog	II
Jemez Mountains salamander	IV
Mountain yellow-legged frog	II
Olympic salamander	II
Oregon ensatina	II
Pine woods treefrog	VII
Red-legged frog	II
Birds	
American swallow-tailed kite	VII
Anhinga	VII
Bay-breasted warbler	V
Bridled titmouse	IV
Brown-crested flycatcher	IV
California quail	II
Cape May warbler	V
Dusky-capped flycatcher	IV
Elegant trogon	IV
Grace's warbler	IV
Hermit warbler	II
Kirtland's warbler	V
Nuttall's woodpecker	II
Prothonotary warbler	V
Red-faced warbler	IV
Siberian tit	I
Strickland's woodpecker	IV
Tennessee warbler	V
Virginia's warbler	III
Mammals	
Abert's squirrel	IV
Brush rabbit	II
Coati	IV
Collared peccary	IV
Douglas' squirrel	II
Gray-collared chipmunk	IV
Mexican woodrat	IV
Mountain beaver	II
Red tree vole	II
Townsend's chipmunk	II
Western gray squirrel	II
Wild boar	VI

Table 5.3. Continued.

Common Name	Region
Reptiles	
California mountain kingsnake	II
Eastern diamondback rattlesnake	VII
Mole skink	VII
Plateau striped whiptail	IV
Sharp-tailed snake	II
Sonoran mountain kingsnake	III
Striped racer	II

Table 5.4. Occurrence of wildlife species by forest region.

Common Name	I	II	III	IV	V	VI	VII
				Region			
Amphibians							
American toad	—	—	—	—	X	X	—
Appalachian woodland salamander	—	—	—	—	—	X	—
Arboreal salamander	—	X	—	—	—	—	—
California newt	—	X	—	—	—	—	—
Carpenter frog	—	—	—	—	—	X	X
Cascades frog	—	X	—	—	—	—	—
Clouded salamander	—	X	—	—	—	—	—
Common gray treefrog	—	—	—	—	X	X	X
Eastern newt	—	—	—	—	X	X	X
Foothill yellow-legged frog	—	X	—	—	—	—	—
Green treefrog	—	—	—	—	—	X	X
Jemez Mountains salamander	—	—	—	X	—	—	—
Long-toed salamander	—	X	X	—	—	—	—
Marbled salamander	—	—	—	—	X	X	X
Mountain dusky salamander	—	—	—	—	X	X	—
Mountain yellow-legged frog	—	X	—	—	—	—	—
*Northern leopard frog	—	X	X	X	X	X	—
Olympic salamander	—	X	—	—	—	—	—
Oregon ensatina	—	X	—	—	—	—	—
Pacific giant salamander	—	X	X	—	—	—	—
Pacific treefrog	—	X	X	X	—	—	—
Pickerel frog	—	—	—	—	X	X	X
Pine woods treefrog	—	—	—	—	—	—	X
Red-legged frog	—	X	—	—	—	—	—
Rough-skined newt	X	X	—	—	—	—	—
Southern chorous frog	—	—	—	—	—	X	X
Southern leopard frog	—	—	—	—	—	X	X
Southern toad	—	—	—	—	—	X	X
Spotted frog	—	X	X	—	—	—	—
Spring peeper	—	—	—	—	X	X	X
Tailed frog	—	X	X	—	—	—	—

*Indicates life history information is included in Appendix C

Table 5.4. Continued.

Common Name	I	II	III	IV	V	VI	VII
				Region			
*Tiger salamander	—	—	X	X	X	X	X
*Western toad	X	X	X	X	—	—	—
Wood frog	—	—	—	—	X	X	—

Birds

Common Name	I	II	III	IV	V	VI	VII
Acadian flycatcher	—	—	—	—	—	X	X
Acorn woodpecker	—	X	—	X	—	—	—
American crow	—	X	X	X	X	X	X
*American dipper	X	X	X	X	—	—	—
American redstart	X	X	X	X	X	X	X
*American robin	X	X	X	X	X	X	X
American swallow-tailed kite	—	—	—	—	—	—	X
American woodcock	—	—	—	—	X	X	X
Anhinga	—	—	—	—	—	—	X
Anna's hummingbird	—	X	X	—	—	—	—
Ash-throated flycatcher	—	X	X	X	—	—	—
*Bald eagle	X	X	X	X	X	X	X
Band-tailed pigeon	—	X	—	X	—	—	—
Barred owl	—	—	—	—	X	X	X
Bay-breasted warbler	—	—	—	—	X	—	—
*Bewick's wren	—	X	—	X	X	X	X
Black-and-white warbler	—	—	—	—	X	X	X
Black-backed woodpecker	X	X	X	—	X	—	—
Black-billed cuckoo	—	—	X	—	X	X	—
*Black-capped chickadee	X	X	X	X	X	X	—
*Black-chinned hummingbird	—	X	X	X	—	X	—
Black-headed grosbeak	—	X	X	X	—	—	—
Black-throated blue warbler	—	—	—	—	X	X	—
Black-throated gray warbler	—	X	X	X	—	—	—
Black-throated green warbler	—	—	—	—	X	X	—
Blackburnian warbler	—	—	—	—	X	X	—
Blackpoll warbler	X	—	—	—	X	—	—
*Blue grouse	X	X	X	X	—	—	—
Blue jay	—	—	—	—	X	X	X
*Blue-gray gnatcatcher	—	X	X	X	X	X	X
Boreal chickadee	X	—	—	—	X	—	—
Bridled titmouse	—	—	—	X	—	—	—
Broad-tailed hummingbird	—	X	X	X	—	—	—
Broad-winged hawk	—	—	—	—	X	X	X
*Brown creeper	X	X	X	X	X	X	X
Brown thrasher	—	—	—	—	X	X	X
Brown towhee	—	X	—	X	—	—	—
Brown-crested flycatcher	—	—	—	—	—	—	—
Brown-headed nuthatch	—	—	—	—	—	X	X
Bushtit	—	X	—	X	—	—	—
California quail	—	X	—	—	—	—	—
Calliope hummingbird	—	X	X	X	—	—	—
Canada warbler	—	—	—	—	X	X	—
Cape May warbler	—	—	—	—	X	—	—

Table 5.4. Continued.

Common Name	I	II	III	IV	V	VI	VII
				Region			
Carolina chickadee	—	—	—	—	—	X	X
Cassins's finch	—	X	X	X	—	—	—
Cedar waxwing	X	X	X	X	X	X	X
Chestnut-backed chickadee	X	X	X	—	—	—	—
Chestnut-sided warbler	—	—	—	—	X	X	—
*Chipping sparrow	X	X	X	X	X	X	X
Chuck-will's-widow	—	—	—	—	—	X	X
Clark's nutcracker	—	X	X	X	—	—	—
Common loon	X	X	X	—	X	X	X
*Common raven	X	X	X	X	X	X	—
*Cooper's hawk	—	X	X	X	X	X	X
*Dark-eyed junco	X	X	X	X	X	X	X
*Downy woodpecker	X	X	X	X	X	X	X
Dusky-capped flycatcher	X	—	—	—	—	—	—
*Eastern screech-owl	—	X	X	X	X	X	X
Eastern wood pewee	—	—	—	—	X	X	X
Elegant trogon	—	—	—	X	—	—	—
*Evening grosbeak	—	X	X	X	X	X	X
Flammulated owl	—	X	X	X	—	—	—
*Fox sparrow	X	X	X	X	X	X	X
*Golden-crowned kinglet	X	X	X	X	X	X	X
Grace's warbler	—	—	—	X	—	—	—
Gray jay	X	X	X	X	X	—	—
Great crested flycatcher	—	—	—	—	X	X	X
Great gray owl	X	X	X	—	X	—	—
*Great horned owl	X	X	X	X	X	X	X
Green-tailed towhee	—	X	X	X	—	—	—
*Hairy woodpecker	X	X	X	X	X	X	X
*Hermit thrush	X	X	X	X	X	X	X
Hermit warbler	—	X	—	—	—	—	—
Hooded warbler	—	—	—	—	—	X	X
Kentucky warbler	—	—	—	—	—	X	X
Kirtland's warbler	—	—	—	—	X	—	—
Lazuli bunting	—	X	X	X	—	—	—
Lewis' woodpecker	—	X	X	X	—	—	—
*Lincoln's sparrow	X	X	X	X	X	—	X
*Long-eared owl	—	X	X	X	X	X	X
MacGillivray's warbler	X	X	X	X	—	—	—
Magnolia warbler	—	—	—	—	X	X	—
Merlin	X	X	X	X	—	—	X
Mississippi kite	—	—	—	—	—	—	X
*Mountain bluebird	X	X	X	X	—	—	—
Mountain chickadee	—	X	X	X	—	—	—
Mountain quail	—	X	X	—	—	—	—
*Mourning dove	—	X	X	X	X	X	X
Nashville warbler	—	X	X	—	X	X	—
*Northern cardinal	—	—	—	X	X	X	X
*Northern flicker	X	X	X	X	X	X	X
*Northern goshawk	X	X	X	X	X	X	—

*Indicates life history information is included in Appendix C

Table 5.4. Continued.

Common Name	I	II	III	IV	V	VI	VII
*Northern oriole	X	X	X	X	X	X	X
Northern parula	—	—	—	—	X	X	X
Northern pygmy owl	X	X	X	X	—	—	—
*Northern saw-whet owl	X	X	X	X	X	X	X
*Northern shrike	X	X	X	X	X	X	—
Nuttal's woodpecker	—	X	—	—	—	—	—
*Olive-sided flycatcher	X	X	X	X	X	X	—
Orange-crowned warbler	X	X	X	X	—	—	X
Ovenbird	—	—	—	—	X	X	—
Painted bunting	—	—	—	—	—	X	X
*Pileated woodpecker	—	X	X	—	X	X	X
*Pine grosbeak	X	—	X	X	X	—	—
*Pine siskin	X	X	X	X	X	X	X
Pine warbler	—	—	—	—	X	X	X
Pinyon jay	—	X	X	X	—	—	—
Plain titmouse	—	X	—	X	—	—	—
Prairie warbler	—	—	—	—	X	X	X
Prothonotary warbler	—	—	—	—	X	—	—
*Purple finch	—	X	X	—	X	X	X
Pygmy nuthatch	—	X	X	X	—	—	—
*Red crossbill	X	X	X	X	X	X	—
Red-bellied woodpecker	—	—	—	—	—	X	X
*Red-breasted nuthatch	X	X	X	X	X	X	X
Red-cockaded woodpecker	—	—	—	—	—	X	X
*Red-eyed vireo	—	—	X	X	X	X	X
Red-faced warbler	—	—	—	X	—	—	—
*Red-headed woodpecker	—	—	—	X	X	X	X
Red-shouldered hawk	—	X	—	—	X	X	X
Red-tailed hawk	X	X	X	X	X	—	X
Rose-breasted grosbeak	—	—	—	—	X	—	X
*Ruby-crowned kinglet	X	X	X	X	X	X	X
Ruby-throated hummingbird	—	—	—	—	—	X	X
*Ruffed grouse	X	X	X	X	X	X	—
Rufous hummingbird	X	X	X	—	—	—	—
*Rufous-sided towhee	—	X	X	X	X	X	X
Scarlet tanager	—	—	—	—	X	X	—
Scrub jay	—	X	X	—	—	—	X
*Sharp-shinned hawk	X	X	X	X	X	X	X
Siberian tit	—	—	—	—	—	—	—
*Solitary vireo	—	X	X	X	X	X	X
*Song sparrow	X	X	X	X	X	X	X
Spotted owl	—	X	—	X	—	—	—
*Spruce grouse	X	X	X	—	X	—	—
*Steller's jay	X	X	X	X	—	—	—
Strickland's woodpecker	—	—	—	X	—	—	—
Summer tanager	—	X	—	X	—	X	X
Swainson's thrush	X	X	X	—	X	X	—
Tennessee warbler	—	—	—	—	X	—	—
*Three-toed woodpecker	X	X	X	X	X	—	—
Townsend's solitaire	X	X	X	X	—	—	—

Table 5.4. Continued.

Common Name	Region						
	I	II	III	IV	V	VI	VII
Townsend's warbler	X	X	X	—	—	—	—
Tree swallow	X	X	X	X	X	X	X
Tufted titmouse	—	—	—	—	X	X	X
Varied thrush	X	X	—	—	—	—	—
Veery	—	—	X	X	X	X	—
*Violet-green swallow	X	X	X	X	—	—	—
Virginia's warbler	—	—	X	—	—	—	—
Warbling vireo	—	X	X	X	X	X	X
Western bluebird	—	X	X	X	—	—	—
*Western flycatcher	X	X	X	X	—	—	—
Western screech-owl	—	—	—	—	—	—	—
Western tanager	—	X	X	X	—	—	—
Western wood pewee	X	X	X	X	—	—	—
*Whip-poor-will	—	X	—	X	X	X	X
White-breasted nuthatch	—	X	X	X	X	X	X
*White-crowned sparrow	X	X	X	X	X	X	X
White-eyed vireo	—	—	—	—	—	X	X
White-headed woodpecker	—	X	X	—	—	—	—
White-throated sparrow	—	X	—	—	—	X	X
White-winged crossbill	X	X	X	—	X	X	—
*Wild turkey	—	X	—	X	—	X	X
Williamson's sapsucker	—	X	X	X	—	—	—
*Wilson's warbler	X	X	X	X	X	—	—
*Winter wren	X	X	X	—	X	X	X
*Wood duck	—	X	X	—	X	X	X
Wood thrush	—	—	—	—	X	X	X
Worm-eating warbler	—	—	—	—	—	X	X
*Yellow-bellied sapsucker	—	X	X	X	X	X	X
Yellow-breasted chat	—	X	X	X	X	X	X
Yellow-rumped warbler	X	X	X	X	X	X	X
Yellow-throated vireo	—	—	—	—	X	X	X
Yellow-throated warbler	—	—	—	—	—	X	X

Fishes

Brook trout	—	X	X	X	X	X	—
*Brown trout	—	X	X	X	X	X	—
*Cutthroat trout	X	X	X	X	—	—	—
*Largemouth bass	—	X	X	X	X	X	X
*Rainbow trout	X	X	X	X	X	X	—
*Smallmouth bass	—	X	X	X	X	X	X

Mammals

Abert's squirrel	—	—	—	X	—	—	—
*Beaver	X	X	X	X	X	X	X
Bighorn sheep	—	X	X	X	—	—	—
*Black bear	X	X	X	X	X	X	X
*Bobcat	—	X	X	X	X	X	X

*Indicates life history information is included in Appendix C

Table 5.4. Continued.

Common Name	I	II	III	IV	V	VI	VII
				Region			
Brush rabbit	—	X	—	—	—	—	—
Bushy-tailed woodrat	—	X	X	X	—	—	—
Coati	—	—	—	X	—	—	—
Collared peccary	—	—	—	X	—	—	—
Cotton mouse	—	—	—	—	—	X	X
*Coyote	X	X	X	X	X	X	X
*Deer mouse	X	X	—	X	X	X	—
Douglas' squirrel	—	X	—	—	—	—	—
Eastern chipmunk	—	—	—	—	X	X	—
*Eastern cottontail	—	—	—	X	X	X	X
Eastern spotted skunk	—	—	—	—	X	X	X
Eastern woodrat	—	—	—	—	—	X	X
*Elk	—	X	X	X	—	—	—
Ermine	X	X	X	X	X	X	—
Fisher	—	X	X	—	X	—	—
Fox squirrel	—	—	—	—	X	X	X
*Gray fox	—	X	—	X	X	X	X
*Gray squirrel	—	—	—	—	X	X	X
Gray wolf	X	X	—	—	X	—	—
Gray-collared chipmunk	—	—	—	X	—	—	—
*Grizzly bear	X	—	X	—	—	—	—
*Hoary bat	—	X	X	X	X	X	X
Hoary marmot	—	X	X	—	—	—	—
Least chipmunk	—	X	X	X	—	—	—
*Long-tailed vole	X	X	X	X	—	—	—
*Long-tailed weasel	—	X	X	X	X	X	X
*Lynx	X	—	X	X	—	—	—
Mantled ground squirrel	—	X	X	X	—	—	—
*Marten	X	X	X	X	X	—	—
Mexican woodrat	—	—	—	X	—	—	—
*Mink	X	X	X	X	X	X	X
*Moose	X	—	X	X	X	—	—
Mountain beaver	—	X	—	—	—	—	—
*Mountain lion	—	X	X	X	—	—	X
*Mule deer	X	X	X	X	—	—	—
New England cottontail	—	—	—	—	X	X	—
Nine-banded armadillo	—	—	—	—	—	X	X
*Northern flying squirrel	X	X	X	X	X	X	—
Pika	—	X	X	X	—	—	—
*Porcupine	X	X	X	X	X	—	—
*Pygmy shrew	X	—	X	—	X	X	—
*Raccoon	—	X	X	X	X	X	X
*Red fox	X	X	X	X	X	X	X
*Red squirrel	X	X	X	X	X	X	—
Red tree vole	—	X	—	—	—	—	—
Ringtail	—	X	—	X	—	—	X
*River otter	X	X	X	X	X	X	X
Short-tailed shrew	—	—	—	—	X	X	X
*Snowshoe hare	X	X	X	X	X	X	—
Southern flying squirrel	—	—	—	—	X	X	X

Table 5.4. Continued.

Common Name	I	II	III	IV	V	VI	VII
*Striped skunk	—	X	X	X	X	X	X
Townsend's chipmunk	—	X	—	—	—	—	—
Uinta chipmunk	—	X	X	X	—	—	—
Vagrant shrew	—	X	X	X	—	—	—
*Virginia opossum	—	X	—	X	X	X	X
Western gray squirrel	—	X	—	—	—	—	—
Western jumping mouse	—	X	X	X	—	—	—
*White-footed mouse	—	—	X	X	X	X	X
*White-tailed deer	—	X	X	X	X	X	X
Wild boar	—	—	—	—	—	X	—
Wolverine	—	X	X	—	—	—	—
*Woodchuck	X	—	—	—	X	X	X
Woodland jumping mouse	—	—	—	—	X	X	—
Woodland vole	—	—	—	—	X	X	X
Yellow-bellied marmot	—	X	X	X	—	—	—
Yellow-pine chipmunk	—	X	X	—	—	—	—

Reptiles

Common Name	I	II	III	IV	V	VI	VII
Broadheaded skink	—	—	—	—	—	X	X
California mountain kingsnake	—	X	—	—	—	—	—
*Coachwhip	—	X	—	X	—	X	X
Coal skink	—	—	—	—	X	X	X
*Common garter snake	—	X	X	—	X	X	X
*Common kingsnake	—	X	—	X	—	X	X
Copperhead	—	—	—	—	—	X	X
Corn snake	—	—	—	—	—	X	X
Eastern box turtle	—	—	—	—	X	X	X
Eastern coral snake	—	—	—	—	—	X	X
Eastern diamondback rattlesnake	—	—	—	—	—	—	X
Five-lined skink	—	—	—	—	X	X	X
*Gopher snake	—	X	X	X	—	X	X
Green anole	—	—	—	—	—	X	X
Ground skink	—	—	—	—	—	X	X
*Milk snake	—	—	X	X	X	X	X
Mole skink	—	—	—	—	—	—	X
Northern alligator lizard	—	X	X	—	—	—	—
*Painted turtle	—	X	X	—	X	X	X
Pigmy rattlesnake	—	—	—	—	—	X	X
Pine woods snake	—	—	—	—	—	X	X
Plateau striped whiptail	—	—	—	X	—	—	—
*Racer	—	X	X	X	X	X	X
Rat snake	—	—	—	—	X	X	X
*Ringneck snake	—	X	X	X	X	X	X
Rough green snake	—	—	—	—	—	X	X
Rubber boa	—	X	X	X	—	—	—
Sharp-tailed snake	—	X	—	—	—	—	—
Short-horned lizard	—	—	X	X	—	—	—
Sonoran mountain kingsnake	—	—	X	—	—	—	—
Striped racer	—	X	—	—	—	—	—

*Indicates life history information is included in Appendix C

Table 5.4. Continued.

Common Name	I	II	III	IV	V	VI	VII
				Region			
Striped whipsnake	—	X	X	X	—	—	
*Timber rattlesnake	—	—	—	—	X	X	X
Western diamondback	—	—	—	X	X	—	—
Western fence lizard	—	X	X	X	—	—	—
*Western rattlesnake	—	X	X	X	—	—	—
Western skink	—	X	X	X	—	—	—
Western terrestrial garter snake	—	X	X	X	—	—	—
Wood turtle	—	—	—	—	X	X	—

5.5 LIFE HISTORY

Life history information describes all the events that happen from the birth to the death of an individual. It includes the specific information on reproduction described under genetics (species characteristics) as well as food and cover requirements that are part of resources. Life history is often separated into two categories—reproduction and habitat requirements—depending on the intended use. However, a complete life history includes distribution data as well as taxonomic classification.

In documenting life history all the subcategories of the six factors (Ha, Pr, Di, Ge, Re, Hu) of Conceptual Equation 3 (Table 2.2) are included but not necessarily in the order of the equation (Table 5.5). The genetics section contains information on taxonomy and there is also a major section for important references. Life history information can be presented in great detail—including information on all the factors, as for elk in the western United States (Appendix D)—or in a summary form providing facts only necessary to become familiar with the species (Appendix C).

5.6 SPECIES CLASSIFICATION

The classification of wildlife into categories for management purposes is difficult because many species can be included in several different groups depending on the purpose of the grouping. Species may be grouped by legal status as:

1. Threatened and endangered
2. Game, such as small game, big game, waterfowl, and fish
3. Furbearer
4. Nongame, such as songbirds, predators, and fish
5. Nonprotected species
6. Introduced, exotic, or feral

Table 5.5. Factors included in a complete life history.

1. GENETIC (SPECIES CHARACTERISTICS AND/OR INFORMATION)

1. GENETIC (SPECIES CHARACTERISTICS AND/OR INFORMATION)
          Taxonomic information
                    Scientific name
                    Common name
                    Family and order
                    Subspecies
          Physical description
                    Weight
                    Color
                    Distinguishing characteristics
          Distribution
                    Geographic range map
          Breeding characteristics
                    Breeding season
                    Age of sexual maturity
                    Number of offspring per breeding
                    Number of breedings per year
                    Gestation period
                    Longevity
          Behavior
                    Territory
                    Migration
                    Niche
                    Mobility (home range)
          Legal status
2. HAZARDS
          Natural
          Man-made
3. PREDATORS
          Primary
          Secondary
4. DISEASES
          Infectious
          Noninfectious
          Herbicides
          Pesticides
          Parasites
5. RESOURCES
          Food
          Water
          Cover Space
          Successful management practices
6. HUMANS
7. MAJOR REFERENCES

Wildlife also may be classed as commercial (for food, fur, or fertilizer) or noncommercial.

A third classification, and one of the most common, is by land use category. This classification usually has five categories:

1. Farm, includes species such as cottontails and quail
2. Forest, includes ruffed grouse, black bear, white-tailed deer, and wild turkey, etc.
3. Wilderness, includes elk, moose, and grizzly bear
4. Wetland, includes species of ducks, geese, and muskrat
5. Rangeland, includes bison, grouse, and pronghorn

A fourth system based on the biological characteristics of a species is very useful and is used when building databases. Biological systems use the life-form of animals for the grouping category as:

1. Amphibians
2. Birds
3. Fishes
4. Mammals
5. Reptiles

Another system in use is that of associating animals with different habitat requirements based on vegetation succession. This system uses three categories:

1. Climax
2. Mid-successional
3. Low-successional

The climax category is close to what was previously identified as wilderness; the mid-successional is similar to forest; and the low-successional species are associated with brushland. Of all the classification systems developed, the species-successional stage comes the closest to being the one wildlife managers can use.

In addition to the five classification systems already mentioned, another is based on an animal's function:

1. Carnivore
2. Herbivore
3. Predator

This classification system is not useful in the decision-making process but is helpful in trying to understand and clarify an animal's role in the functioning of an ecosystem.

No single means of classifying wildlife species into a group for management purposes is beneficial in all situations. In selecting a scheme, *the objectives for using a given classification must be the deciding factor.* In most cases, some combination based on life-forms, legal status, and ecological association is used to convey the desired information such as in management plans.

5.7 SUMMARY

A forest contains vegetation whose dominant life-form is trees. The trees are 10.2 cm (4 in) in diameter and occupy at least 25% of the forest area. About one-third of all terrestrial wildlife species are found only on forested land in North America. Trees provide these species with food items leaves, bark, twigs, and associated insects—as well as with environments for reproduction activities and protection from the weather and humans. Because humans use trees to produce over 5,000 commercial products, conflict arises between humans and wildlife which use trees for food and cover.

About one-third of the total land area in the United States is forest land with at least 10% of that area stocked with trees. In 21 states more than 50% of the area is forested, private land constituting over twice the commercial acreage of federal lands. Within the United States there are 7 general forest regions, 4 west and 3 east of the 100th meridian. Forests in the West are primarily coniferous while those in the East are mostly hardwoods or mixtures of both. Each forest region contains one or more of 24 major or minor forest cover types that indicate on a broad scale the vegetation characteristics of that region. The seven geographical regions are distinguished on the basis of characteristic combinations of these forest cover types. In general, the types contain a dominant tree species or group of species along with several minor components.

These forest types provide food and cover for a variety of wildlife species of amphibians, birds, mammals, and reptiles. The number of forest regions where a particular species is found gives an indication of the ecological amplitude of that species. All these species have a life history that includes species characteristics and habitat requirements. For quick reference an abbreviated form of life history is useful.

There are several ways in which wildlife can be classified including legal, land-use, commercial, biological, or ecological. The choice of a particular way depends on the need for a classification in the decision-making process.

5.8 RECOMMENDED READING

Eyre, F. H., ed. 1980. *Forest Cover Types of the United States and Canada*. Soc. of Amer. Foresters, Wash., DC.

McCormick, J. 1966. *The Life of a Forest*. McGraw-Hill, New York.

Sutton, A., and M. Sutton. 1987. *Eastern Forests*. Audubon Soc. Nature Guides. Alfred A. Knopf, New York, NY.

U.S. Dept. of Agriculture. 1973. *Silvicultural Systems for the Major Forest*

Types of the United States. Hdbk. No. 445, For. Serv., Div. of Timber Manage., Wash., DC.

U.S. Dept. of Agriculture. 1977. *Vegetation and Environmental Features of Forest and Range Ecosystems.* Hdbk. No. 475, For. Serv., Wash., DC.

Wenger, K. F., ed. 1984. *Forestry Handbook.* John Wiley and Sons, New York.

Whiting, S. 1985. *Western Forests.* Audubon Soc. Nature Guide. Alfred A. Knopf, New York.

Chapter 6

MANAGING THE CENTRUM OF FOREST WILDLIFE

Wildlife was defined in Chapter 1, and environment and habitat in Chapter 2. However, a definition of wildlife management is needed to complete the background to discuss the topics that follow. *Wildlife management* is the art and science of manipulating the *centrum* of wild animal populations to meet specific objectives. This definition does not require a specific reason for managing wildlife, for society typically decides which species are important. It is the job of management then to increase, decrease, or to maintain wildlife populations stable in keeping with the desires of society.

Manipulating populations requires a basic knowledge of all the factors affecting the ability of an animal to survive and reproduce. These factors are defined in Conceptual Equation 3 (Table 2.2) and each will be reviewed in this chapter to provide an understanding of how it contributes to managing a particular wildlife species.

6.1 MANAGEMENT OF HAZARDS

The modification of *natural hazards* such as rain, snow, and wind can only be accomplished on a small scale and with limited success so in most cases their management is not a realistic consideration. Rain can be produced through cloud seeding but the technique is expensive and the results are too unpredictable at the present time to be useful. One aspect of producing rain, should it become practical in the future, is that large amounts of precipitation at one time may be detrimental to many wildlife species. For this reason the effects of cloud seeding must be known before it becomes a practical tool.

Snow movement can be controlled to some degree in open areas by installing *drift fences* or *wind rowing* logging debris to provide open or lightly covered areas on the downwind side making forage easier to obtain by deer and elk. *Wind breaks* as have been developed for agricultural shelterbelts can also provide protection for wildlife.

Fire as a natural hazard is much more easily managed than rain, snow, or wind. Fire management involves planning the creation of fuel breaks in vegetation important for a particular species. In using fire the identification of the areas to be protected is the first priority. The second job is to determine how large the fuel breaks must be and where they will be located to provide the most protection. The third task is to develop a schedule of maintenance, for if maintenance is not done the fuel break will lose its protective value over time.

Fire as a man-made hazard results from accidents or prescribed burns. Before initiating a prescribed burn, the plant and animal species native to the area, along with any endangered and threatened species existing there, must be thoroughly documented. Special areas such as nest and roost trees, water holes, and browse concentrations are identified for protection by fire lines. Prescribed fires can be just as hazardous to wildlife as natural fires, but hazards can be reduced if wildlife requirements are considered in the planning and implementation of a prescribed burn.

In general, the effects of fire as a management technique can be summarized as follows (from Lyon et al. 1978; Wright and Bailey 1982; Kramp, Brady, and Patton 1983):

1. Some browse species used by herbivores proliferate in early postfire succession and their nutritional quality is high for several years after burning.

2. In coniferous forests fire stimulates plants which produce fruits and berries. A parallel increase in seed availability favors an increase in use of burned habitat by birds and small mammals.

3. Wood boring insects, some of which are important foods for woodpeckers and other insectivorous birds, may increase. Fires also may decrease populations of wildlife pest insects and parasites.

4. Fire affects the scale and pattern of the vegetation mosaic through fire size, intensity, and frequency. These factors influence the relative abundance of plants and subsequent successional stages. The resulting patterns in turn affect herbivore populations.

5. Carnivores dependent on certain herbivore populations benefit from a fire-created habitat mosaic because it increases their food base—herbivores.

Hazards created by humans are marginally easier to control than natural hazards. *Man-made* hazards are primarily roads and fences.

Road hazards can be reduced in the planning process by avoiding alignment through areas critical for survival and reproduction of wildlife species. In addition, roads can be fenced and underpasses built to provide for free movement where a road separates two use areas or crosses a migratory route.

Road hazards can be further reduced on high speed highways by selecting unpalatable cover plants for roadside and median plantings. The opposite holds true for unpaved forest or other infrequently used roads. Also, road kills along major highways can be reduced by placing 90° light reflectors about 10 yards apart along both sides and in the median.

Fences, depending on the type, are obstacles to free movement of wild animals. Barbed wire is especially hazardous for juvenile antelope, deer, and elk. This problem can be reduced by using smooth wire on the top and bottom strands of the fence. High fences of wood or hog wire can be installed to direct animals to desirable locations or away from undesirable locations.

6.2 MANAGEMENT OF DISEASES

The control of wildlife diseases is a subject that has not been given the attention devoted to other decimating factors. In part this lack of attention indicates a widely held belief that wild animals cannot be treated for diseases so why waste time and effort in trying. Disease is a natural way of controlling animal populations by eliminating unhealthy animals. Disease-weakened animals are easy prey for predators that otherwise would try to kill healthy individuals. Leopold (1933) believed that the effect of disease on wildlife was underestimated and this is probably the case even today. Diseases which clearly kill wild animals are difficult to identify except in cases where there is a large die-off or symptoms are discovered during routine checking of animals killed in a hunting season.

Not all diseases and parasites kill animals directly. Some cause side effects that may be objectionable to hunters. For example, botfly larvae parasitize the Eastern gray squirrel causing nonlethal wounds which hunters find so unpleasant they discard the squirrel carcass (Jacobson, Guynn, and Hackett 1979). The manager's remedy for this problem is to postpone the squirrel hunting season from September to October after the botfly larvae have left their host and the objectionable wounds are healed. Cottontail rabbits not only suffer from the same botfly problem but in addition carry ticks which cause tularemia in rabbits and humans (Yeatter and Thompson 1952). Again the managers can reduce the incidence of the disease in humans by having a hunting season during cold months when ticks have left the host and shortly after all infected rabbits have died.

At one time there was great concern about meningeal worm infections of white-tailed deer. Research results later showed that the parasite had little effect on white-tailed deer but it is known to cause death in other ungulates such as mule deer, moose, reindeer, and elk (Anderson 1972; Anderson and Priestwood 1981; Davidson et al. 1981). This particular parasite is transmitted by snails as the intermediate host. The solution to the problem is to find a way to control the infected snails.

In the case of highly prized trophy species, managers have found it desirable to capture animals and treat them with drugs to ward off disease. This was done in the case of bighorn sheep in Colorado (Schmidt et al. 1979). This technique is expensive, time consuming, and the results are difficult to evaluate, but benefits to management through increased survival may make it worthwhile. However, it is hoped that through research various diseases might be associated with habitat quality and thereafter techniques developed to prevent many diseases through habitat manipulation. In this regard, manipulating habitat with the objective of increasing a population may cause concentrations of animals which the transmission of disease is enhanced. As can be seen the control of diseases and parasites in wild populations is at best a difficult but mostly an impossible task.

Herbicides and pesticides (insecticides, rodenticides, etc.) are included with diseases because like diseases they affect the physiology of animals. But unlike diseases, they can be controlled because they are manufactured and sold by humans. One of the problems involved with these two groups of compounds is the attitude by some users that if one kilogram per hectare will do the job, then a higher concentration will do it better.

Herbicides and pesticides are now licensed for use by government agencies (Environmental Protection Agency in the United States) after a period of testing and documentation of the toxic effects. When using herbicides and pesticides it is incumbent on the user to follow the prescribed precautions listed on the label. The trend now is to develop herbicides and pesticides which biodegrade in a short time and to develop natural (biological) techniques to control the target organism or vegetation.

6.3 MANAGEMENT OF PREDATORS

Predator control as a wildlife management technique is largely oriented toward those animals which prey on game species. However, modern ecological thinking which focuses on the food chains and webs involved in nutrient cycling and energy flow in ecosystems strongly opposes the killing of every predator. As a consequence predator control has awakened public concern and criticism making it desirable to conduct hearings before any large-scale control program is undertaken.

The history of wildlife management contains many examples of failed and successful predator control programs. This past experience has led to the development of some general principles useful to managers in the decision-making process.

1. As a management tool, predator control is generally not feasible over a wide area for both biological and economical reasons. From an economic point of view, the cost-benefit ratio must favor predator control if it is to be a feasible management tool.

2. Predator control is best used only to deal with a local and temporary problem, such as unusually heavy predation in a refuge, fish hatchery, or other area where the prey species is found in unusually high numbers (poultry farms).

3. In many cases predation results from the behavior of an individual animal, not the species as a group.

4. Intensive predator control programs do not result in an increase in game commensurate with the control effort and expense in most wildlife management areas.

5. A predator control program often inadvertently increases the populations of other species such as small rodents.

6. Predator control must on occasion be initiated for health reasons such as control of rabies by reducing populations of foxes, skunks, or raccoons.

Methods to control predators are varied, but programs usually involve one of three methods:

1. Professional hunters and trappers
2. Bounty systems open to the general public
3. Poisons

There are advantages and disadvantages of all three methods. Poisons are not selective to a species, bounty systems are open to fraud, and professional hunters and trappers are very expensive. Most game and fish departments now use their own employees to control predators after intensive review and documentation of the environmental effects.

6.4 MANAGEMENT OF SPECIES

In the centrum, innate genetic characteristics, while centered in the individual animal, influence the development of populations whereby managers view the two matters together. Management of species includes the topics of stocking, refuges, and harvesting.

Stocking and Game Ranching

One of the first techniques used in wildlife management in the early 1930s and 40s was stocking of game animals. Stocking includes both the introduction and reintroduction of a species. As with predator control there have been many successes in stocking animals but wildlife literature also contains examples of unsuccessful attempts. Stocking is not a substitute for habitat improvement and the need for stocking must be examined very carefully before initiating this expensive management technique. Some of the reasons for stocking are:

1. To populate an unoccupied habitat or reintroduce an extirpated species. *Extirpated* means driven out or eliminated in a specific area. Such stocking programs using animals trapped in other areas have been very successful with some species such as deer, turkey, elk, and bighorn sheep. Needless to say, the unoccupied habitat must be suitable for the introduced or reintroduced species.

2. To bolster a declining population. Stocking has rarely been a successful means of accomplishing this objective as the deterioration of habitat quality usually accounts for the decline in native populations.

3. To increase the number of animals available to hunters. The cost of this method in areas other than a hunting preserve make it impractical. However, some government agencies have found that "put-and-take" fishing can be cost effective.

4. To introduce a new species, usually an exotic. An *exotic* is any animal not native to the area where it is being introduced. Several exotics have been successfully introduced in North American including the ringneck pheasant, chukar partridge, Hungarian partridge, and rainbow trout.

The last objective differs from the first three because of the greater biological risk involved in introducing exotics which may possibly compete with native species. Many exotics, as rabbits in Australia, English sparrows in North America, and deer in New Zealand, have caused serious management problems; in these three cases the introduced species had no natural predators so populations are now out of control with quite disastrous results particularly in the case of rabbits and deer.

Before introducing exotics managers should ask these questions:

1. Is the niche occupied by a native species?

2. Will it become a pest?

3. Will it bring in disease?

4. Does it have recreational value?

5. Is there habitat of suitable quantity and quality to meet the life requirements of the species?

An outgrowth of the introduction of exotics has been *game ranching* where animals are raised for food. This method has been tried in several African countries because of the need to provide protein for undernourished people. One of the advantages of using exotic animals in such a program is their ability to reproduce and survive much better than domestic cattle.

When a decision has been made to introduce an animal into unoccupied habitat careful thought and attention must be given to the source of animals for stocking. This should include information about where the animals will be obtained. Will they come from game farms? In most cases, these animals are unable to take care of themselves when released. Will the animals be wild trapped? Trapping wild animals is expensive but the animals are better able to survive and reproduce. In the case of birds there is also the possibility of using *foster parents* or *double clutching*. For example, Texas bobwhite have been used to hatch masked bobwhite eggs. In double clutching, eggs are taken from the nests of wild birds to get them to lay another egg. The egg that is removed is hatched artificially and the chick raised without human contact to be returned to the wild. This technique has been used successfully with endangered species such as the peregrine falcon.

Refuges

The notion of establishing refuges was originally based on the premise that a protected population would increase and overflow into adjacent areas. Refuges are used as a management area to protect one or several species while a *hunting preserve* is used to raise animals for sport hunting. The refuge system has worked well for highly mobile species and for species needing special protection. Refuges are generally established for the following reasons:

1. Nesting, feeding, or loafing areas

2. Escape areas

3. Protection of an endangered species

4. Research purposes

Refuges established in the waterfowl flyways have been the most successful. Income from the sale of waterfowl stamps is used to acquire wetlands. In Canada and the northern United States, waterfowl refuges are established on these lands; in the central and southern United States, waterfowl nesting areas provide protection within wintering areas.

Harvesting

The principle management tool in the control of game animals has traditionally been the regulation of hunting seasons designed to kill excess populations created by management. While harvesting of animals is often a controversial topic it is still done in every state in the United States, every Canadian province, and in most other countries. Some of the objectives of regulating hunting seasons are:

1. To structure the harvest in such a way that the maximum number of animals is provided to the maximum number of hunters.

2. To regulate the number of animals killed so as to ensure the continued productivity of the hunted species. The regulation of numbers killed in a hunting season involves controlling:
   a. The timing and duration of the hunting season
   b. Daily and seasonal limits
   c. Licenses or tags sold, rationed, or auctioned to hunters
   d. Types of hunting equipment permitted

All the mechanisms indicated have the effect of reducing hunting pressure by applying restrictions at different degrees of intensity. The wildlife manager therefore has a variety of tools to limit the number of animals killed in a hunting season. In addition, some of the controls, such as types of equipment and restricting the number of hunters by licenses or tags, also contribute to hunter safety.

When determining hunting regulations, the following general principles of harvesting have proved useful in guiding the control of hunting:

1. A given amount of hunting pressure takes a larger proportion (%) of a population when it is high than when the same population is low. For example, a 10% harvest from 1,000 animals is 100, but a 10% harvest from a population of 100,000 is 10,000. The 100,000 animals may be able to withstand the 10% harvest, but the 1,000 animals may not.

2. Any game species characterized by a high rate of increase can normally sustain a high rate of kill. Examples of population percentages which can be harvested are shown in Table 6.1.

3. In some instances, a high kill rate stimulates production in a given game species.

4. The law of diminishing returns often sets the harvest for some species. As more effort is needed to hunt to find an animal, the number of days hunted and the number of hunters declines.

5. Unhunted populations will reach a dynamic equilibrium (ecological carrying capacity) between animal density and vegetation.

6. Hunted populations will reach a dynamic equilibrium below the ecological carrying capacity.

Table 6.1. Percentage of a population that can be removed by hunting.

Species	Percent	Season
Wild turkey	25–35	
Deer	5–10	Bucks only season
	20–30	To hold population stationary, any choice season
	30–40	To reduce population, any choice season
Rabbit	60–70	
Squirrel	25–50	
Bear	5–15	
Quail	30–40	

In recent years hunting has come under attack by many environmental, ecological, and animal rights groups primarily trying to stop the killing of big game animals by legal action or by active protest. The question is: What would happen if all hunting were stopped? Initially there may be an increase in the density of game animals, but there also would be a loss of millions of dollars in license fees which now contribute to maintaining habitat of many nongame wildlife species. Eventually, without hunting there probably would be some steady state reached where game populations would be lower as a result of the directly acting factors in Conceptual Equation 3 (Table 2.2).

Humans, as a result of an ancient need to hunt for food and fur persisting over thousands of years, may have developed a psychological need or at least a predisposition to hunt. Although most hunters in developed countries do not need to hunt for food, the compulsion to do so remains ingrained in the psyche of many. The hunting experience also seems to lend to a sense of becoming a part of nature where the role of predators is acted out to achieve some psychological benefit that is difficult to explain.

Life Equation

Before a hunting season can be set some accounting must be made of the yearly population cycle of the animal hunted so that the numbers to be harvested can be estimated. The life history of the species provides the background information but it is necessary to condense and reduce this information to develop a useful management tool. This synthesis of information yields a *life equation*. While life history deals with events over the entire life of an animal, a life equation accounts for changes in a population as a result of yearly life events.

The best way to understand a life equation is to construct an example, but first several points must be made. A life equation should be flexible and generally applicable to a wide range of habitats, all of which on the average are capable of supporting a population. These populations should, in the absence of hunting, continue to increase in density until hazards, disease, predators, and resource shortages bring the population into equilibrium with its environment.

Second, these generalized life equations must be modified by the manager to reflect the actual conditions found in individual management units. Third, life equations should represent average conditions, especially for weather which exerts a tremendous short-time effect on total populations of some animals such as quail and rabbits.

Finally, life equations do not generally take into account cyclical phenomena exhibited by some species. Nor do they take account of random ecological changes by floods, timber harvesting, and/or vegetation pattern changes even though in long-range planning such factors are important and must be considered. This difference arises because the life equation is recomputed on an annual basis to set hunting seasons while the time frame of long-term planning extends over a number of years.

A review of a hypothetical life equation example based on the wild turkey using generalized and not actual data can be found in Figure 6.1. The wild turkey life equation is typical of the annual cycles of virtually every animal population. If 2 males mate with 8 females and the females lay an average of 15 eggs, ideally up to 120 eggs could be hatched. But for several reasons, only 96 (80%) hatch and of these, 16 chicks die for various reasons. Another 30 chicks are lost to hazards. Of the 120 eggs laid, 50 chicks live to become adults in the fall which, added to the original 10 birds, yields a total of 60 in the population. Of these, 10 birds are killed by poaching or predators. Assuming a 50:50 sex ratio, 25 male and 25 female adult turkeys remain.

TIME	ACTIVITY	YOUNG	ADULTS
SPRING	1. 2 MALES MATE 8 FEMALES.		10
	2. CLUTCHES AVERAGE 15 EGGS. (8 x 15 = 120)		
	3. 80% HATCH.	96	
	4. 16 DIE FOR VARIOUS REASONS.	80	
SUMMER	5. 30 DIE FROM HAZARDS.	50	
FALL	6. 50 YOUNG BECOME ADULTS.		60
	7. 10 ARE LOST TO POACHERS AND PREDATORS.		50
	ASSUMING A 50-50 SEX RATIO, THERE ARE 25 MALES AND 25 FEMALES REMAINING.		
	BASE SPRING POPULATION DESIRED IS 2 MALES AND 8 FEMALES.		
WINTER	8. POPULATION CONSISTS OF 25 MALES AND 25 FEMALES.		50
	9. 30 BIRDS WILL DIE OVER THE WINTER SO THE LEGAL HARVEST CAN BE 30 IN THE FALL, WITH 10 LEFT AS A SAFETY FACTOR.		
SPRING	10. 2 MALES MATE WITH 8 FEMALES.		10

Figure 6.1. Hypothetical life equation for wild turkey.

Experience has shown that a spring population of 2 males to 8 females will maintain the base population as a sustainable population and that over the winter there will be a loss of about 30 more turkeys from the earlier population of 50 birds. These 30 turkeys at the time of the fall hunting season are surplus to the management unit's capacity so they can be killed by hunters, leaving 10 birds for spring breeding and 10 as a safety factor. This example does not take into account a common practice of a spring gobbler season. Even though the life equation is hypothetical and simple, it does illustrate the way in which a population is analyzed to determine the surplus which management has developed for the benefit of hunters.

Sustained Yield

In an unhunted population the difference between the uninhibited growth model and the inhibited growth model is the number of animals that are lost due to diseases, hazards, and other factors. When a population has reached ecological carrying capacity there is still a seasonal fluctuation in animal density. In the spring population density is low because of mortality, but in the fall it is high as a result of natality. The mortality occurring from fall to spring is a *huntable surplus*.

The idea is that a surplus can be removed which if left would be lost to winter mortality. The surplus of game animals killed each year is referred to as *yield*. *Sustained yield* is the number of animals that can be removed from a population over a period of years leaving not only a breeding population sufficient to maintain both the equilibrium number of animals but also a harvestable surplus. This conclusion assumes that the number of animals killed by hunters is *compensatory*—that is, it does not add to natural mortality. The concept is that the total effect of mortality does not vary, but different types of mortality such as those caused by disease, predation, or resource shortages can each vary and one will compensate for the other. Hunting is hypothesized to replace the mortality caused by other decimating factors and seems to hold true for species with high natality such as small game, but may not hold for big game species.

In the logistics model (inhibited growth) of population growth the maximum rate of production occurs at the inflection point (Fig. 3.4) which is about 50% of the carrying capacity of the habitat for the species. It is the point where growth is increasing but at a decreasing rate. If the population is reduced each year to this level then the recovery rate will be at a maximum which is called *maximum sustained yield* (MSY). Several problems make it difficult to apply MSY in a field situation. One is that K is almost never known and a statistically valid estimate for P is difficult to obtain even under the best of conditions.

Maximum sustained yield has also been criticized as not being directed at maintaining ecosystems (Holt and Talbot 1978). Further deficiencies of MSY include:

1. Focusing only on the population of a single species.

2. Assuming dependence on a species' ability to adjust that may not exist.

3. Emphasizing yield without regard to the effort needed to produce the yield.

4. Not taking into account fluctuating yearly conditions that produce variable vegetation growth.

5. Not including an insurance factor to assure that a viable population will always exist.

To compensate for the shortcomings of MSY, Holt and Talbot (1978) offer four suggestions:

1. Ecosystems maintenance must be optimized so that consumptive (hunting, fishing) and nonconsumptive (photographing, viewing) values can be maximized on a long-term basis, present and future options are ensured, and the risk of irreversible change or long-time adverse effects resulting from human use is minimized.

2. Management decisions must include a safety factor to allow for the fact that human knowledge is limited and human institutions are imperfect.

3. Measures to conserve all wild living resources should be formulated and applied so as to avoid wasteful use of other resources.

4. Survey or monitoring, analysis, and assessment need to precede planned use and accompany actual use of wild living resources.

The fact of the matter is that public desires and opinions may control the annual yield regardless of the intent of science. Public desires and opinions translate into lobbying of elected officials. This mixing of biology and politics is appropriately called *Bio-politics* and is of major importance in the decision-making process of all wildlife managers.

6.5 MANAGEMENT OF RESOURCES

Resources are categorized into terrestrial and aquatic and because of the physical differences between them they will be considered separately. Aquatic resources as used in this text refers primarily to the habitat of fish and not to wildlife species such as mink, muskrat, or waterfowl that use both land and water for food and cover.

Terrestrial Resources

The abundance of wildlife in a habitat depends heavily upon the adequacy of food, cover, water, and space and the desires of people (bio-politics). Habitat management for a given species involves the manipulation of vegetation as the most important factor contributing to maintaining a viable population. Even carnivores are indirectly dependent on plants for energy since most of the animals they eat are themselves herbivores.

Each successional stage provides different types of food and cover which satisfy the minimum requirements of some but not all species and which provide optimum conditions for yet others. Thus, several species may have all their requirements met in a single stage while some require one stage for food and another for cover. Because seral stages progress over time to climax, the vegetation cycle of food and cover for different wildlife species occurs on the same tract of land.

Vegetation is controlled by nature but can be influenced by humans. In the absence of human interference, vegetation will naturally develop in a balanced way that will always contain habitat components for a number of species. A balanced system is what one would expect in undisturbed wilderness areas, but in reality vegetation is always changing even though the rate of change may be difficult to detect.

Under human influence management can be directed to maintain vegetation for a single species or a group of species with overlapping habitat components. Using this approach vegetation can be kept in or directed to a seral stage producing the combinations of food and cover required by one or more wildlife species. This is done by directly interfering with succession to return stages to a lower level of secondary succession. Another alternative, if the environmental factors are suitable, is to convert the area to a completely different vegetation type such as forest to grassland or hardwoods to conifers.

Control of wildlife populations is directly the responsibility of game and fish agencies and is usually accomplished by regulating hunting seasons and bag limits. However, wildlife populations can be controlled indirectly through habitat manipulation—activities which are the responsibility of state, federal, or private land management agencies. Direct techniques used to modify succession in forested areas are:

1. Prescribed fire

2. Controlled grazing

3. Timber harvesting

4. Herbicides

5. Seeding and planting

In addition to the techniques employed by mankind, nature has from time immemorial controlled succession by wildfires, such as those in Yellowstone National Park in 1988, or by insect and disease outbreaks. While humans may not manipulate vegetation for brief periods of time, natural forces are always at work throughout the world.

As was indicated at the beginning of this chapter initially wildlife managers were largely concerned with population regulation of a small number of game species by controlling the length of hunting seasons and the *bag limit*. But with the rapid increase in understanding of ecological processes and a growing public and professional concern for maintaining the integrity of ecological systems, wildlife managers are increasingly turning to cooperative efforts with other natural resource managers to meet the complex planning and management requirements of a large number of species. Wildlife managers can no longer focus exclusively on those animals which hunters find desirable or others find undesirable. They increasingly must plan for and manage all wildlife species many of which are a matter of indifference to a segment of our human population.

In view of this approach to management which seeks to benefit all forest wildlife, I have furnished a summary listing of habitat management practices and functional measures in the following two sections. These practices and measures are stated only in a general way as the details must be left to the managers of specific forest types. These details include such considerations as size of openings to create, how much vegetation to leave as cover for a particular species, the amount and species of food plants to seed or plant for a desired wildlife species, and the types and sizes of buffer strips to leave along streams. These and other factors have to be tailored to a particular vegetation type and given species. Any management practice or coordination measure used must be closely aligned to the silvicultural requirements of tree species in the forest type and the life history requirements of the wildlife species associated with the type.

Habitat Management Practices

Most of the direct habitat management practices listed below are the result of research or many years of practical experience. The list is not all-inclusive, but it does contain a number of generic measures suitable for a variety of forest types. Detailed descriptions of each practice can be found in many state and federal publications dating from the 1930s to the present time. The practices listed are for the most part purposely undertaken to benefit wildlife and so all the costs are usually paid from wildlife funds.

There are many practices that can be used to improve wildlife habitat in forested ecosystems, but *CAUTION* is urged: before transporting technology from one forest region to another, first document that the practice actually benefits the targeted species or group of species.

1. Maintain (plant, cultivate, and fertilize) permanent openings (food plots) with wildlife food plants. This practice benefits many species but generally is used to attract deer and turkey to specific areas.

2. Cut, burn, or disc concentrations of shrub species to stimulate growth of browse. Used primarily to produce food for deer and elk.

3. Create snags by directly killing trees. Benefits many species of cavity nesting birds and provides hunting perches for raptors.

4. Fence riparian areas to exclude cattle. Benefits many species but cost is high.

5. Restrict access to critical habitat of threatened and endangered species. Proven technique, but cost may exclude its use in many areas. Good public involvement project for volunteers.

6. Plant food and cover species along streams to stabilize bank erosion and provide shade. Generally good conservation technique. Benefits aquatic species.

7. Create water sources by building small ponds, guzzlers, and trick tanks. Benefits many species. Cost can be high and maintenance is required.

8. Improve existing water sources such as seeps and springs to create permanent water sources. May need to be fenced to exclude livestock.

9. Build dams to create shallow lakes for waterfowl habitat. Good technique but can be very expensive. Best situation is for a multiresource structure as in flood control.

10. Build small check dams in eroded wet meadows. Generally gets quick results. Provides conditions to establish riparian vegetation.

11. Remove invading trees from wet meadows. Expensive but effective in maintaining high forage producing areas for deer and elk, and "bugging areas" for turkey poults.

Coordination Measures

The opportunity to improve conditions for wildlife through indirect techniques or coordination measures with other agencies can often be accomplished at greatly reduced costs and with a minimum expenditure of wildlife funds. Indirect techniques often depend upon coordination of activities with other land management agencies including those dealing with timber, range, engineering (roads, etc.), watershed, and recreation as well as management of other special land-use projects.

Coordination measures differ from direct habitat management practices only in the way they are financed and brought to completion. These measures generally use funds from severalresource agencies or functions to accomplish the job. Coordination measures are the bread and butter of wildlife management and their contribution should not be underestimated even though results in terms of increased wildlife populations may not be immediately evident.

Timber Management: Timber management activities affect a large number of wildlife species either positively or negatively depending on the requirements of the individual species. For this reason the removal of tree overstory can be used as a management tool to meet specific wildlife objectives. The coordination measures listed here are intuitive but are based on sound ecological principles:

1. Locate timber roads and skid trails to eliminate or minimize siltation of streams.

2. Reserve, in connection with tree planting and regeneration, natural openings, access ways, and brush areas for food and cover. Benefits deer, elk, and turkey.

3. Plant or reserve groups of trees for cover (thermal, escape, hiding, etc.) for targeted species.

4. Reserve and release fruit and nut trees and shrubs to increase food production. Benefits many species.

5. Retain den and roost trees. Generally species specific for squirrels, raccoons, turkeys, band-tailed pigeons and some raptors.

6. Create openings by harvesting in dense timber stands. One of the most common techniques to benefit deer, elk, and turkey as well as many species of small mammals and birds.

7. Plan timber harvest roads to provide access for those who hunt and fish.

8. Retain tree buffer areas or riparian vegetation along streams and lakeshores to maintain cold water temperatures (Fig. 6.2).

9. Include provisions in timber sale contracts to prevent entry of large amounts of small-sized debris into perennial streams. Debris of large size may be desirable in some streams.

10. Favor aspen reproduction in timber sales where it is a seral stage. Management objectives must be clearly defined especially where beaver or deer could cause damage to a watershed.

11. Locate logging roads away from live streams to prevent destruction of natural stream channels. Often requires persistence to keep roads away from water.

12. Prohibit log landings in live streams in timber sale contracts.

13. Bridge water courses on log haul roads.

14. Limit the number of stream crossings in logging operations. May increase cost of road building to an unacceptable level.

15. Eliminate tree-felling into stream channels or dragging trees along or across channels. Enforcement and supervision a must where sale affects an important fishery.

16. Prevent stream pollution from sawmill waste.

17. Adjust slash disposal plans to include requirements of wildlife species; slash piles benefit small mammals and birds.

18. Prevent log skidding across meadows, along stream-banks or through food, escape, nesting, or roosting cover. Areas must be identified before the sale.

19. Reserve cover adjacent to seeps and stringer meadows in cut areas. Areas must be identified before timber sale.

20. Promote sales to regenerate aspen. A very effective technique to increase forage production for deer and elk where the type occurs.

21. Withhold from tree planting special areas needed for food and cover plantings. Designate species to benefit in management plans.

22. In releasing coniferous timber species, leave hardwoods in strips or patches to provide a mast crop. Benefits squirrel, deer, and turkey.

23. Suspend logging during spawning seasons to avoid interference with upstream movement, redds, and fry.

24. Scatter the location of small timber sales to break up large areas of a single age class. Provides diversity of food and cover areas for many species.

25. In large clearcuts, leave small scattered plots uncut for cover. Increases use by deer and elk of large cut areas.

26. Seed skidways, roadsides, and landings on sale areas with wildlife food mixtures of forbs, grasses, and shrubs. Provides food for many species.

27. Burn slash away from water so ash will not enter streams or lakes.

28. Provide for juxtaposition of stands in different vegetation types.

29. Provide for interspersion of habitat units throughout the vegetation type.

30. Thin dense stands of trees to stimulate understory growth. Good technique for deer and elk.

31. Close permanent and seasonal logging roads when necessary to protect a species during all or part of a year.

32. Mark and save a given number of snags per acre for cavity nesting birds.

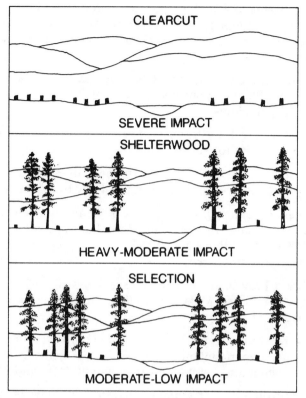

Figure 6.2. Effects of timber harvesting on riparian habitat. (Modified from Thomas, Maser, and Rodiek 1979a).

Range Management: The coordination measures undertaken with range management agencies are largely directed toward including wildlife requirements for grazing and browsing species such as deer, elk, and bighorn in allotment analysis plans and reducing the livestock-wildlife competition for food and space. The following coordination measures are those which range conservationists and wildlife biologists have generally agreed upon, due to their wide application. They are practical, beneficial, and can be accomplished with little difficulty.

1. Allocate a percentage of forage to wildlife species in range allotment plans to keep total livestock/wildlife numbers within carrying capacity.

2. Manage livestock to protect key wildlife areas. This is especially important in overlapping winter use areas.

3. In placing salt on the range, distribute it to allow use by big game animals. While this technique was used extensively in the past, there is some question about its value for wildlife.

4. Reseed depleted grass ranges with palatable species to reduce the use of shrubs by livestock. Plant according to local guidelines.

5. Reduce shrub use by livestock through grazing of moderate intensity for shorter periods.

6. Install escape ramps for small game in livestock watering tanks to keep water from being polluted by dead animals.

7. Include legumes, browse, and other important wildlife foods in range reseeding mixtures. Plant species adapted to local conditions and palatable to target species such as deer, elk, and turkey.

8. Leave browse for wildlife in reseeding projects involving brush removal. Make certain that browse species are not decadent.

9. Defer spring livestock grazing until young wildlife are no longer dependent on green feed. Very effective in important gamebird habitat and also benefits deer and elk.

10. Control, fence, or eliminate livestock driveways across winter game ranges. Fencing is expensive.

11. When initiating brush control on winter ranges, consider wildlife needs in the early planning stages of a project in relation to the surrounding area.

12. Consider the possible competition effects on wildlife when changing the kind of livestock in an allotment. Competition between livestock and wildlife must be considered early in the planning process.

13. Reserve forage at heavily used hunting and fishing camps for recreational pack and saddle stock.

14. Do not build range improvements that concentrate or encourage use of key wildlife habitat by domestic livestock. Plan for a good distribution of improvements.

15. Locate stock watering improvements to benefit wildlife. Water improvements have proved an important factor in increasing deer, elk, and turkey where food and cover but not water were available.

16. Fence springs at water developments to provide cover and nesting sites for upland birds. An effective technique in heavily stocked livestock allotments.

17. Consider the effects on other wildlife when initiating predator control, particularly if poison is used.

18. Where there is direct competition for forage between livestock and big game, select a reasonable balance. Must be provided for in the planning stage.

Engineering: Engineering concerns deal primarily with cultural and structural features including roads. The placement of buildings, fences, recreational areas, electrical lines, and water structures can have highly detrimental effects on many wildlife species. As a consequence coordination measures undertaken with engineering agencies are largely targeted at mitigating negative impacts on habitat and populations. The following measures are among those wildlife biologists have found to be most acceptable and successful:

1. Limit borrow from gravel bars to above water level. Borrow pits below high-water stage should be leveled and provided with drainage to avoid trapping fish when high water recedes.

2. Install culverts in streams at the grade of the stream and below stream level. Drop-offs at outlet are impossible for fish to negotiate. Drop-off may undercut culvert and require expensive maintenance procedures.

3. Eliminate or minimize use of heavy equipment in streambeds to prevent silting of stream.

4. Consider the access needs for those who hunt and fish when planning roads and trails.

5. In locating roads, avoid streambanks, meadows, water courses, and other areas used by wildlife. Locating roads away from water or wet areas is also a good soil conservation technique.

6. Plant wildlife food on cut-and-fill slopes. Good conservation practice to reduce erosion. Plant species palatable to targeted wildlife.

7. In constructing roads along a stream leave fringes of timber and brush to provide shade and bank protection.

8. Keep construction debris out of streams to reduce siltation of spawning areas.

9. Provide for spreading of drainage water from roads to prevent silting of streams. Reduces siltation and pollution.

10. Consider designing road fills to serve as dams which can provide additional water. Must be planned carefully so that dirt dam is stable.

Watershed Management: Considerable watershed restoration work is done under state and federal programs to provide jobs for unemployed people. Such projects are often dependent on the political climate, so when a favorable climate exists, a backlog of projects can often be completed in a short time.

Some of the measures listed below grew out of the experience of the Civilian Conservation Corp of the 1930s, but they remain just as effective today. In the 1980s these coordination measures are often programmed for specific projects in small watersheds with small work crews.

1. Use browse and legumes in erosion-control plantings to benefit wildlife species.

2. Include wildlife food plants in bank stabilization projects.

3. Integrate watershed plans and big game hunting to provide for game management in municipal watersheds to increase recreational opportunities for a local area.

4. Include the protection and improvement of wildlife habitat in water impoundment programs. If planned properly, can benefit many locally important species.

5. Safeguard or improve aquatic habitats when using structures to alter stream courses, streambanks, or for channel and gully control. Common sense practice.

6. In gully and other erosion control plantings, use fruit-bearing and thicket-forming species to provide food and cover for wildlife.

Recreation: Recreation coordination measures are aimed at controlling the behavior of people not just hikers, picnickers, etc. but hunters and fishermen as well. Considerable effort is required to mitigate the effects of recreational activities on both habitat and populations particularly on sites having fragile soils or containing sensitive plants and animals. Most recreation control measures require long-range planning involving public review and acceptance.

1. Restrict cross-country use of four-wheel drive vehicles or all-terrain vehicles to designated roads in sensitive soil areas. A controversial technique always requiring public review.

2. Locate hunting and fishing camps and enforce camping regulations so that stream pollution or damage to habitat is held to a minimum.

3. Corral horses used at hunting camps located within deer and elk winter ranges. Enforcement may be difficult in remote areas.

4. Restrict camping and grazing of pack and saddle animals at heavily used hunting and fishing sites. The terrain immediately adjacent to these areas may need to be closed to grazing.

5. Restrict speed boating in high-quality fishing waters.

6. Consider camping needs of those who hunt and fish when selecting areas for campground development.

7. Through hunting and fishing clubs and other organizations, develop programs of cooperative camp cleanup, sanitation, and fire prevention. An effective way to educate sports enthusiasts on just how much trash is left by their activities.

8. Plan for campsite areas for those who hunt and fish non-commercially before issuing special permits to commercial guides.

Special Land Uses: These coordination measures grow out of public and private projects not normally part of the day-to-day management of wildlife. Such projects often result from state or federal undertakings that create or maintain large bodies of water or extensive land areas, or that provide benefits to large numbers of people at a distance from the management area. As with the other coordination measures the list below is not inclusive:

1. Install fish screens at headgates of irrigation ditches.

2. Maintain minimum pool levels in artificial reservoirs as a mitigating procedure. Must be provided for early in the planning stage.

3. Keep minimum and maximum rates of discharge from impoundments below levels hazardous to fish and wild-

life. Determining such rates must be done early in the planning process.

4.   Provide vegetative cover on ditchbanks to reduce soil erosion.

5.   Consider needs of wildlife when acting on applications for agricultural use of land.

6.   Provide for protection of wildlife and habitat, including the control of organic and inorganic pollution, when issuing occupancy permits. Enforcement of existing regulations is generally the major problem.

7.   Plant forb, grass, and browse species along power lines and similar clearings to produce wildlife food and cover and at the same time reduce the cost of periodic clearing. If properly coordinated with utility companies this technique can be very effective in providing food and cover for many wildlife species.

Wildlife Food Plants

Many wildlife habitat practices and coordination measures require the seeding, planting, or protection of areas containing certain trees, shrubs, forbs, and so forth as wildlife food or cover plants. Plant species possess nutritional elements in differing quantities and their value for one wildlife species may be different than that for another. On the other hand, a single plant species may possess food or cover value for a number of wildlife species. Clearly such plants are those that managers should preferentially employ in habitat improvement projects.

Trees and Wildlife

Live Trees: Single live trees provide the physical structure needed for a nest by a squirrel or a hawk, for overnight roosting by wild turkeys, or for perching and resting by doves and pigeons. When single trees accumulate to make a small group and can be identified as a stand, then additional protection is provided to many forms of wildlife. In their early developmental stage, sprouts, limbs or twigs of trees are food for browsing animals such as deer and elk. The importance of live trees in providing wildlife food and cover cannot be overemphasized. Approximately 28 genera make up the groups of tree species across North America that are commonly used by some wildlife species (Table 6.2).

Table 6.2. Trees used by a number of wildlife species for food and cover.

Common Name	Scientific Name (Genus)
Alder	*Alnus*
Ash	*Fraxinus*
Aspen	*Populus*
Beech	*Fagus*
Birch	*Betula*
Blackgum	*Nyssa*
Juniper	*Juniperus*
Cherry	*Prunus*
Chinaberry	*Melia*
Baldcypress	*Taxodium*
Dogwood	*Cornus*
Douglas-fir	*Pseudotsuga*
Fir	*Abies*
Hemlock	*Tsuga*
Hickory	*Carya*
Incense-cedar	*Libocedrus*
Larch	*Larix*
Magnolia	*Magnolia*
Maple	*Acer*
Mulberry	*Morus*
Oak	*Quercus*
Pine	*Pinus*
Sassafras	*Sassafras*
Sourwood	*Oxydendrum*
Spruce	*Picea*
Cedar	*Thuja*
Willow	*Salix*
Tuliptree	*Liriodendron*

Dead Trees and Snags: Forests are more than live, straight trees used for sawlog production (Fig. 6.3). Dead trees, snags, hollow trees, decaying logs and stumps are used as hunting perches for raptors as den trees for raccoons, squirrels, and bears; as feeding sites for woodpeckers; and as nests for cavity nesting birds. On occasion, rabbits use rotted tree root canals for tunnels and dens, as do snakes, lizards, and small rodents. The cavity nesting birds are a particularly important class of forest birds because the majority of them are insectivorous and help control endemic forest insects that damage valuable timber trees.

Recognition that snags and dead or dying trees are an important component of the forest is not new. In 1848 Richard Jefferies wrote:

Those hollow trees, according to woodcraft, ought to come down by the axe without further loss of time. Yet it is fortunate that we are not all of us, even in this prosaic age, imbued with the stern utilitarian spirit; for a decaying tree is perhaps more interesting than one in full vigour of growth. The starlings make their nests in the upper knot-holes; or, lower down, the owl feeds her young; and if you chance to pass near, and are not aware of the ways of the owls, you may fancy that a legion of serpents are in the bushes, so loud and threatening is the hissing noise made by the brood. The woodpecker comes for the insects that flourish on the dying giant; so does the curious little tree-climber, running up the trunk like a mouse; and in winter, when insect-life is scarce, it is amusing to watch there the busy tomtit. He hangs underneath a dead branch, head downwards, as if walking on a ceiling, and with his tiny but strong bill chips off a fragment of the loose dead bark. Under this bark, as he well knows, woodlice and all kinds of creeping things make their home. With the fragment he flies to an adjacent twig, small enough to be grasped by his claws and so give him a firm foothold. There he pecks his morsel into minute pieces and lunches on the living contents. Then, with a saucy chuckle of delight in his own cleverness, he returns to the larger bough for a fresh supply. As the bough decays the bark loosens, and is invaded by insects which when it was green could not touch it.

For the acorns the old oak still yields some rooks, pigeons, and stately pheasants, with their glossy feathers shining in the autumn sun. Thrushes carry wild hedge-fruit up on the broad platform formed by the trunk where the great

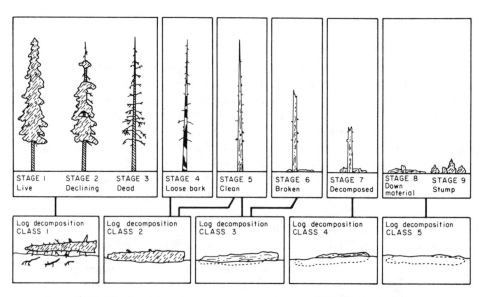

Figure 6.3. Dead trees and snags are important wildlife habitat components (Thomas, Anderson, et al. 1979).

limbs divide, and, pecking it to pieces, leave the seeds. These take root in the crevices which widen out underneath into a mass of soft decaying "touchwood"; and so from the crown of the tree there presently stream downwards long trailing briars, bearing in June the sweet wild roses and in winter red oval fruit. Ivy comes creeping up, and in its thick warm coverts nests are built. Below, among the powdery "touch-wood" which lines the floor of this living hut, great fungi push their coloured [sic] heads up to the light. And here you may take shelter when the rain comes unexpectedly pattering on the leaves, and listen as it rises to a roar within the forest. Sometimes wild bees take up their residence in the hollow, slowly filling it with comb, buzzing busily to and fro; and then it is not to be approached so carelessly, though so ready are all creatures to acknowledge kindness that ere now I have even made friends with the inhabitants of a wasp's nest.

A thick carpet of dark green moss grows upon one side of the tree, and over it the tall brake fern rears its yellow stem. In the evening the goat-sucker or nightjar comes with a whirling phantom-like flight, wheeling round and round: a strange bird, which will roost all day on a rail, blinking or sleeping in the daylight, and seeming to prefer a rail or branch without leaves to one that affords cover. Here also the smaller bats flit in the twilight, and, if you stand still, will pursue their prey close to your head, wheeling about it so that you may knock them down with your hand if you wish. The labouring people call the bat "batmouse." Here also come many beetles; and sometimes on a summer's day the swallows will rest from their endless flight on the dying upper branches, for they too like a bough clear or nearly clear of leaves. All the year through the hollow tree is haunted by every kind of living creature, and therefore let us hope it may yet be permitted to linger awhile safe from the axe.

The number of dead trees, snags, logs, and stumps to leave as a component of wildlife habitat is variable depending on the species and forest region. For example, in Arizona and New Mexico the U.S. Forest Service policy is to leave 7.4 snags per ha (3 per ac) within 152 m (500 ft) of forest openings and water, with 5 per ha (2 per ac) over the remaining forest (U.S. Dept. of Agric. 1976). In the Southwest good snags are considered to be over six years old, more than 46 cm (18 in) dbh, and with more than 40% bark cover. In the Pacific Northwest the suggested snag density is 2.5–3.0 per ha (6–7 per ac) for ponderosa pine (Thomas, Anderson, et al. 1979). The wildlife species using dead trees and snags in the 7 forest regions include more than cavity nesting birds, as Table 6.3 indicates.

Table 6.3. Some wildlife species that use dead trees and snags for food and cover.

Common Name	Life-form*	Use**
Abert's squirrel	Ma	DEN
Acorn woodpecker	Bi	PCN
Ash-throated flycatcher	Bi	SCN
Barn owl	Bi	SCN
Barn swallow	Bi	SCN
Black bear	Ma	DEN
Black-backed woodpecker	Bi	PCN
Black-capped chickadee	Bi	SCN
Boreal chickadee	Bi	SCN
Bridled titmouse	Bi	SCN
Brown-crested flycatcher	Bi	SCN
Brown-headed nuthatch	Bi	SCN
Carolina chickadee	Bi	SCN
Carolina wren	Bi	SCN
Chestnut-backed chickadee	Bi	SCN
Chimney swift	Bi	SCN
Douglas' squirrel	Ma	DEN
Downy woodpecker	Bi	PCN
Dusky-capped flycatcher	Bi	SCN
Eastern screech-owl	Bi	SCN
Fox squirrel	Ma	DEN
Gila woodpecker	Bi	PCN
Golden-fronted woodpecker	Bi	PCN
Gray squirrel	Ma	DEN
Great-crested flycatcher	Bi	SCN
Hairy woodpecker	Bi	PCN
House wren	Bi	SCN
Ladder-backer woodpecker	Bi	PCN
Lewis' woodpecker	Bi	PCN
Mountain bluebird	Bi	SCN
Northern flicker	Bi	PCN
Northern Flying squirrel	Ma	DEN
Nuttall's woodpecker	Bi	PCN
Pileated woodpecker	Bi	PCN
Plain titmouse	Bi	SCN
Pygmy nuthatch	Bi	SCN
Porcupine	Ma	DEN
Prothonotary warbler	Bi	SCN
Raccoon	Ma	DEN
Red squirrel	Ma	DEN
Red-bellied woodpecker	Bi	PCN
Red-breasted nuthatch	Bi	SCN
Red-cockaded woodpecker	Bi	PCN
Red-headed woodpecker	Bi	PCN
Ringtail	Ma	DEN
Siberian tit	Bi	SCN
Southern flying squirrel	Ma	DEN
Strickland's woodpecker	Bi	PCN
Three-toed woodpecker	Bi	PCN
Tree swallow	Bi	SCN

Table 6.3. Continued.

Common Name	Life-form*	Use**
Tufted titmouse	Bi	SCN
Vaux's swift	Bi	SCN
Western bluebird	Bi	SCN
Western gray squirrel	Ma	DEN
Western screech-owl	Bi	SCN
White-breasted nuthatch	Bi	SCN
White-headed woodpecker	Bi	PCN
Williamson's sapsucker	Bi	PCN
Winter wren	Bi	SCN
Yellow-bellied sapsucker	Bi	PCN

*Ma = Mammal
 Bi = Bird
**PCN = Primary cavity nester (excavates cavity)
 SCN = Secondary cavity nester (uses cavity excavated by PCNS)
 DEN = A cavity used by mammals

Shrubs and Wildlife

The planting and production of shrubs was emphasized in the past, particularly in the West, due to their importance in maintaining big game populations. More recent ecological thinking has encouraged their cultivation for a variety of other life-forms as well. Many shrubs are browse food plants for deer and elk but also provide seeds and fruits for birds and small mammals and, in many cases, also provide cover for the same groups of species (Table 6.4).

Approximately 1,000 species of shrubs, semishrubs, and woody vines grow in United States national forests and adjacent lands (Dayton 1931). Martin, Zim, and Nelson (1951), in their classic book, account for over 100 wildlife species using shrubs for food. It is probably not an exaggeration to assert that every shrub species provides food or cover for some animal species. The association of specific animal species with specific shrub species is recognized in their respective common names, such as bearberry, rabbitbrush, deerbrush, and antelope bitterbrush.

Shrubs grow in pure stands in the absence of an overstory or in scattered single or small clumps as understory plants. Production of shrubs follows the well-documented inverse relationship of high percent tree canopy/low production; low percent tree canopy/high production. However, total productivity is modified by soil and rainfall so there is great variation from location to location. Furthermore, production is commonly lower in shaded stands.

Table 6.4. Shrubs used by a number of wildlife species for food and cover.

Common Name	Scientific Name (Genus)
Apache-plume	*Fallugia*
Azalea	*Rhododendron*
Barberry	*Berberis*
Bearmat	*Chamaebatia*
Blueberry	*Vaccinium*
Buckthorn	*Rhamnus*
Bush-honeysuckle	*Diervilla*
Buttonbrush	*Cephalanthus*
Calliandra	*Calliandra*
Ceanothus	*Ceanothus*
Cherry	*Prunus*
Chokecherry	*Aronia*
Cinquefoil	*Potentilla*
Cliffrose	*Cowania*
Condalia	*Condalia*
Cyrilla	*Cyrilla*
Dalea	*Dalea*
Elder	*Sambucus*
Euonymus	*Euonymus*
Gooseberry	*Ribes*
Hackberry	*Celtis*
Hawthorn	*Crataegus*
Hazelnut	*Corylus*
Holly	*Ilex*
Huckelberry	*Gaylussacia*
Juniper	*Juniperus*
Kalmia	*Kalmia*
Manzanita	*Arctostaphylos*
Menziesia	*Menziesia*
Mountainmahogany	*Cercocarpus*
Oak	*Quercus*
Palmetto	*Sabal*
Paperflower	*Psilostrophe*
Porliera	*Porliera*
Rabbitbrush	*Chrysothamnus*
Raspberry	*Rubus*
Rose	*Rosa*
Sagebrush	*Artemesia*
Saw-palmetto	*Serenoa*
Serviceberry	*Amelanchier*
Silktassel	*Garrya*
Snowberry	*Symphoricarpos*
Sumac	*Rhus*
Viburnum	*Viburnum*
Waxmyrtle	*Myrica*
Willow	*Salix*
Winterfat	*Eurotia*
Witchhazel	*Hamamelis*
Wolfberry	*Lycium*

Herbaceous Plants and Wildlife

Herbaceous plants include the many species of grasses such as wheatgrasses, bluegrasses, and bromes, and the low, wide-leaved flowering plants such as buttercups, cinquefoils, and clovers, often referred to as forbs. Grasses, grass-like, and herbaceous plants are important sources of wildlife food and provide cover for a range of animals from small mammals—for example, chipmunks, mice, and ground squirrels—to large grazing animals including elk, deer, and pronghorn. Given limits of space it is not possible here to list and describe all the grass or forb species of value to all wildlife species, but an appreciation of their importance can be obtained by reading current issues of the *Wildlife Society Bulletin* or *Journal of Wildlife Management*. Articles in these journals contain information on herbaceous plants used by various wildlife species either for food or cover. Some of the most common genera are listed in Tables 6.5. and 6.6.

Table 6.5. Forb species used by a number of wildlife species for food and cover.

Common Name	Scientific Name (Genus)
Agoseris	*Agoseris*
Alumroot	*Heuchera*
Arnica	*Arnica*
Balsamroot	*Balsamorhiza*
Brodiaea	*Dichelostemma*
Bundleflower	*Desmanthus*
Buttercup	*Ranunculus*
Cinquefoil	*Potentilla*
Clover	*Trifolium*
Croton	*Croton*
Dandelion	*Taraxacum*
Deervetch	*Lotus*
Dock	*Rumex*
Eriophyllum	*Eriophyllum*
Eupatorium	*Eupatorium*
Filaree	*Erodium*
Fleabane	*Erigeron*
Galax	*Galax*
Goldenrod	*Solidago*
Groundcherry	*Physalis*
Groundsmoke	*Gayophytum*
Hawkweed	*Hieracium*
Hollyfern	*Polystichum*
Knotweed	*Polygonum*
Lespedeza	*Lespedeza*
Lettuce	*Lactuca*
Lupine	*Lupinus*
Marshpurslane	*Ludwigia*
Medic	*Medicago*

Table 6.5. Continued.

Common Name	Scientific Name (Genus)
Oxalis	*Oxalis*
Parthenium	*Parthenium*
Partridgeberry	*Mitchella*
Peavine	*Lathyrus*
Penstemon	*Penstemon*
Plantain	*Plantago*
Pussytoes	*Antennaria*
Ragweed	*Ambrosia*
Royal fern	*Osmunda*
Senna	*Cassia*
Snapweed	*Impatiens*
Strawberry	*Fragaria*
Sweetclover	*Melilotus*
Tansymustard	*Descurainia*
Thistle	*Cirsium*
Tickclover	*Desmodium*
Trailing-arbutus	*Epigaea*
Turkey-mullein	*Eremocarpus*
Vetch	*Vicia*
Violet	*Viola*
Waterlily	*Nymphaea*
Wild-buckwheat	*Eriogonum*
Willow-herb	*Epilobium*
Wintergreen	*Gaultheria*
Wyethia	*Wyethia*
Yarrow	*Achillea*

Aquatic Resources

Within the large land areas classified as a forest a great many smaller communities and ecosystems can be identified. Among them are marshes, riparian zones, and aquatic habitats of emergent and submergent vegetation, all of which add to the total diversity and wildlife productivity of the forest type.

The management and protection of *watersheds*, as the source of water for streams and lakes, have always been important functions of forest management. Watersheds control water quality. When maintained in the proper condition they prevent soil erosion and serve as filters for the streams within the watershed boundary. Foresters and wildlife biologists are concerned with the effects of various watershed uses and treatments on the natural character of the stream environment providing fish habitat. In streams running through forested areas trout and salmon are considered the most important game fish so the discussion which follows focuses on these two groups.

Table 6.6. Grass and grasslike plants used by a number of wildlife species for food and cover.

Common Name	Scientific Name (Genus)
Grass	
Saltgrass	*Distichlis*
Bluegrass	*Poa*
Bristlegrass	*Setaria*
Brome	*Bromus*
Cupscale	*Sacciolepis*
Cutgrass	*Leersia*
Fescue	*Festuca*
Cane	*Arundinaria*
Junegrass	*Koleria*
Muhly	*Muhlenbergia*
Needlegrass	*Stipa*
Panicum	*Panicum*
Paspalum	*Paspalum*
Squirreltail	*Sitanion*
Wheatgrass	*Agropyron*
Wildrye	*Elymus*
Grasslike	
Bulrush	*Scirpus*
Rush	*Juncus*
Sedge	*Carex*
Spikerush	*Eleocharis*

Most spiny-rayed fish are tolerant of a wide range of water conditions, but soft-rayed trout and salmon are very intolerant. Trout and salmon are most often found in streams characterized by the following conditions which are favorable to their survival and ability to reproduce (Forbs 1961).

1. A good steam is a relatively fast-moving body of clear water averaging about 14°C (57°F) in the summer months, with deep pools at intervals along its length.

2. The stream must have frequent spawning areas of coarse, clean gravel in moderately fast water.

3. The streambanks must be well-forested and -vegetated for food production and temperature control. For streams 1.5–6.0 m (5–20 ft) in width, a cover of 50–75% of tall trees and shrubs is a necessity. Cover along wider and deeper streams is not as important since water depth itself furnishes cover.

4. The stream should have about 50% pools and 50% riffles, well spaced. The riffles are food-producing areas while the pools provide hiding spots.

5. The stream should not be subject to flooding sufficient to cause lateral course changes each year. Normal spring flooding may not be a cause for concern.

The factors initiated by humans which directly affect trout streams are mainly logging, grazing, and in some cases, high recreational use. The measures necessary to maintain viable trout streams during logging operations were outlined earlier in this chapter in the section "Timber Management" under "Coordination Measures." Where forests are removed in large clearcuts the volume of water is greatly increased. Additionally, spring runoffs are regularly larger while summer water flows are typically lower following a heavy timber harvest.

The key to protecting forest streams is to harvest small areas over longer time intervals to even out water flows. However, if larger permanent water flows are needed and would benefit trout waters, then heavier harvesting, if properly done, may improve water temperature and depth.

Livestock, deer, elk, and moose tend to concentrate much of their use close to or in riparian zones. Overuse of streamside vegetation by grazing animals contributes to the destruction of stream banks and streamside vegetation, thereby increasing erosion. Fencing has been used to restrict livestock access to streams, but deer and elk can easily jump a fence. Successful management under these circumstances requires a balanced multiple-use approach which includes limiting livestock and wildlife numbers together with routine watershed maintenance.

Recreational campground use is often the cause of erosion and streambank deterioration. In addition, water quality in the absence of sanitation facilities is adversely affected. Considerable progress has been made in monitoring water quality as a result of the National Environmental Policy Act. However, an increasing human population, resulting in more use of outdoor areas, necessitates constant monitoring of water quality in heavily used areas. In some situations erosion can be controlled by building structures to stabilize the banks or by plantings aimed at the same goal. Other control measures include limits on campground use.

Coordination measures are just as important to maintaining fish species as they are to terrestrial species. As with birds, mammals, and reptiles, habitat improvement is best done through coordination measures where funds can be combined from several sources to accomplish multiple management objectives. Some direct measures that have been successful in improving fishery resources are:

1. Fertilize ponds to increase phytoplankton as a food source with a 5–10–5 fertilizer.

2. Plant trees or tall shrubs along stream banks to maintain a cold water temperature.

3. Build small dams in streams to create pools. An excellent stream improvement technique. Dam maintenance must be included in future work plans.

4. Create gravel beds in lakes for spawning areas. A cost-effective technique but requires review and planning to verify what is needed.

5. Create cover in lakes by sinking brush piles or old car bodies. Very effective technique for providing cover at a low cost.

6. Build small fishing lakes within the forest environment. Requires planning for location and degree of use.

7. Remove undesirable vegetation from lakes and ponds. Heavy concentrations of aquatic vegetation can destroy a good local fishery.

Habitat improvement to maintain a good trout stream is very expensive. Improvements in the stream must be preceded by improvements in watershed conditions to achieve long-term results.

6.6 MANAGEMENT OF HUMANS

The management of humans as a directly acting factor of Conceptual Equation 3 centers on education and legal regulation. Education provides information to understand the role and utility of animals in the functioning of ecosystems. Education is also involved with regulations aimed at protecting resources necessary for the survival of both humans and wild animals.

Humans are ecological dominants so have to be controlled because a few are given to exploiting the wildlife resource. With modern tools and weapons humans can do great damage in a short time. Conversely, others choose to protect everything. In such circumstances—with vocal partisans at both extremes—conflict inevitably arises. Managing humans, as most managers soon realize, is generally more difficult than managing the wildlife resource. In the past natural resource managers have not received the training required to deal with the human dimension. But in recent years managers have placed greater emphasis on understanding human behavior and the values of wildlife as perceived by the citizenry (Kellert 1980).

Colleges and universities are also doing a better job in educating resource managers in psychology and operations research. Further, government agencies are now providing training in conflict resolution to field managers and enforcement staff. As a result, managers are more aware of the human dimension in affecting changes in management strategies which in turn will effect wildlife populations, habitat, and human experiences.

A large number of resource-oriented organizations influence their members and others through educational programs and involvement in the formulation of conservation and regulatory laws. They include professional societies such as the Society of American Foresters, The Wildlife Society, The Society for Range Management, and conservation organizations such as the National Wildlife Federation, National Audubon Society, Ducks Unlimited, Wild Turkey Federation, Rocky Mountain Elk Foundation, and Sierra Club. In many cases the members and officers influence resource management agencies in setting goals and objectives and in monitoring situations and conflicts which they consider important to their interests.

Government management agencies depend on conservation organizations to support their objectives because it is often necessary to enlist their help in securing funds for research and management purposes. Indirectly this is a form of people management since the agencies must convince the organizations that their goals are sound and will benefit the public. The manager or biologist demonstrating a good understanding of wildlife and how to deal with the people involved with wildlife, either directly or indirectly (biopolitics), will be the successful manager.

6.7 SUMMARY

Wildlife management is the art and science of manipulating the centrum of wild animal populations to meet specific objectives. The modification of natural hazards such as rain, snow, and climate is not practical, but control of man-made hazards such as roads, fences, and fire is within the scope of management. Diseases of wild populations have not been given as much attention as the other directly acting factors affecting an animal's chance to survive and reproduce. Herbicides and pesticides are included in the disease category because they work by affecting the physiology of an animal.

The management of predators is usually directed only to animals which prey on game species. However, predators are part of ecosystem functioning and so should be recognized before considering control measures. Species management includes techniques of stocking, game ranching, refuges, hunting, and manipulating habitat factors. Management of terrestrial resources deals primarily with the manipulation of food, cover, and water, either directly or through coordination measures. Aquatic resources are influenced by the surrounding watershed, so watershed manipulation or protection affects the associated water resource.

Humans are an important part of Conceptual Equation 3 because they are ecological dominants with the ability to drastically change the environment of any wildlife species. Humans can be managed by increasing their awareness of the fragile nature of wildlife habitat and populations through education and regulations that enforce the desires of society and the control of human activities.

6.8 RECOMMENDED READING

Brown, E. R., ed. 1985. *Management of Wildlife and Fish Habitats in Forests of Western Oregon and Washington.* Part I. USDA For. Serv., Pacific Northwest Region in cooperation with USDI Bureau of Land Management, Portland, OR.

Hoover, R. L., and D. L. Wills, ed. 1984. *Managing Forested Lands for Wildlife.* Colorado Div. of Wildl. in cooperation with USDA For. Serv., Rocky Mountain Region, Denver, CO.

Martin, A. C., H. S. Zim, and A. L. Nelson. 1951. *American Wildlife and Plants: A Guide to Wildlife Food Habits.* Dover Publications, New York.

Thomas, J. W., ed. 1979b. *Wildlife Habitats in Managed Forests: The Blue Mountains of Oregon and Washington.* Hdbk. 553, U.S. Dept. of Agric., Wash., DC.

Chapter 7

MANAGEMENT SYSTEMS

In the past 10 years biologists have had to devise new ways to manage wildlife populations and habitat because recent legislation has introduced terms and concepts, such as diversity and management indicator species, into the planning process. As a result several new approaches to management are being tried by resource agencies and this experimentation will continue until there is some agreement on which one best meets the various legislative intents.

In the past five years there have been significant advances in the systems available to biologists and foresters in meeting goals and objectives of management plans. Because of the diversity of conceptual approaches to a variety of systems, confusion often arises as to which concepts to adopt or which techniques to use to formulate and meet a given plan's requirements. Many administrators are bothered by the fact that resource agencies do not use standard techniques in the decision-making process across an array of land-use plans. There can be different but equally correct approaches to a problem and we need to look no further than the field of physics for an example.

Neils Bohr, an atomic physicist, concluded that information obtained under one set of experimental conditions might not be the same as or even consistent with information obtained under a different set of procedures (Weaver 1975). In such cases the information obtained from the first experiment must be viewed as complementary to the information obtained in the second. Even if the two sets of data appear contradictory both must be accepted as equally valid if no errors exist in the data. For example, under some experimental conditions electrons behave as particles, but under others they behave as waves. The question becomes: Can electrons be both particles and waves? The *principle of complementarity* asserts that they can.

By analogy, Bohr's principle can mean that different approaches to solving a common problem are equally valid. The point is that there can be several correct answers to managing wildlife populations and habitat so the manager is free to select from a variety of approaches to accomplish a set of objectives. The choice depends on the degree of complexity in the data needed to make sound management decisions, the amount of data available, and the amount of money needed to meet the objectives of the management plan. The realization that managers have several systems from which to choose provides them with a richer set of opportunities to devise a successful plan and more alternatives by which to explain decisions.

In general, the current systems of managing wildlife are either single species– or multispecies-oriented. However, there is now a trend to integrate these into one system based on habitat relationships. In the following sections we will explore both approaches to management with a look to what will be possible in the future as new and more precise information becomes available.

7.1 SINGLE SPECIES

Excellent references are available documenting the food and cover requirements of single species, especially game animals. Publications such as the *Black-tailed Deer of the Chaparral* (Taber and Dasmann 1958), *The Wild Turkey in Alabama* (Wheeler 1948), *The Bobwhite Quail, Its Habits, Preservation and Increase* (Stoddard 1931), *The Pronghorn Antelope and Its Management* (Einarsen 1948), and the books on deer (Wallmo 1981) and elk (Thomas and Toweill 1982) are only a few of the hundreds of publications contributing to the understanding of a single species. For years game and fish departments have been using single species management for hunted or fished species. The experience gained from single species management has resulted in an accumulation of a large amount of data for such species as deer, elk, and turkey and has generally led to successful programs by management agencies.

In the early stages, information on a single species tends to be at a broad scale concerned with distribution and general food and cover requirements. As information accumulates, data is refined and becomes more specific, pertaining, for example, to a particular mountain range or management unit. Single species management has largely been guided by obtaining information on food and cover requirements to use in developing written habitat guidelines. The data are often further compiled into a rating system for habitat, as has been done for big game in Montana (Cole 1958), California (Dasmann 1948), and New Mexico (New Mexico Game and Fish Dept. 1973). These guidelines have proved to be a real asset to field biologists. But their very success has led to recognition of the need to quantify requirements to better predict the effects of land treatments on wildlife habitat.

The simulation of forage for elk in Montana (Giles and Snyder 1970) and for deer in Colorado (Wallmo et al. 1977) are examples of techniques used for quantifying data and extending its utility for management. These investigators used forage as the critical component of habitat; research on other species has indicated that food and cover variations can successfully be integrated to develop habitat capability models (HCM) (Fig. 7.1) or habitat suitability indices (HSI). Both model types use an integrative approach incorporating several habitat variables to arrive at a composite rating. For the Abert squirrel, the habitat quality rating is based on tree density, size, and grouping to provide information on the habitat requirements for a given number of squirrels.

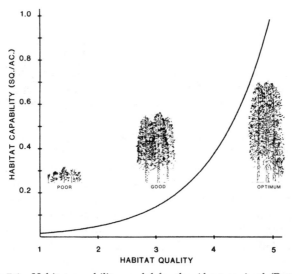

Figure 7.1. Habitat capability model for the Abert squirrel (Patton 1984).

Featured Species

A modification of single species management is *featured species* whereby one animal is selected for an area and management efforts are concentrated to fulfill its needs through coordination of timber and wildlife habitat activities (Holbrook 1974). Small areas within the habitat of the featured species are selected to meet the needs of other species (Fig. 7.2). If properly planned and implemented, featured species management provides a variety of habitats for other species that overlap the requirements of the featured species. For many years featured species management has been used successfully by state

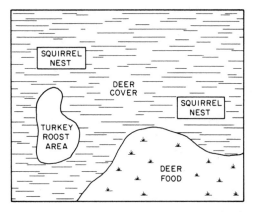

Figure 7.2. Deer is the featured species, but habitat is provided for other species as well.

agencies primarily for game animals such as deer, elk, and turkey.

The data used in single species models and guidelines provides the basis for future multispecies habitat models, but at present such models are only in the conceptual and early developmental stages. Progress has been made in developing simulation techniques that identify single species as a component of a main program. For example, the ECOSIM model developed for ponderosa pine in Arizona includes information and projections for the Abert squirrel, as well as for deer and elk forage (Rogers, Prosser, and Garrett, 1981). There is little doubt that single species management is currently the leading approach and will continue to be important in the management of game animals and wildlife classified as threatened and endangered.

7.2 MULTISPECIES

The concept of *multispecies management*, while not new, received little attention until recent legislation required consideration of all species; it now has become a favored management approach in land-use plans. Multispecies management involves grouping species with common habitat requirements and then managing for identified, overlapping critical habitat components. In the hypothetical model presented in Figure 2.7, the vertical axis represents the individual species while their broad scale association with different forest successional stages is shown along the horizontal axis. A vertical line drawn through the pole stage crosses the habitat requirements of species "E" through "N", suggesting that managing for mid-successional vegetation provides habitat of varying degrees of utility for a group of species. Multispecies management in this form is very similar to featured species. A good example of multispecies management is the management of snags for cavity nesting birds (Scott and Patton 1989).

Multispecies management techniques are presented in the form of guilds, life-forms, habitat profiles, and ecological factors. Managers using these various approaches are all trying to achieve the same goal which is to manage the forest as a *commons* for many species. This is an appealing approach because it reduces the cost of providing habitats for many single species by manipulating habitat through coordination measures with other uses of timber, range, watershed, and so forth.

Guilds

Guilds are defined as groups of species exploiting the same class of environmental resources in a similar way (Root 1967). A guild groups species together for habitat reasons, without regard to taxonomic criteria. The important consideration when developing guilds or related systems of species grouping is whether the result provides information that can be used in decision-making. Classifying animals into carnivores, herbivores, gleaners, and peckers, for example, does not provide adequate detail for management purposes. Such classifications, however, do provide a broad hierarchical level as a starting place for developing databases, models, and relationships.

PRIMARY FEEDING ZONE

PRIMARY NESTING ZONE		GROUND	SHRUBS	TREE BOLES AND LIMBS	TREE CANOPIES	SNAGS
	GROUND	RCSP WEME CORA CAQU TUVU BRTO ROAD BHCO LEGO COFL STAR				
	SHRUBS	BRTO	CATH WREN BEWR			
	TREE BOLES AND LIMBS	STAR AMKE SCOW COFL GHOW	HOWR BEWR	NUWO WBNU	ACWO COHA ATFL ANHU PLTI WEKI WEBL BGGN PHAI HUVI NOOR	VGSW
	TREE CANOPIES	RTHA MODO GHOW SCJA HOFI	BUSH			
	SNAGS					
	BREEDS ELSEWHERE	GCSP RSTO AMRO		HETH RCKI WCSP FOSP LISP	YBSA	SSHA BTPI YRWA DEJU

Figure 7.3. Guild matrix for birds using the pine-oak woodlands in California (Verner 1984).

Typically, guilds are contained in a matrix table with one axis representing the feeding habitat, and the other axis, the reproducing habitat. In the past 10 years much useful research on the theory of guilds. with an emphasis on birds, has been completed (Severinghaus 1981; Short and Burnham 1982; Mannan, Morrison, and Meslow 1984; Verner 1984). A good example of an easy-to-understand guild matrix for birds was developed for the pine-oak woodlands of California (Fig. 7.3) (Verner 1984). The primary feeding and nesting zones form the two axes of the matrix. Within the matrix, birds are assigned to each cell according to their feeding and nesting habits. Birds are identified by using a four letter code derived from their scientific name (two from genus and two from species). The matrix has a potential of 30 guilds if all cells are filled.

Life-form

Thomas, Miller, et al. (1979) have successfully used a life-form approach of the multispecies concept of management in the Blue Mountains of Oregon, and the concept has been extended to other areas (Brown 1985). A *life-form* is a single category synthesizing

LIFE FORM	REPRODUCES	FEEDS	NO. SPECIES	EXAMPLES
1.	IN WATER	IN WATER	1	BULLFROG
2.	IN WATER	ON THE GROUND, IN BUSHES, AND/OR IN TREES	9	LONG-TOED SALAMANDER, WESTERN TOAD, PACIFIC TREEFROG
3.	ON THE GROUND AROUND WATER	ON THE GROUND; AND IN BUSHES, TREES, AND WATER	45	COMMON GARTER SNAKE, KILLDEER, WESTERN JUMPING MOUSE
4.	IN CLIFFS, CAVES, RIMROCK, AND/OR TALUS	ON THE GROUND OR IN THE AIR	32	SIDE-BLOTCHED LIZARD, COMMON RAVEN, PIKA
5.	ON THE GROUND WITHOUT SPECIFIC WATER, CLIFF, RIMROCK, OR TALUS ASSOCIATION	ON THE GROUND	48	WESTERN FENCE LIZARD, DARK-EYED JUNCO, ELK
6.	ON THE GROUND	IN BUSHES, TREES, OR THE AIR	7	COMMON NIGHTHAWK, LINCOLN'S SPARROW, PORCUPINE
7.	IN BUSHES	ON THE GROUND, IN WATER, OR THE AIR	30	AMERICAN ROBIN, SWAINSON'S THRUSH, CHIPPING SPARROW
8.	IN BUSHES	IN TREES, BUSHES, OR THE AIR	6	DUSKY FLYCATCHER, YELLOW-BREASTED CHAT, AMERICAN GOLDFINCH
9.	PRIMARILY IN DECIDUOUS TREES	IN TREES, BUSHES, OR THE AIR	4	CEDAR WAXWING, NORTHERN ORIOLE, HOUSE FINCH
10.	PRIMARILY IN CONIFERS	IN TREES, BUSHES, OR THE AIR	14	GOLDEN-CROWNED KINGLET, YELLOW-RUMPED WARBLER, RED SQUIRREL
11.	IN CONIFERS OR DECIDUOUS TREES	IN TREES, IN BUSHES, ON THE GROUND, OR IN THE AIR	24	GOSHAWK, EVENING GROSBEAK, HOARY BAT
12.	ON VERY THICK BRANCHES	ON THE GROUND OR IN WATER	7	GREAT BLUE HERON, RED-TAILED HAWK, GREAT HORNED OWL
13.	IN OWN HOLE EXCAVATED IN TREE	IN TREES, IN BUSHES, ON THE GROUND, OR IN THE AIR	13	COMMON FLICKER, PILEATED WOODPECKER, RED-BREASTED NUTHATCH
14.	IN A HOLE MADE BY ANOTHER SPECIES OR IN A NATURAL HOLE	ON THE GROUND, IN WATER, OR THE AIR	37	WOOD DUCK, AMERICAN KESTREL, NORTHERN FLYING SQUIRREL
15.	IN A BURROW UNDERGROUND	ON THE GROUND OR UNDER IT	40	RUBBER BOA, BURROWING OWL, COLUMBIAN GROUND SQUIRREL
16.	IN A BURROW UNDERGROUND	IN THE AIR OR IN THE WATER	10	BANK SWALLOW, MUSKRAT, RIVER OTTER
		TOTAL:	327	

Figure 7.4. Life-forms for Blue Mountains, OR (Thomas, Miller, et al. 1979).

information on where an animal feeds and breeds (Haapanen 1965). Although the life-form is a specific category, like the guild it is represented by a matrix. Each life-form combination has an identification number and forms a simple but easy-to-use table (Fig. 7.4). All the vertebrates (26 amphibians and reptiles, 263 birds, and 90 mammals) known to occur in the Blue Mountains were ultimately grouped into 16 life-forms (Thomas, Miller, et al. 1979). The most obvious advantage of both the guild and life-form concepts is that a large amount of information can be condensed into tabular form for quick reference.

Species Habitat Profiles

Species habitat profiles (Patton 1978) were developed for the Southwest to meet the need to categorize large amounts of wildlife species information quickly to comply with the *Resource Planning Act Assessment*. Species habitat profiles combine life-forms and guilds into a hierarchical system with different levels of detail from general to very specific habitat requirements.

Each habitat profile contains six digits (Fig. 7.5). Each digit represents a level in the hierarchy and the system works like a dichotomous key used in plant and animal taxonomy with two major sections: one for reproduction and the other for feeding. Two hundred forty-six profiles containing six digits were identified for southwestern species ranging from high-elevation alpine vegetation to low-altitude deserts (Fig. 7.6). The system is open-ended so additional digits can be added, thereby providing more detailed information than the levels currently used. Information underlying the

BIOLOGICAL FACTOR	HIERARCHICAL CODE
I. REPRODUCES	100 000
I. IN WATER	110 000
I. RUNNING WATER	111 000
I. SPRINGS	111 100
I. COLD	111 110
2. INTERMEDIATE	111 120
3. WARM	111 130
I. CURRENT	111 131
2. POOLS	111 132
2. FEEDS	200 000
I. IN WATER	210 000
I. RUNNING WATER	211 000
I. SPRINGS	211 100
I. COLD	211 110
I. CURRENT	211 111

Figure 7.5. Species habitat profiles for the Southwest (Patton 1978).

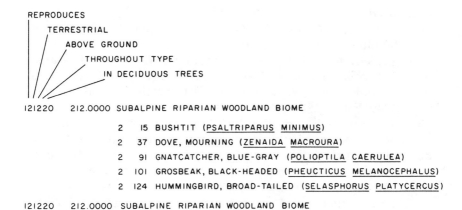

Figure 7.6. Profiles by vegetation type.

species habitat profiles was developed for use on a mainframe computer and was made available to forest planners through remote terminals. In cases where terminals were not available biologists and planners still could access the data through microfiche (Patton 1979a).

Ecological Factors

Management plans must clearly provide an array of ecological factors which have to be taken into account, manipulated, and monitored to realize the wildlife objectives set by a plan. Several different concepts for understanding and providing for ecological factors have been developed. Guilds, life-forms, and species habitat profiles are direct ways to group species with overlapping physical habitat requirements. There are, however, indirect ways to approach multispecies management through the notion of ecological factors, which are categorized as indicators, diversity (biodiversity), edge effect, habitat extent, and habitat relationships.

Indicators

Indicators were first used in resource management by plant ecologists working with soil productivity and agricultural crops (Schantz 1911). In the 1930s, vegetation condition and trend techniques were developed for range management using plant indicators, (Talbott 1937). Modified forms of these techniques are still in use today. Recently, the term *management indicator species* (MIS) was introduced to the wildlife profession by the National Forest Management Act (NFMA). This type of indicator is different from those described in ecological literature.

An *ecological indicator* is a plant or animal so closely associated with a particular environmental factor that the presence of the species is clearly indicative of the presence or absence of the factor (Clements 1916, 1949; Patton 1987). In most cases a carefully selected indicator identifies a cause-effect relationship; though the cause cannot always be easily detected, identified, or quantified, the effect is clearly desirable or undesirable. Every plant or animal is a product and measure of its environment, and therefore an indicator of some habitat factor because of its inherent need for that factor.

Ecological theory and experience suggest some general criteria for selecting indicators. First, indicators should have a narrow and specific tolerance to the factor being indicated, so that by definition the relationship has a high degree of certainty. Species with wide ecological amplitudes, which can use a combination of factors none of which is limiting and many of which can be substituted one for another, are not suitable indicators.

Second, in most cases plants are the best indicators of change since they are directly influenced by environmental factors. Plants are nonmobile, easy to count, and indicate change with a high degree of certainty. Finally, because single plants or particular species may indicate the successional stages of plant communities, the community as a whole is the best indicator of the total environment. Whether this community approach is equally valid for groups of animals has not been substantiated.

Ecological literature contains a large number of examples of the use of indicators, so it is not surprising that the framers of environmental resource legislation extended the concept to be to monitor progress in land management plans. Regulations implementing the NFMA define management indicator species as

1. Endangered and threatened plant and animal species identified on state and federal lists for the planning area.

2. Species with special habitat needs that may be influenced significantly by planned management programs.

3. Species commonly hunted, fished, or trapped.

4. Nongame species of special interest.

5. Plant or animal species selected because changes in their population are believed to indicate effects of management activities on other species of a major biological community or on water quality.

Each alternative in National Forest plans must establish objectives for the maintenance and improvement of habitat for MIS (U.S. Dept. of Agric. 1980). The five categories listed above provide great latitude in what can be selected as an indicator. Category 5 suggests that an indicator species can be used to monitor the effects of management activities on the population trend of a group of species, but research shows that this may not be a valid use of indicators (Mannan Morrison, and Meslow 1984; Verner 1984; Szaro 1985). The important consideration for using any indicator, whether it is ecological or management, is to decide what is to be indicated and then determine if there are cause-effect relationships that can be indicated with a high degree of certainty.

Biological Diversity

A term receiving much attention from forest biologists is biological diversity (also termed biodiversity or simply diversity), but the topic is not new to the wildlife profession as detailed in Pimlot's (1969) review. The National Forest Management Act (NFMA) states that *diversity* is the distribution and abundance of different plant and animal communities and species within the area covered by a land or resource management plan. A large number of species coupled with a variety of natural biological communities in an area indicates a high degree of ecological diversity. The concept of diversity in recent times has been expanded to imply that the more complex a community is, the more stable it is and the more readily it can adjust to perturbations from within without major effects on the structure of the community as a whole. In a report to the U.S. Congress, diversity has been referred to as

> the variety and variability among living organisms and the ecological complexes in which they occur. Diversity can be defined as the number of different items and their relative frequency. For biological diversity, these items are organized at many levels, ranging from complete ecosystems to the chemical structures that are the molecular basis of heredity. Thus, the term encompasses different ecosystems, species, genes, and their relative abundance (Office of Tech. Assessment 1987).

The many food chains and webs, energy pathways, and interactions in complex communities create buffers which minimize the effects of mankind's and nature's destructive forces on the community. Intuitively, diversity is thought to be "good" because complex interactions are harder to destroy than simple ones. But the view that diversity promotes stability has been challenged (May 1973).

Diversity in forest ecosystems implies an increase in the diversity of habitats and consequently an increase in potential habitat to support diverse kinds of organisms (Boyce and Cost 1978). Habitat diversity to increase species richness has been used in a practical way to manage forests in the Eastern Region of the Forest Service (Siderits and Radtke 1977). *Species richness* is the number of species occurring in a defined area and is often referred to as *alpha diversity* (MacArthur 1965). The number of species between areas is *beta diversity*. It is a measure of heterogeneity between plant or animal communities. *Gamma diversity* includes both alpha and beta in the same unit but separated from another unit by geography (Cody and Diamond 1975). In addition to alpha, beta, and gamma diversity, there is also horizontal and vertical diversity.

Horizontal diversity results from the arrangement of plant life-forms in different successional stages along a horizontal plane (Fig. 7.7). An aerial view of diversity created by openings and forest stands is depicted in Figure 7.8. In this figure a homogeneous opening can become diverse to a point, then revert back to homogeneity over ecological time. *Vertical diversity* (contrast) is the degree of divergence in height between two vegetation types or stages in a stand. It portrays the effect of layering of plant life-forms (Fig. 7.9).

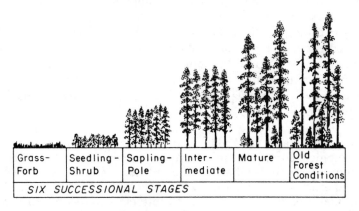

Grass-Forb	Seedling-Shrub	Sapling-Pole	Inter-mediate	Mature	Old Forest Conditions

SIX SUCCESSIONAL STAGES

Figure 7.7. Diversity along a horizontal plane.

A grass-forb successional stage adjacent to a seedling-shrub successional stage is low in contrast and thus low in diversity, but a grass-forb stage adjacent to a mature tree stage is high in contrast. The greater the contrast (high diversity), the greater the difference in structure and the number of species supported (Thomas, Maser, and Rodiek 1979b). In low-diversity situations edge cannot or can hardly be detected, but as the contrast increases the outline of the edge becomes recognizable and is fully identifiable when a high degree of

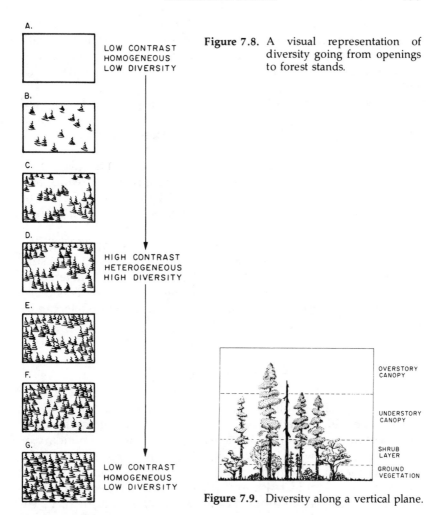

A.

LOW CONTRAST
HOMOGENEOUS
LOW DIVERSITY

Figure 7.8. A visual representation of diversity going from openings to forest stands.

B.

C.

D.

HIGH CONTRAST
HETEROGENEOUS
HIGH DIVERSITY

E.

F.

G.

LOW CONTRAST
HOMOGENEOUS
LOW DIVERSITY

OVERSTORY
CANOPY

UNDERSTORY
CANOPY

SHRUB
LAYER

GROUND
VEGETATION

Figure 7.9. Diversity along a vertical plane.

contrast is reached. Because *biological diversity* is a complex subject, it is useful to consider the different levels of diversity from the narrowest to the broadest perspective (U.S. Dept. of Agric. 1990). Four general categories—genetic, species, ecosystem, and landscape—illustrate the range of complexities contained in the term biological diversity.

Genetic diversity refers to the inherited characteristics found among individual representatives of a species which are critical to the survival of the species. A variety of genes provides for resilience under environmental stress which allows a species to adapt to changing conditions.

Species diversity is the level of biodiversity that receives the most attention. It is the variety of living organisms found in a particular place—for example, the hundreds of different species inhabiting a ponderosa pine forest including all life-forms of plants and animals. To address species diversity, one must consider how the species changes from place to place as well as over time at the same place (Fig. 1.6).

Ecosystem diversity includes the variety of species and ecological processes (nutrient cycling and energy flow) that occurs in different physical settings, such as in an old-growth forest, a riparian zone, or a desert habitat.

Landscape diversity (gamma diversity) deals with the dispersion of ecosystems—such as grasslands, wetlands, ponds, streams, and forest—in a geographical setting. A broad expanse of forests or grasslands is not as diverse as the same expanse containing a mixture of ecosystems.

By considering all the definitions and explanations discussed in this section it is easy to understand why biodiversity is a complex subject. However, one should not forget that understanding diversity is no different than understanding ecosystem functioning. This really should be the objective so that ultimately practical guidelines can be developed for management purposes.

The Edge Effect

A concept closely allied to diversity is that of the "edge effect." When two distinct vegetative life-forms (trees, shrubs, grass) or structures within life-forms (mature trees, saplings) meet, a boundary or edge is created between the two (Fig. 7.10). This change from one life-form to another or from one vegetative structure to another intrinsically produces an increase in variety or diversity. Leopold (1933) referred to this change or increase in diversity as "edge effect". He viewed and animal habitat as a phenomenon of an edge, from which developed his *Law of dispersion*:

> the potential density of game of low radius requiring two or more types is, within ordinary limits, proportional to the sum of type peripheries.

An edge quite commonly provides food and cover in proximity for low-mobility species (Fig. 7.11). Leopold's law is based on the premise that home range is fixed for a species as an innate characteristic and determines how far an animal will go in search of food and cover. While a rabbit may be restricted by instinct to remain within a 0.15-km (quarter-mile) radius, a deer is not. Yet, both rabbit and deer use edges to meet their life needs, but at different scales. Although

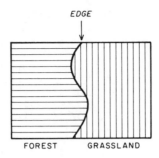

Figure 7.10. An edge is created where two vegetation types meet.

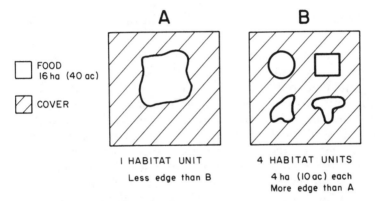

Figure 7.11. For low-mobile species the creation of edge places food and cover in close proximity.

Leopold's law of dispersion was developed in relation to small game and nongame species, his edge principle applies in varying degrees to all wildlife. Efforts to understand edge and how it can be used in wildlife management have evolved into more profound meanings and techniques for quantification of the concept.

Edge can be of two types: inherent and induced (Thomas, Maser, and Rodiek 1979b). An *inherent edge* is the natural outcome of the evolution of successional and structural stages of plant communities. *Tension zones* arise in cases where the position of the naturally evolved inherent edge is in constant change (Fig. 7.12). The factors influencing inherent edge are those identified in Conceptual Equation 1 (Table 1.1).

Induced edges result from management practices such as fire, timber harvesting, grazing, or seeding and planting. All are direct but short-term activities undertaken to manipulate succession. Both types of edges can be either abrupt or meandering depending upon the nature of the vegetation types or successional stages. Techniques

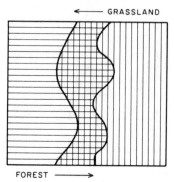

Figure 7.12. A tension zone is created by changes in edge.

have been proposed to measure differences between the two, not only in terms of diversity but in terms of wildlife as well.

Edge can be quantified by relating it to area (Patton 1975). The geometric figure possessing the greatest area and least perimeter or edge is a circle. If the ratio of circumference to area of a circle is given a value of 1, a formula can be derived to compute an index for any area to compare with a circle. A value larger than 1 is a measure of irregularity and can be used as a diversity index (DI). A 1 ha circle (2.47 ac) has a circumference of 354.49 m (1162.87 ft) and an area of 10,000 m² (107,639 ft²). The formula to set the ratio equal to 1 is:

$$\frac{C}{2\ (A \times 3.1416)^{.5}} = 1 \qquad [7.1]$$

where C is the circumference, A is the area, and 3.1416 is Pi. This same formula is often used by limnologists to express shoreline irregularity of a lake. The next step is to restate the formula for habitat diversity as:

$$DI = \frac{TP}{2\ (A \times 3.1416)^{.5}} \qquad [7.2]$$

where TP is the total perimeter around the area plus any linear edge within the area.

Several examples demonstrate how DI is computed and what it means. A 1-ha square is 100 m (328 ft) on a side, so its perimeter is 400 m (Fig 7.13A). Solving for these values:

$$DI = \frac{400}{2\ (10,000 \times 3.1416)^{.5}} = 1.13 \qquad [7.3]$$

The result indicates that a square of 1 ha (2.47 ac) has 0.13 times more edge than a circle of 1 ha.

By rewriting the formula, DI can be expressed as a percent where:

$$\% = (DI-1) \times 100$$

Thus, the computation for a 1-ha square with a DI of 1.13 is:

$$(1.13-1)*100 = 13\%$$

This figure simply means that a 1-ha square has 13% more perimeter than a 1-ha circle. A square kilometer (or square mile) by analogy also has 13% more perimeter than a circle of the same area.

Dividing the 1-ha block of the original example into 4 units of different vegetation types increases the DI to 1.69 (Fig. 7.13B). A block of the same size but long and narrow has a DI of 1.41 (Fig. 7.13C). If the long and narrow 1 ha block is divided into 4 smaller units, then the DI is increased to 1.83 (Fig. 7.13D). In Figure 7.13D, the TP of 650 m (2,132 ft) is computed by adding the outside perimeter 500 m (1,640 ft) to the 3 inside edges of 150 m (492 ft).

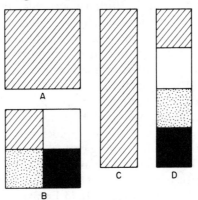

Figure 7.13. Comparison of edges created by different block patterns.

The diversity index assumes that the total perimeter of an area is actually edge. In the case of artificial unit boundaries this assumption is not valid, but the index can be derived from induced or inherent edge (Thomas, Maser, and Rodiek 1979b).

The creation of edges has been one of the major tools used by wildlife managers to increase game species. However, as is usually the case when more information about a principle accumulates from research and experience, a need to review and revise the principle develops and this is probably true for edge (Harris 1988c).

Harris has suggested that while many game species such as the white-tailed deer benefit from edge, nongame species or species needing large blocks of land with interior habitats often do not. The wildlife manager's adage that what is done to benefit one species will in turn be detrimental to another still holds.

Habitat Extent

Another concept in approaches to multispecies management involves the size of habitat needed to maintain viable populations of both alpha and gamma species using the same area. Habitat extent results from the theory of island biogeography, which seeks to explain why the number of species on one island differs from that found on another.

Darwin was the first to pursue this question in connection with his findings in the Galapagos Islands. MacArthur and Wilson (1967) integrated the concepts of ecology (including genetics) into the island theory of biogeography. Harris (1984) later translated the ideas of biogeography to isolated stands of old growth (islands) surrounded by younger stands of trees (sea).

The idea of habitat extent, however, is not new. Elton (1966) employed this same concept in developing his image of the inverse pyramid of habitat—that is, tertiary animals, those carnivorous species at the top of the food chain, are always low in numbers (gamma species) relative to the large areas of habitat (high mobility, Fig. 2.4) which they occupy, while small herbivores maintain populations of large numbers occupying small habitat areas (alpha species). Managers cannot discuss critical habitat or viable populations of gamma species such as mountain lions and grizzly bears at the same scale as say for squirrels and rabbits. Habitat extent has recently received more attention due to large blocks of public forest land which are now scheduled for cutting or conversion to other uses and thereby affect the habitat of some gamma species.

One of the major tenets of the theory of forest fragmentation is that the number of species in an island habitat is directly proportional to the size of the habitat area. This simply means that as the size of the island increases, the number of species increases. On the other hand, when larger areas of habitat are reduced in size, the number of animal species in the reduced habitat is proportionally reduced. The idea is not different from the species-area curve used by plant ecologists to determine plot size for sampling (Phillips 1959), but in the case of wildlife management, the plot size can be a block of land no smaller than is necessary to preserve a species.

In many areas, such as the Pacific Northwest, large blocks of old growth timber have been removed so the problem now is one of providing corridors so wildlife species can move between habitat islands (Fig. 7.14). Harris (1984) has termed this the "archipelago approach." Evidence for the use of corridors has been developed; thus MacClintock, R. Whitcomb, and B. Whitcomb (1977) found that small areas connected by corridors to larger areas supported a species composition comparable to that of the larger area.

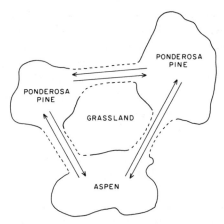

Figure 7.14. Corridors provide pathways for species moving between habitat islands.

In the southwestern United States, riparian vegetation forms corridors traversing otherwise inhospitable habitat from high-elevation spruce-fir and mixed conifer forests to the lower pinyon-juniper woodlands. In addition, stringers of ponderosa pine and pinyon-juniper, resulting from differences in soils and exposure (inherent edge), penetrate the surrounding vegetation increasing not only the edge but food and cover areas. Corridors of these types exist in many regions and must be protected and maintained as critical wildlife habitat (Thomas, Maser, and Rodiek 1979a).

The minimum size of forest blocks which must be maintained is not yet well defined, but there is some indication that a reasonable area is 30–40 ha (75–100 ac) for maximum species diversity for birds (Galli, Leck, and Forman 1976; Thomas, Maser, and Rodiek 1979b). An important consideration is that an increase in the number of species does not mean an increase in the density of species.

Species of low to moderate mobility requiring several edge types to fully meet their food and cover requirements do not benefit from large blocks of old growth forest. However, as stated earlier, some species need large blocks of mature and old growth because of the presence of specialized conditions in the interior of the forest stand (Galli, Leck, and Forman 1976). Wildlife managers must consider the size of the management unit together with forest habitat island size because the management unit is where the resources are provided for species to survive and reproduce.

Harris (1984) has suggested a system to maintain old-growth forest conditions by maintaining habitat islands in long rotations surrounded by normal rotation patches. The key to the system is scheduling stands to be harvested so as to maintain the island and corridor effect necessary for species diversity. In the process, care must be taken not to increase the number and size of the habitat islands in a combination that will in turn lead to homogeneity over a large area. For example, a large number of small openings of the same size when increased in number reduces the size of the remaining forest stands so that both openings and stands are the same size. In such circumstances homogeneity is created by the evenness of the spacing of stands and openings.

An Expanded Definition of Habitat

The definition of habitat presented in Chapter 2 stated that habitat is the sum of all the factors affecting an animal's chance to survive and reproduce in a specific place. There is, however, a scale factor of space and time that must be considered as a part of the habitat definition. There is *primary* habitat, which is all the combined habitat areas and environmental factors necessary to support a viable population of the species, and there is *secondary* habitat, the area in which an organism may spend part of its time but which does not meet all its life requirements (Harris 1984).

The genetics factor (species characteristics) outlined in Chapter 2 includes behavior traits of migration and reproduction activities. These innate behavioral characteristics have a distinct bearing on primary habitat. For example, elk migrate from summer use of high-elevation, mixed conifer forest habitat to a low-elevation, conifer-shrub habitat during the winter. Winter and summer habitat together must be delineated as primary elk habitat, but individually they are secondary habitats.

7.3 WILDLIFE HABITAT RELATIONSHIPS

Wildlife habitat relationships (WHR) is a systems approach growing out of the theory of systems analysis developed during and after World War II. This approach as used in wildlife management is an effort to comprehend all the associations with and uses of habitat by wildlife. It is a way of thinking about the complexities of plant-animal interactions. Information from both single and multispecies management systems is combined into an integrated system to meet the objectives of managing the total ecological system on a sustainable basis. The process depends upon integrating all linkages (relationships) and interactions between plant and animal species in a given ecosystem.

WHR as a process is receiving increased emphasis as a result of legislation requiring land management agencies to put in place management plans with goals, objectives, and alternatives along with a monitoring system to track progress. WHR can either be applied in general to a way of carrying on the work of a resource management agency or it can be applied in a strictly ecological way as a means of modeling plant-animal interactions. The first approach was adopted as policy by the U.S. Forest Service (Nelson and Salwasser 1982). Other land management agencies followed but modified to respond to their set of laws and regulations governing their practices.

The ecological approach to using WHR in a systematic way has been documented by Patton (1978), Thomas (1979a), Verner and Boss (1980), and Hoover and Wills (1984), but the study of the association of animals with their habitat has been in progress for many years. For example, Reynolds and Johnson (1964) provided a matrix to associate vegetation with different life-forms of animals in the Southwest (Fig. 7.15). Attempts at a functional classification for species was in progress in Texas as early as the 1940s, resulting in a

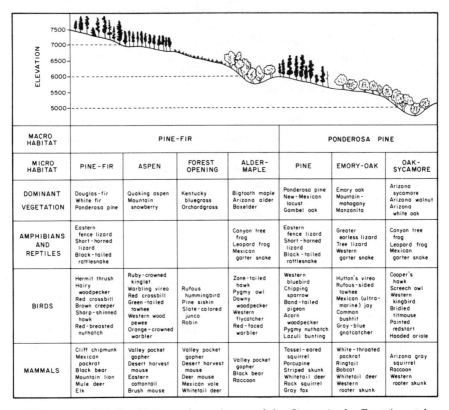

MACRO HABITAT	PINE–FIR				PONDEROSA PINE		
MICRO HABITAT	PINE–FIR	ASPEN	FOREST OPENING	ALDER–MAPLE	PINE	EMORY–OAK	OAK–SYCAMORE
DOMINANT VEGETATION	Douglas-fir White fir Ponderosa pine	Quaking aspen Mountain snowberry	Kentucky bluegrass Orchardgrass	Bigtooth maple Arizona alder Boxelder	Ponderosa pine New-Mexican locust Gambel oak	Emory oak Mountain-mahogany Manzanita	Arizona sycamore Arizona walnut Arizona white oak
AMPHIBIANS AND REPTILES	Eastern fence lizard Short-horned lizard Black-tailed rattlesnake			Canyon tree frog Leopard frog Mexican garter snake	Eastern fence lizard Short-horned lizard Black-tailed rattlesnake	Greater earless lizard Tree lizard Western garter snake	Canyon tree frog Leopard frog Mexican garter snake
BIRDS	Hermit thrush Hairy woodpecker Red crossbill Brown creeper Sharp-shinned hawk Red-breasted nuthatch	Ruby-crowned kinglet Warbling vireo Red crossbill Green-tailed towhee Western wood pewee Orange-crowned warbler	Rufous hummingbird Pine siskin Slate-colored junco Robin	Zone-tailed hawk Pygmy owl Downy woodpecker Western flycatcher Red-faced warbler	Western bluebird Chipping sparrow Band-tailed pigeon Acorn woodpecker Pygmy nuthatch Lazuli bunting	Hutton's vireo Rufous-sided towhee Mexican (ultra-marine) jay Common bushtit Gray-blue gnatcatcher	Cooper's hawk Screech owl Western kingbird Bridled titmouse Painted redstart Hooded oriole
MAMMALS	Cliff chipmunk Mexican packrat Black bear Mountain lion Mule deer Elk	Valley pocket gopher Desert harvest mouse Eastern cottontail Brush mouse	Valley pocket gopher Desert harvest mouse Deer mouse Mexican vole Whitetail deer	Valley pocket gopher Black bear Raccoon	Tassel-eared squirrel Porcupine Striped skunk Whitetail deer Rock squirrel Gray fox	White-throated packrat Ringtail Bobcat Whitetail deer Western rooter skunk	Arizona gray squirrel Raccoon Western rooter skunk

Figure 7.15. Habitat relations of vertebrates of the Sierra Ancha Experimental Forest, Arizona (Reynolds and Johnson 1964).

scheme classifying 49 mammals and 171 birds into nine categories from "conspicuously insectivorous birds" to "tree protectors" (Taylor 1940). The intent of the classification was to show how nongame species habitat was ecologically interrelated to the habitat of game animals.

A modification of Taylor's efforts resulted in a classification for North American mammals and birds by forest habitat preference (Yeager 1961). In this scheme, birds and mammals were placed into three groups according to whether they were primarily of forest or brushland, whether they associated with forest or brushland, and lastly whether they only rarely, if ever, used woody cover. The significance of this form of synthesis is greater today than it was 30 years ago, though Yeager's summary is still appropriate:

> The ecologist capable of arranging all species of mammals and birds, or all vertebrates or animals, into a habitat classification fully considerate of vegetation, terrain, water, elevation, exposure, latitude, and climatic factors would indeed provide a vast fund of useful knowledge.

Because of the large amount of data accumulated since the 1940s present day wildlife managers understand habitat relationships in more detail and so are better able to develop a systematic scheme for use of plant and animal data.

The relationships hierarchy, when included in a relational database, can combine the multispecies management systems of guilds, life-forms, habitat profiles, and ecological factors into a detailed but easy-to-use system where information can be obtained from the level needed to meet management objectives or for decision-making. The different levels of abstraction allow "nested questions" to be asked about the relationships. This approach will be dealt with in greater detail in Chapter 9.

7.4 SUMMARY

Two systems have evolved for managing wildlife: single species and multispecies. The development of both systems started with single species largely because information on hunted or fished species was needed. The featured species system, which directs management to one species but in the process maintains habitat for other species, grew out of single species management. Environmental laws enacted in the 1970s required management information on all species of wildlife not simply game animals. This requirement led to the development of new ways of combining single species information into a multispecies format.

Guilds, life-forms, species habitat profiles, and ecological factor theories were developed and then improved to make this integrated information available to managers for decision-making. The first three of these multispecies techniques uses hierarchies and matrices to classify species into groups having various habitat requirements in common. The idea is to manage for the common habitat factors to meet the needs of all the species in the group.

As a consequence of new ways of thinking about wildlife resulting from the rapid accumulation of scientific information and legislatively driven changes in planning for and management of wildlife, ecological factor theory became the fourth approach to multispecies management. Ecological factors emphasize indicators, diversity, edge, and habitat extent to set management direction.

Both single species and multispecies management systems are of value and provide a range of management tools from which to select in making decisions. However, the need to build on past efforts and combine information into more complexly related and better understood systems involving plant and animal interactions as attributes and relations in a habitat relationships system remains to be completed.

7.5 RECOMMENDED READING

Boyce, S. G., and N. D. Cost. 1978. *Forest Diversity: New Concepts and Application*. Res. Paper SE-194, USDA For. Serv., Southeastern For. and Range Exp. Sta., Asheville, NC.

Harris, L. D. 1984. *The Fragmented Forest: Island Biogeography Theory and the Preservation of Biological Diversity*. Univ. of Chicago Press, Chicago, IL.

Harris, L. D. 1988a. Edge effects and conservation of biotic diversity. *Conservation Biology* 2:330–332.

Harris, L. D. 1988b. *Landscape Linkages*. A Florida Films production. Video tape. 25 min. Tele. 904-377-3456

Nelson, R. D., and H. Salwasser. 1982. The Forest Service wildlife and fish habitat relationships program. *Trans. N. Amer. Wildl. and Nat. Resour. Con.* 47:174–182.

Pimlot, D. H. 1969. The value of diversity. *Trans. N. Amer. Wildl. and Nat. Resour. Conf.* 34:265–280.

Yeager, L. E. 1961. Classification of North American mammals and birds according to forest habitat preference. *J. For.* 59:671–674.

Chapter 8

TIMBER HARVESTING, LIVESTOCK GRAZING, AND WILDLIFE

Wildlife management has often been in conflict with other resource management practices and uses, but the two principal areas of contention, both in the past and present, are the harvesting of trees and the grazing of forested areas by livestock. The issues of the past are still with us today but at a more refined level. In the 1930s the *Journal of Forestry* included articles discussing the relationship of forestry and wildlife (Chapman 1936; Gabrielson 1936; Cahalane 1939), and that of timber versus wildlife (Davenport 1940). In recent years that journal has included articles on biological diversity (Blockstein 1990), wildlife management (Wood 1990), the new focus for wildlife management (Giles and Nielsen 1990), redesigning the forest (Lee 1990), and spotted owls (Fox and Rhea 1989).

The concerns of the 1980s will continue to be the basis of conflict in the 1990s for such matters as the cutting of old-growth forests, snags, and riparian trees, the reduction of ecological diversity, and the consumption of forage by livestock in riparian and other forested areas. These issues have created the controversies involving the spotted owl, cavity nesting birds, bald eagles, and other species classified as threatened or endangered. Yet timber harvesting to provide wood for human use while maintaining the forest ecosystem for wildlife can be complementary if the objectives for a management unit are clearly defined to include both.

Livestock and wildlife conflicts are the direct result of both groups of animals using the same forest land for food and cover. Ranchers lease range lands from state and federal governments for livestock grazing. These same lands, however, are also used by native wildlife species. Some ranchers therefore claim that deer and elk are using forage that is reserved for their livestock. State wildlife agencies hold the opposite view.

A precedent was set early in U.S. land policy to allow livestock grazing in forested areas. Today, where public lands are managed under a multiple-use concept, livestock grazing is permitted but is more or less carefully controlled. However, in many areas the livestock/wildlife conflict is as intense as the timber/wildlife conflict. An example of the type of problem that can occur is the 1989 incident of an Arizona rancher killing elk because he claimed they were using forage that belonged to his cattle.

8.1. SILVICULTURE

In the forest ecosystem, timber harvesting is the primary means available to managers to set back succession for producing food and cover for some wildlife species now and for the future. Given a set of goals which remain constant over a sufficiently long period of time, the silviculturist can prescribe treatments for stands of trees to meet multiresource goals.

Reproducing, growing, and improving stands of trees in a forest is referred to as *silviculture*. It can be used not just to produce timber but to meet multiresource objectives of wildlife, watershed, recreation, visual quality, and range management as well. Ideally silviculture begins with the regeneration of a forest stand following harvesting by one of four basic cutting methods: selection, shelterwood, seed-tree, and clearcutting. However, silvicultural treatments can begin anywhere in the life cycle of a forest stand. Once a stand has been regenerated and has progressed to the sapling, pole, or sawtimber stage, thinning then can be used to influence subsequent tree growth.

Tree Species Tolerance

An understanding of a species light tolerance is critical in selecting a harvest method to regenerate and manage a forest. For example, loblolly pine cannot grow in stands of larger and older trees that have a high degree of canopy closure. To establish the species it is necessary to cut all the trees in a designated area to remove the shade cast by overstory trees. Northern hardwoods, such as sugar maple and yellow birch, however, grow under a relatively closed canopy and thus can be regenerated in shade.

Every tree species has a different requirement for the amount of light needed for vigorous, healthy growth. Species needing large amounts of sunlight—ponderosa pine, yellow poplar, and Virginia pine—are categorized as *intolerant*—that is, they cannot grow in shade. Conversely, sugar maple, Engelmann spruce, redwood, and western hemlock can survive and grow in reduced sunlight, and thus are termed *tolerant* species (Baker 1949). Ranging between the two categories is a spectrum of species which make up an *intermediate class*—for example, Douglas-fir, white oak, and sycamore. Species tolerance is related to successional development; intolerant species are the first to germinate or sprout and grow following disturbance, while the tolerant species establish later, closer to the climax stage.

Even-aged Management

Tree species requiring considerable sunlight to germinate and grow benefit from even-aged management. Three of the four harvest methods—clearcutting, selection, and seed-tree—are used to produce even-aged stands (Fig. 8.1A). Even-aged management is directed to establishing a cohort of trees that can be harvested in the future. At the time of harvesting all the trees will be in the same age or size class.

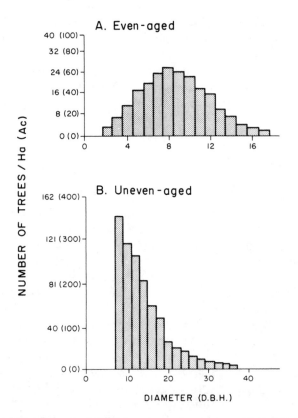

Figure 8.1. Characteristics of even- and uneven-aged stands by diameter and number of trees per hectare.

Uneven-aged Management

Theoretically, trees of all ages and sizes can be found on the same acre in uneven-aged forests. In practice, all ages or sizes of trees are not distributed equally, even though all are represented in a management unit (Fig. 8.2). Uneven-aged forests can also be managed as small units of even-aged stands. There are differences between even- and uneven-aged management, such as species composition and frequency and intensity of harvest on a particular acre, but the fundamental difference between the two is the arrangement of the growing stock.

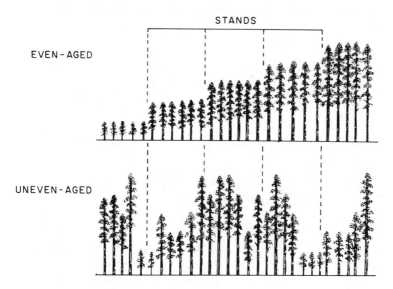

Figure 8.2. Visual representation of even- and uneven-aged stands.

Old Growth

The National Forest Management Act of 1976 focused attention on the need to maintain and provide for a diversity of plant and animal communities including old-growth forests. Forest communities aged 175–1,000 years provide habitat for certain associated wildlife species, such as the spotted owl. Under the law all the values found in these old forest communities must be considered, not simply their value for timber production. The old-growth forests in the Pacific Northwest became the center of attention in the 1980s due to the impact of this legislation on the increased demand for wood products from that region.

One of the major problems with old-growth issues is definition. Foresters usually consider old growth as over-mature or decadent stands exceeding rotation age based on the criterion f culmination of mean annual increment (MAI). Environmental groups conceive old-growth stands as possessing pristine or virgin conditions. As a result of the issues developing in the Pacific Northwest and because no common understanding of old growth existed, the Society of American Foresters appointed a Task Force to formulate an ecological definition of an average old-growth acre. The definition adopted (Heinrichs 1983) states that old growth forest:

1. Consists of 2 or more tree species of a wide range of sizes and ages. Often a long-lived seral dominant such as Douglas-fir is associated with a shade-tolerant species such as western hemlock.

2. Contains a deep, multilayered crown canopy.

3. Contains more than 10 trees at least 200 years or 40 inches dbh.

4. Contains 10 snags over 20 feet tall, and more than 20 tons of down logs.

5. Contains at least 4 large snags and an equal number of logs 25 inches dbh and 50 feet long.

The definition applies specifically to Douglas-fir forests of the Pacific Northwest, but many of the characteristics cited are applicable to other western vegetation types. The items included relate not just to growth of a single tree but to stands and their structure and composition as well.

A more recent definition of old growth is a broader one based on the premise that old-growth forests are relatively old and undisturbed by humans (Hunter 1989). Seven questions are asked to determine old-growth characteristics:

1. Has the forest reached an age at which the species composition has stabilized?

2. Has the forest reached an age at which average net annual growth is close to zero?

3. Is the forest significantly older than the average interval between natural disturbances that are severe enough to lead to succession?

4. Have the dominant trees reached the average life expectancy for that species on that type of site?

5. Is the forest's current annual growth rate below the lifetime average annual growth rate?

6. Has the forest ever been extensively or intensively cut?

7. Has the forest ever been converted by people to another type of ecosystem?

Clearly, old growth is an ecological entity, and humans have placed a high value on preserving part or all of it for its ecological significance, wildlife habitat, or recreational and aesthetic values. However, several issues petaining to old growth remain and will have to be addressed by land management agencies:

1. How many acres should be conserved?

2. Can silvicultural techniques duplicate existing conditions in future stands?

3. At what point in time should old growth be regenerated?

The first question will take several years to answer since it involves hotly debated social and political issues. The second question cannot be addressed as easily and it may take several generations before an answer is found. The third question will also take time because it deals with the problem of when does old growth cease to be old growth and become a decadent stand which may lose the value for which it was protected in the first place.

The third question raises another question: Is old growth a renewable resource? Kimmins (1987) makes the point that the notion of a renewable resource should not be based on whether it is living or dead, but whether it can be restored to a point of reuse after a period of time. The period of time must either be within our current economic or social planning scale, or it must be renewed at a rate that renders investment in its renewal economically attractive. Resources not meeting these criteria are classed as nonrenewable. Kimmins further observes that with the general antipathy of the public towards cutting 1,000-year-old redwoods in California or a 300 year-old oak tree in France these trees are nonrenewable on the basis of his conditions: Once cut the social value the trees provided is gone, never to be renewed.

8.2 HARVEST METHODS

The harvesting of all trees in a given area at one time is called *clearcutting* (Fig. 8.3). The objective is to create an even-aged stand. A new stand regenerates naturally by seed from the surrounding trees, from root or stump sprouts, or artifically by planting nursery-grown seedlings or by direct seeding. The size of a clearcut area is dictated by the requirements of the tree species, economics, state and federal laws, and constraints imposed to protect other resources, e.g., scenic beauty and wildlife. Sizes of clearcuts can range from a small stand of 4 ha (10 ac) or less to a larger stand containing 40 or more hectares (100 ac), depending on the tree species, and in some areas clearcuts have been as large as 2470 ha (1000 ac). A point to be made here is that large homogeneous areas of thousands of hectares (acres) can accomodate larger clearcuts than smaller areas. No matter what the size, the primary objective is to regenerate the stand in a short time interval. Clearcutting is done in strips, patches, or blocks and can create large scale landscape diversity.

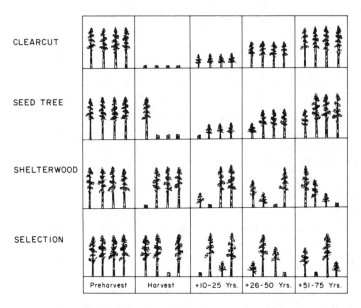

Figure 8.3. Diagrammatic concept of harvest methods.

If trees on a site are removed over time with the intent of creating an even-aged stand in a series of cuts, the method is referred to as *shelterwood* harvesting. Mature trees left following the initial cut provide both seed and shelter for seeds to germinate and for seedlings to get established. Several intermediate cuts are common but the last one always removes the remaining shelter trees.

The first entry is a preparatory cut to open the forest canopy by about 50% to stimulate seed production and provide space for seedlings. A second cut, the seed cut, can be made to remove another 25% of the original stand. Seed cuts can be bypassed if adequate seedling establishment occurred following the preparatory cut. A third or final cut is made to remove the remaining trees after seedlings are established. Shelterwood cuts can be made in strips, patches, or blocks.

Removing almost all the trees but leaving a few scattered mature trees of good genetic stock to produce seed to regenerate the harvested area is called *seed-tree* harvesting. The seed source can be a single tree or a small group of 2–3 trees. After adequate regeneration has been established the remaining trees are removed. Like clearcutting and shelterwood harvesting, the seed-tree cut is used for regenerating intolerant species of even-aged stands.

Removing single trees or small groups at each cutting is called *selection harvest*. The objective of this method is to create or maintain an uneven-aged stand in which regeneration goes forward on a continuous basis. If the groups of trees harvested are large, the result resembles a small clearcut. Trees selected for cutting are of all ages and diameters, including the oldest or largest in the stand. In the selection method, cuttings are made more frequently in a given stand than in the other three harvest methods. The selection method is designed to duplicate forest succession in which individual trees die and are replaced. It resembles gap phase replacement (Chapter 1) which creates small scale diversity within a stand.

Harvesting selected trees before a final or regeneration cut is made is called *thinning*. If the trees are usable then it is called a *commercial thinning*. A stipulated percentage of trees are removed from a stand to improve the growth or form of the remaining trees. If the cut is made to remove trees damaged by fire, insects, or disease, it is called a *salvage cut*. The number and intensity of thinnings made before the final harvest varies according to management objectives and how well the stand responded to prior thinnings.

Spacing of Trees

Thinning dense stands of unevenly spaced trees promotes timber growth and increases the quality of wood but can have a negative effect on some wildlife species by virtue of reducing protective cover. At the same time, thinning can have a positive effect on the production of herbaceous vegetation. The cover value of a stand depends on whether the trees are evenly or unevenly spaced. Thinning as a forestry practice is based on the hypothesis that trees in fully stocked stands follow an identical rate of self-thinning. The implication is that if the growing space is fully occupied, an increase in average stand diameter can only occur if there is a reduction in the density of trees.

A loss of cover for wildlife is not due solely to the reduction in the number of trees but to the *plantation effect* (even distances between trees) as well. In evenly spaced stands the cover value is greatly reduced as compared to the same stand where trees are randomly spaced. Groups of trees in stands provide greater densities by creating microsites for cover. Thinnings, if done with wildlife constraints in mind, can produce a stand arrangement which provides future benefits for deer, elk, turkey, squirrels, and other species.

8.3 FOREST REGULATION

The process of bringing a forest into balance to produce a sustained yield of wood volume is called *forest regulation*. Forest regulation is accomplished by either area or volume control. In *area regulation* a given size of area is cut each year, or at other established time periods, but the volume between areas is variable. *Volume* refers to the amount of wood removed. *Volume regulation* is realized by removing a stipulated volume of wood each year, regardless of the size of area cut.

Regulation also involves rotation and cutting cycle. *Rotation* is the time interval between one regeneration cut and the next on the same area of land to perpetuate an even-aged forest. Rotation age varies by tree species but is a function of tree growth and management objectives. It is also guided by the age at which a particular tree species reaches the culmination of mean annual increment (CMAI). MAI is the age at which wood fiber is being added at the highest average annual rate the tree can achieve. It is not the greatest age or largest size a given tree species can attain.

Cutting cycle refers to the interval of time between cuts in an uneven-aged stand. The idea is to cut the growth that has accumulated over the interval and then return later to the stand to repeat the cycle. Examples using rotation and cutting cycle for even-aged and uneven-aged stands managed by area regulation will illustrate the ideas presented.

Given the following conditions:

$$Area \ (A) = 960 \ ha \ (2371 \ ac)$$
$$Rotation \ (R) = 120 \ years$$

then the area to be cut per each is $A/R = 960 \ ha/120 \ yrs = 8 \ ha/yr$ (Fig. 8.4A). Eight-hectare stands in 120 age classes from 1 to 120 years (120 yrs × 8 ha/yr = 960 ha) are designated. The 8 ha can be from one stand or from any combination of stands—as large as the total equals 8 hectares. For example, 4 stands of 2 hectares each could be cut each year (Fig 8.4B). The result of the configuration is 480 stands (4 in each of 120 age classes) equalling 960 hectares (4 stands/age class × 2 ha

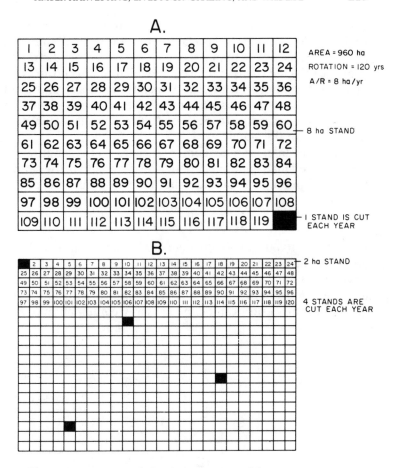

Figure 8.4. Area regulation in an even-aged forest.

stands × 120 age classes). In either case the number of hectares (acres) treated annually is identical. This of course assumes an all-aged stand with the same growth rate. In reality stands are never all-aged. By cutting a smaller sized stand, harvesting and transportation costs may increase to an unfavorable cost-benefit ratio.

Conforming to a strict stand size to meet area requirements does not provide the diversity that is needed to maintain habitat for a large number of wildlife species. The question revolves around maintaining a large number of small stands, a small number of large stands, or a variety of stand sizes from large to small in a management unit. Landscape diversity is greatest with the third option.

The problem then become one of how many large and how many small stands (or openings) to create or perpetuate. One suggestion is that the distribution of sizes should follow a log normal curve (Harris

1984) because it occurs frequently in nature (there are many more small species of animals than large ones, more small fires than large fires, more small trees than large trees, etc.). Another distribution that could be selected is the normal curve which centers on an average but includes deviation both large and small from the average. In any case, the number and size distribution of openings and stands will vary considerably according to local conditions, species requirements, and the desires of the public.

While even-aged management focuses on the length of rotation, uneven-aged management depends on the length of the cutting cycle. Given the previous area and using following information:

$$\text{Area (A)} = 960 \text{ ha (2371 ac)}$$
$$\text{Cutting cycle (CC)} = 20 \text{ years}$$

then A/CC (960/20) = 48 hectares in one stand to be entered annually (Fig. 8.5A). The diameters of trees in each of the 20 stands have a frequency distribution as shown in Figure 8.1B. All age classes are represented in the forest, but each stand varies by 20 years (Table 8.1). The yearly entry can be made in any combination of stand size

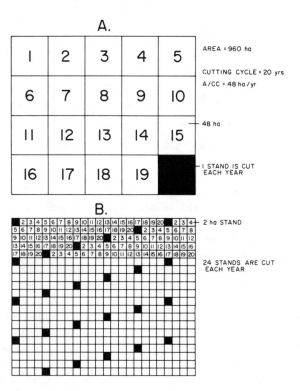

Figure 8.5. Area regulation in an uneven-aged forest.

Table 8.1. Stands and years scheduled for harvesting in an uneven-aged forest with a cutting cycle of 20 years (see Figure 8.5).

Stand	Years harvested					
1	1	21	41	61	81	101
2	2	22	42	62	82	102
3	3	23	43	63	83	103
4	4	24	44	64	84	104
5	5	25	45	65	85	105
6	6	26	46	66	86	106
7	7	27	47	67	87	107
8	8	28	48	68	88	108
9	9	29	49	69	89	109
10	10	30	50	70	90	110
11	11	31	51	71	91	111
12	12	32	52	72	92	112
13	13	33	53	73	93	113
14	14	34	54	74	94	114
15	15	35	55	75	95	115
16	16	36	56	76	96	116
17	17	37	57	77	97	117
18	18	38	58	78	98	118
19	19	39	59	79	99	119
20	20	40	60	80	100	120

and number to equal 48 hectares. For example, 24 stands of 2 hectares each could be entered every year to equal 48 hectares, resulting in 480 stands of 2 hectares each (24 stands/yr × 2 ha stands × 20 years = 960 ha) (Fig 8.5B).

Removing old or mature trees in a stand every 20 years does not provide habitat for animals needing "wolf trees," snags, or old forest conditions. Therefore, several stands considered for harvesting and some old trees in each of the other stands must be saved. These stands and individual trees should be allowed to go through a natural growth and decay process which may be as long as 300–400 years. However, as previously discussed, age of trees by itself is not necessarily indicative of old-growth conditions.

By preserving old-growth stands and decadent trees in other stands, conditions can be perpetuated to meet the needs of species depending on older, mature trees. The trees saved ultimately become the wolf trees and snags used by many nongame bird species and small mammals. The same principle can be applied to even-aged management systems in that some stands and trees are reserved to go through their normal life cycle, which is beyond rotation age.

Because factors such as topography, site quality, and climate lead to deviations from what is possible, the examples presented are ideal and will probably never occur. However, they are the consequences of the theory of silviculture and forest regulation which provides the foundation for practical application of forest science. The manager's job is to consider all these factors in designing a system that will perpetuate the plant species being manipulated to meet the needs of both humans and native wildlife species for an indefinite period of time involving hundreds of years into the future.

8.4 HARVEST SCHEDULING

An examination of the patterns of harvesting in the examples for even-aged (Fig. 8.4A) and uneven-aged management (Fig. 8.5A, Table 8.1) indicates a continuum of adjacent age/size classes. While this may be a good arrangement for timber management it eliminates much of the physical juxtaposition and interspersion of trees important to maintaining viable populations of many wildlife species.

The legal requirements for habitat distribution on National Forests in the United States are contained in the National Forest Management Act which Mealey, Lipscomb, and Johnson (1982) summarize as follows:

> The NFMA implicitly establishes the legal requirement for habitat dispersion by setting maximum size limits for areas to be regeneration harvested in one operation (Sec. 6(g)(3) (F)(iv)) and by requiring that such cuts be carried out in a manner consistent with the protection of soil, watershed, fish, wildlife, recreation and esthetic resources, and the regeneration of the timber resource. Maximum size limits on cuts require that some portions of some harvestable stands remain uncut. This imposes some degree of scattering of harvest blocks among uncut areas. Compatibility of such cuts with the protection of wildlife resources demands a certain amount of edge and retention of cover which are necessary for wildlife. Effective edge and cover in timber harvest areas result from adequately scattering cuts through uncut areas.

The distribution problem can be remedied by either randomly selecting stands to be cut in each harvest or by identifying the stands to be cut based on the constraint that adjacent stands must be of different age/size classes. This form of restriction is referred to as a *nonadjacency constraint* (NAC). By randomly selecting stands for cutting there is some level of probability that two adjacent stands of the same age or size class might result, thereby creating a larger habitat unit than desired.

Harvests can be arranged by visual adjustments on maps so that no two stands of the same age/size class are adjacent. By this process the stands in Figure 8.5A can be rearranged to create considerable contrast between stands (Fig. 8.6). Other more restrictive constraints

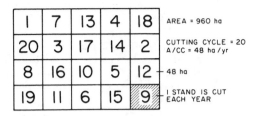

Figure 8.6. Rearrangement of stands so that no two of the same age/size class are adjacent (each number represents a unique age/size class).

can be adopted, such as requiring arbitrary differences between adjacent stands in age class (20–30 years), average stand diameter, basal area, or vegetation height.

When restrictions are applied the arrangement of the stands on the ground becomes more complicated and may be impossible to realize in one rotation without violating the restriction sometime during the rotation (Gross and Dykstra 1989). Restrictions can reduce outputs and value for other resources in addition to wildlife habitat.

The shape of the boundary of adjacent stands imposes constraints upon the planning and creation of a mosaic of stands. Squares are the easiest to consider since mathematically the minimum number of age/size classes needed to assure that no 2 are adjacent is 2. The same mathematical theory requires only 2 colors to create a checkerboard. The square pattern assumes that the touching corners do not violate the nonadjacency constraints. While there is no perfect arrangement of stands which will benefit all species of wildlife, allowing some stands of the same age/size class to be adjacent during a rotation is not catastrophic since there will be some wildlife species that benefit from this spatial pattern.

8.5 TIMBER HARVESTING

Much information has now accumulated on the effects of timber harvesting on various species of wildlife in different geographical areas. Intuitively, it has long been understood that timber harvesting can have positive, negative, or neutral effects on wildlife habitat depending upon the life requirements of the species inhabiting the area. The kinds of data necessary to support intuition and field experience are illustrated by an example of plant/animal associations needed to build habitat relationships (Table 8.2). These associations illustrate the accepted ecological law that *plant succession determines animal succession*. While the species might differ in any particular geographical area the process is the same in every forest ecosystem.

Table 8.2. Generalized plant-animal associations resulting from succession setback by clearcutting and burning (Ream and Gruell 1980).

Clearcut and Burn	Herbaceous Growth	Brush Seedlings Saplings	Mature Trees
*************************** Deer mouse ***************************			
********************* Microtines *********************			
**************** Ground squirrels ******************			
*********** Chipmunks ***********			
*************** Rabbits ***************			
********** Red squirrel **********			
********* Flying squirrel ********			

The effects of timber harvesting on wildlife are best judged by what actually happens to an animal's habitat as compared to its habitat requirements. The main concern is not what harvest method will be used or how many trees will be removed but what age and size classes will remain and where will they be located.

Both the professional wildlife literature and the news media commonly invoke the words "effect on wildlife" when discussing forest management practices such as cutting, burning, and spraying. In general, this association without naming the specific species being affected commonly results in a misunderstanding of the problem by other disciplines and the general public. The examples in Table 8.3 demonstrate that the effects of timber harvesting are clearly species specific and do not affect all species in the same way. This point must be emphasized when discussing timber harvesting and wildlife management.

Table 8.3. Examples of the effects of timber harvesting on wildlife habitat by geographical region in the United States.

EFFECTS OF CLEARCUTTING

Southeast

Cuts provided good deer forage, which was quickly eliminated by over-browsing on cuts <20 hectares. Large cuts became too dense and unattractive to deer (Harlow and Downing 1969).

Initial reduction in diversity of breeding bird species and number of birds. Increase in 3-, 7-, and 12-year cuts. Peak was at 7 years (Conner and Adkisson 1975).

Many clearcuts fragment habitat to a degree that none of the stands are large enough to fulfill wildlife species food and cover requirements (Harris and Smith 1978).

Northeast

Reduction in gray squirrel numbers; small clearcuts have least impact (Nixon, Haver, and Hansen 1980b).

Minor effect on small mammals (Krull 1970).

Table 8.3. Continued.

Lake States

Clearcutting swamp conifers increased the relative abundance of small mammals (Verme and Ozoga 1981)

Southwest

Clearcuts 8–16 ha (20–40 ac) in spruce-fir and ponderosa pine increased forage and deer and elk use (Reynolds 1966, 1969).

Dear and elk use increased when ponderosa pine in Arizona was clearcut in small patches next to stands of saplings, poles, and sawtimber (Patton 1974).

Rocky Mountains

Increases snow depth and wind; reduces cover; increases animal damage; simplifies ecosystem (Pengelly 1972).

Alternating widths of clearcut strips doubled the use by mule deer 10 years after cutting in lodgepole pine and spruce-fir in Colorado (Wallmo 1969).

Number of bird species increased after clearcutting aspen in patches of 3–20 acres in Colorado. Bird density was not significantly different from uncut areas (Scott and Crouch 1987).

Deer obtained over 2 times as much crude protein and digestible dry matter in clearcut areas as in uncut areas (Reglin et al. 1974).

Clearcutting of aspen in Utah resulted in a decline or loss of 5 bird species and an increase or invasion of 3 others (DeByle 1981).

Pacific Northwest

Bird populations increased and reached a peak 3–6 years after cutting (Hagar 1960).

Clearcut openings distributed in patchwork fashion in Douglas-fir offered a range of optimum habitat for wildlife (Hooven 1973).

Red squirrel populations were eliminated in clearcuts and reduced in density in shelterwood cuts compared to adjacent uncut areas in interior Alaska (Wolff 1975).

Clearcutting in Douglas-fir led to an increase in several small mammal species and a decrease or elimination of others depending on their specific habitat requirements (Gashwiler 1970).

Elk use was higher in clearcuts than in partial cuts in mixed conifer forests of Oregon. Deer used clearcuts but use was lower than for elk (Edgerton 1972).

EFFECTS OF SELECTION

Northeast

Variable effects on birds when northern hardwoods are cut at intensities of 25–75%: 11 species were not affected, habitat for 8 species improved, and 7 species responded negatively to tree crown removal (Webb, Behrand, and Saisorn 1977).

Vertical development maximized. Produces continuous cover of many ages and heights. Bird populations remained more stable. Not many standing dead trees (Titterington, Crawford, and Burgason 1979).

Removal of 37–55% basal area (not including cull trees) had no effect on survival, reproduction, or density of gray squirrels (Nixon, Haver, and Hansen 1980a)

Table 8.3. Continued.

Southeast

Thinning increased bird diversity by stimulating understory growth (Hooper 1967).

Southwest

The number of bird species increased when mixed conifer in Arizona was cut by single tree and group selection but the total bird density was lowered slightly (Scott and Gottfried 1983).

Pacific Northwest

Decreases edge; lack of variety of successional stages; old snags and large trees lacking; forest diversity lacking (Hall and Thomas 1979).

EFFECTS OF VARIABLE HARVESTING

Northeast

Long rotations enhance habitat for more bird species than short rotations regardless of the harvest system used in even-aged management (Crawford and Titterington 1979).

Number of species and diversity indices for songbirds was higher in logged areas and was positively correlated with increased logging intensity (Webb, Behrand, and Saisorn 1977).

The Scale Factor

As indicated earlier, the most important aspect of timber harvesting is not how many trees are removed but *how much vegetation remains* for food and cover for the single species or group of species selected for management. Practical evidence is presented in Figures 8.4 and 8.5 to demonstrate that as stand size is reduced, a point is reached where the scale relating to silvicultural and regulation methods start to converge sufficiently so that contrasts between the two become hard to distinguish. Therefore, it does not matter which harvest method is used to regenerate a stand as long as the *scale of the factors involved (time, management unit size, and size and intensity of harvest, etc.) meet the needs of the species to be managed.*

Since different vegetation types are found in diverse soil, rainfall, and temperature conditions, it is not practical here to prescribe silvicultural treatments appropriate for all forest areas and all wildlife species. However, some general guidelines that can be followed in selecting a suitable harvest method to benefit wildlife in any region can be stated in terms of scale:

1. The size of the harvest area in the management unit.

2. The habitat requirements of the wildlife species in the management unit.

3. Successional patterns of forest stands during a rotation.

4. Size and intensity of a harvest, including the physical structure, size, and juxtaposition of stands and size and juxtaposition of openings remaining after the timber harvest.

Size of Management Unit

The size of a management unit which includes all native wildlife species is determined primarily by the area needed to maintain a viable population of each species or, alternatively, whether it is to include only selected species (featured species). To illustrate, consider the approximate home range (mobility) of the following animals:

Abert squirrel	40	ha	(100 ac)
Bear	4,000	ha	(10,000 ac)
Cottontail	2	ha	(5 ac)
Mountain lion	12,900	ha	(32,000 ac)
Mule deer	600	ha	(1,500 ac)
Skunk	240	ha	(600 ac)
Grace's warbler	0.6	ha	(1.5 ac)

One mountain lion, a gamma or landscape species, may require a minimum of 12,900 hectares (32,000 ac) for a home range. If 12,900 hectares are all that are available in a forest, the mountain lion may not be able to survive and reproduce. The area is too small to maintain a viable population. Thus, it is evident that more than the home range of a single individual needs to be considered. By contrast, it is not realistic to manage units in the 2-hectare (5 ac) size necessary for low-mobility cottontails, an alpha species. A mountain lion confined to the home range of a rabbit could not survive, but several thousand rabbits can exist within the home range of a mountain lion. However, the controlling factor in determining management unit size is often not biologic but economic.

The advantages of maintaining a small unit—for example, <200 hectares (500 ac)—may be outweighed by the cost of administration, except in the case of habitat for an endangered species. Conversely, planning for and maintaining wildlife habitat in areas from 12,000 to 20,000 hectares (30–50,000 ac) is difficult because it may not be possible to provide information at the detail required to maintain viable populations of all native species.

One solution to the problem of management unit size is to select a size that includes the home range of a majority of species in the forest and then combine management units by providing connecting corridors for species of high mobility. A management unit in the range of 3–4,000 hectares (7–10,000 ac) seems to be a reasonable compromise for alpha species, as it includes about 80–90% of the species inhabiting most forest types.

Time

The successive seral stages occurring during a rotation progress through time from either primary or secondary succession to climax. The result is a change in wildlife food and cover accompanied by a change in wildlife species on the same tract of land. A resource manager thoroughly familiar with the time requirements for development of the seral stages of a particular vegetation type can simulate the location of food and cover, and therefore, the presence of wildlife species for any future time period.

Because humans tend to think in terms of the here and now, difficulties arise in visualizing the concept that timber harvesting is a process which resets the renewal cycle for vegetation and animals back to another ecological period. Theoretically, present forest conditions can be duplicated in the future if the correct silviculture prescriptions are applied when harvesting and regenerating a stand of trees.

Size and Intensity of Harvest

Populations of animals of low-mobility and specific habitat requirements—amphibians, reptiles, small birds, and mammals—can be adversely affected at the time of a timber harvest even if the cut is limited to a small area or to a single tree. The Abert squirrel, a moderately mobile animal, builds a nest in a single tree so the removal of that tree is a direct loss of a habitat component. A loss of a nest tree does not imply that an animal using the tree is destroyed, but it must use energy in selecting a new location for building a nest. This energy could otherwise be used for body maintenance and reproduction.

Highly mobile animals—large birds, mule deer, and elk—are least affected by a small cut or the removal of a single tree in one stand. As the size of cut areas increases, the food-cover ratio changes, thereby providing a positive benefit, but only to the point where the amount of cover decreases to the minimum required to maintain a viable population. The gain in forage for deer and elk from the removal of one tree or a small group of trees is not significant unless similar cuts are repeated many times in the management unit.

What Scale is Best for Wildlife

Figure 8.7 portrays three sizes of stands with three distributions to pose the question: Which distribution and size is best for wildlife? Both parts of the question should elicit another question from the reader: What species of wildlife are of concern? All the sizes of trees from seedlings to mature sawtimber are represented in each stand arrangement. As a mental exercise, select a species to associate with the smallest stand size and then ask another question: What habitat factors will change for the species as I increase stand size in the three different distributions? While this is a mental exercise, it demonstrates a major type of decision that wildlife biologists are faced with when reviewing a timber sales prescription.

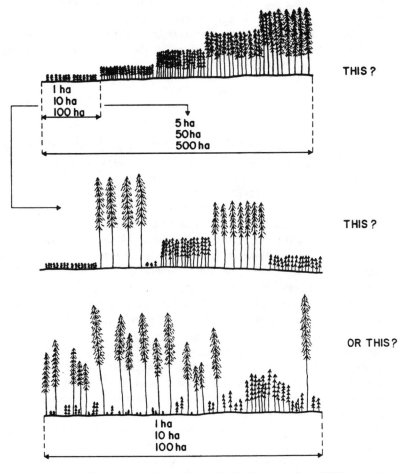

Figure 8.7. Which scale (size and distribution) is best for wildlife?

8.6 LIVESTOCK GRAZING

It is common to broadly group forests and rangelands for statistical reporting because the same resource agencies often manage both. A forest was defined in Chapter 1, but what is a *rangeland*? The U.S. Department of Agriculture (1967) defines it as land on which the climax plant community is composed principally of grasses, grasslike plants, forbs, and shrubs valuable for grazing and in sufficient quantity to justify grazing use.

Rangelands are typically regions of low and erratic precipitation, rough topography, poor drainage, and cold temperatures, which are unsuited to cultivation, but are a source of forage for free-ranging native and domestic animals as well as a source of wood products and water (Stoddart, Smith, and Box 1975). Under this definition forest types containing a sufficient quantity of forage to graze animals are rangelands. As a result practically all of the national forests and Bureau of Land Management lands in the western United States have grazing *allotments*—bounded units of land on which grazing is permitted.

Grazing Systems

Due to the ecological and economic problems of managing livestock on public lands a variety of systems have been developed to reduce soil degradation, to reduce competition with wild herbivores, and to provide sustained economic returns. These various systems are termed continuous, deferred, rotation, deferred-rotation, and rest-rotation (Stoddart, smith, and Box 1975).

The year-long use of an area by livestock in consecutive years is referred to as *continuous grazing*. This system reduces livestock handling to a minimum, which in turn reduces labor costs. Moving livestock between different subdivisions of a range is called *rotation grazing*. The livestock are moved between pastures on a planned schedule. *Deferred grazing* involves delaying grazing for a period of time during seasons when herbaceous plant growth is highest. The idea is to allow time for plants to mature and produce seed before being consumed as forage. In the combination system of *rotation-deferment*, grazing is delayed in each pasture, the deferment being rotated on a planned schedule between pastures. *Rest-rotation grazing* is similar to deferred-rotation with the exception that the period of deferment for a pasture is an entire year. Other pastures are in turn rested on a rotating basis.

The grazing systems outlined are used as the basis for many combinations of deferment, rotation, and rest specifically tailored to special rangeland conditions. In addition other systems are available (e.g., high-intensity–short duration grazing) (Holechek 1983). Each system has certain advantages and disadvantages relative to local conditions which must be evaluated before a grazing system can be selected.

Livestock-Wildlife Interactions

In the western United States large numbers of livestock, under the concept of multiple-use management, are permitted to roam free in allotments in the same forests and rangelands occupied by many species of wildlife. Conflicts arise because the diet of some species of livestock overlaps the diet of big game—mule deer, elk, bighorn sheep, and pronghorn—thereby creating direct competition for food.

The range of overlap of plant use between wildlife and domestic species, clearly demonstrates the potential for competition (Table 8.4). In general, the food preferences of deer and cattle are sufficiently different that their natural occupancy of a unit is not entirely competitive, while the diets of cattle and elk are almost identical. Discretion must be exercised when comparing overlapping diets so that season as well as location are taken into account.

Table 8.4. Generalized overlap of food habits for wildlife and livestock.

	Vegetation		
Species	Grass :	Forbs :	Woody species
Cattle	****************************		
Bighorn sheep	***************************		
Horse	***		
Elk	**		
Sheep	**		
Pronghorn	***************************		
Goat	***		
Deer	**********************************		

When several species of grazing animals use the same area a method has to be devised to compare the use of that ecosystem in equivalency of animals to determine carrying capacity and to provide a basis for charging the allotment holder for the forage eaten by domestic animals. The method that has been adopted is the *animal unit*—the weight of one mature cow or 455 kg (1,000 lbs) (Soc. for Range Manage. 1974). An animal unit is converted to an *animal unit month* (AUM), the amount of forage required for one animal for one month (30 days). To compare the forage consumption between livestock and wildlife an animal unit equivalent is computed using a 1,000-pound (455 kg) cow consuming 20 pounds (9.1 kg) of forage

Table 8.5. Animal unit equivalents for livestock and wildlife (Holechek 1988).

Animal (mature)	Weight lb (kg)	Forage Consumed lb (kg)	Animal Unit Equivalents[1]
Bison	1800 (818)	36.0 (16.4)	1.80
Bighorn sheep	180 (82)	3.6 (1.6)	0.18
Cattle	1000 (455)	20.0 (9.1)	1.00
Elk	700 (318)	14.0 (6.4)	0.70
Goat	100 (45)	2.0 (0.9)	0.10
Horse	1200 (545)	36.0 (16.4)	1.80
Moose	1200 (545)	24.0 (10.9)	1.20
Mule deer	150 (68)	3.0 (1.4)	0.15
Pronghorn	120 (55)	2.4 (1.1)	0.12
Sheep	150 (68)	3.0 (1.4)	0.15
White-tailed deer	100 (45)	2.0 (0.9)	0.10

[1]The comparison is to cattle with a value of 1.00.

Table 8.6. Examples of the effects of grazing on wildlife populations and habitat by geographic region.

Northeast

Breeding birds are 4 times more abundant on ungrazed area as compared to grazed area (Dambach 1944).

Reduced tree and shrub bird nesters; bird populations rose 35–40% when grazing was eliminated (Good and Dambach 1943).

Rocky Mountains

Songbird and raptor use increased in enclosures protected from grazing (Duff 1979).

Number of all game species observed was significantly greater on ungrazed areas versus grazed areas on the South Platte River in Colorado (Crouch 1961).

Southwest

Distribution of mule deer and elk changed when cattle were introduced after being ungrazed for 20 years. Use of habitat by elk shifted from open mesic and silviculturally disturbed areas to more closed forests. (Wallace and Krausman 1987).

White-tailed deer preferred a rangeland grazed under a system of rotational deferment, and the more frequent the deferment, the higher the preference (Reardon, Merrill, and Taylor 1978).

Pacific Coast

Moderate grazing maintained habitat for California quail in humid areas where brush or tree growth was dense. (Leopold 1977).

per day for 30 days, or 600 pounds (273 kg) (Table 8.5).

The livestock-wildlife problem is not confined simply to direct competition for food plants but also includes overgrazing of forest and rangeland habitats. *Overgrazing* or overuse refers to the degree of deterioration from a set of original conditions usually comparable to climax grassland. Overgrazing does not always result from too many livestock on a unit, but also can result from too many deer and elk. However, control of livestock grazing through animal distribution and manipulation of density is more easily done than is control of wildlife. Some species of amphibians, birds, small mammals, reptiles, and fishes are also affected by grazing through a loss of cover or special use areas. For example, ground-nesting birds, such as the wild turkey and blue grouse, lose nesting cover and poult- and chick-rearing cover when tall grasses are grazed to stubble in forested areas.

Grazing can have a positive effect on wildlife species that depend on brushland for food and cover. The heavy consumption of grasses by livestock leads to the invasion of shrubby species as has happened in the western expansion of the bobwhite quail (Edminister 1954), deer (Holechek 1982), and California quail (Leopold 1977). Studies in the United States to determine the effects of grazing on wildlife populations and habitat have produced varying results (Table 8.6).

8.7 THE FUTURE

Forestry and Wildlife

While a conflict between the harvesting of trees and the provision of habitat for forest wildlife species is not only often perceived but repeatedly occurs, there are also opportunities to improve habitat for many species. Thomas (1979a) captured the substance of the opportunities for forest-wildlife management in the following statement:

The basic assumption about wildlife habitat management in forests that are managed under the policy of multiple-use is that it must be carried out in cooperation with timber management. On public lands .... timber management is the dominant land management activity. Large-scale wildlife management usually results from the manipulation of forest vegetation primarily for wood production. Timber management is wildlife management. The degree to which it is good wildlife management depends on how well the wildlife biologist can explain the relationship of wildlife to habitat and how well the forester can manipulate habitat to achieve wildlife goals. . . .

Wildlife habitat management in forests require manipulation of tree cover (Trippensee 1948), but this is usually too expensive solely for wildlife purposes. Forest management practices undertaken to enhance wood production, however, cause dramatic changes in wildlife habitat. If correctly planned and executed, timber management practices are potentially the most practical way to achieve wildlife habitat goals.

In most situations, the wildlife biologist is responsible for making the forest manager aware of the ramifications of proposed forest management activities on wildlife habitats. The forest manager considers advice from many staff specialists and selects a course of action. But it is the field forester who actually manipulates the vegetation and alters habitat. It is essential, therefore, that the forest manager, field forester, and wildlife biologist work closely together.

The habitat management practices and coordination measures listed in Chapter 6 include over 30 associated with timber management. These coordination measures are the basic practices to review when looking for habitat improvement opportunities in timber sales. Understanding these coordination measures and the four factors involved in the scale of timber harvesting will strengthen the forest biologist's position in making recommendations for mitigation of negative effects or for habitat improvement when reviewing timber management plans.

Tradition and Other Considerations

In the earlier sections on harvest methods, forest regulation, and harvest scheduling (see 8.2–8.4), the ideas presented derive from traditional forestry practices. There is a growing concern and awareness by the public, as well as by professional foresters and wildlife managers, that some of the traditional approaches to management, which were initiated or at least influenced by Pinchot and Leopold, may no longer be adequate for today's more sophisticated professional managers.

It is now widely accepted that many combinations of cutting cycles and rotations in even-aged and uneven-aged forest management as well as many combinations of silvicultural systems can be used together. Thus, it is no longer thought necessary to be bound by the tradition of removing every tree from a clearcut, nor is it necessary to remove all down and dead material, as well as all dead standing trees, when using the other harvest systems. Wildlife managers are coming to perceive that creating edges need not always be the sole guiding principle of management, because some exceptions to the rule provide a better balance between vegetation heterogeneity and homogeneity and because both relate to sizes and shapes of stands.

The joint practice of forest and wildlife management must concentrate on maintaining ecosystems at the landscape or macro-level. This approach necessarily calls for adjustments at the micro- or stand-level to encourage diversity of plant and animals species within the landscape to insure the functioning of the whole system. Such a view demands that any given stand cannot be managed in isolation but must include an assessment of the impacts of management on adjacent stands. By the same token one animal species must not be managed in isolation but rather in connection with the other species inhabiting the ecosystem.

Wildlife managers must adopt a multispecies philosophy, taking into account ground squirrels and chipmunks as well as small passerines when considering habitat improvement projects for game animals such as deer, elk, and turkey. Fortunately this orientation has already become established and in some state and federal agencies genuine progress in multispecies ecosystem thinking is now evident in many habitat practices.

As I have suggested here and elsewhere in this volume, the traditional approaches to forest and wildlife management can no longer be thought of as the only means to manage trees and animals. However, the older approaches must not be discarded as useless but rather used as building blocks to create new and different combinations of management systems that respond to the more sophisticated and more profound understanding of forest ecosystems which has grown out of the intensive biological research of the last decade.

Wildlife and Livestock

The conflicts between wildlife and domestic livestock had their historical beginning in the development of this continent, and the problems created in the early years remain with us today. Overcoming these problems requires the careful integration of social, economic, and ecologic factors. Social factors involve how past customs have been carried forward to the present. Economic factors include the consumption of red meat as food and how much red meat the public is willing to forego to recover or maintain existing forest and rangeland conditions. Ecologic factors focus on nutrient cycling and energy flow which scientists are only now beginning to understand.

No general solutions to the livestock/wildlife problems exist today, other than complete removal of cattle from public lands. This is not an acceptable solution to ranchers and the cattle industry at this time, however. A better alternative is to evaluate each management unit in terms of its capacity to produce goods and services with the objective of protecting the soil base.

8.8 SUMMARY

Silviculture is that body of knowledge gained about the growing, reproduction, and improvement of stands of trees in a forest setting. Clearcutting, seed-tree, and shelterwood are three methods used to produce even-aged stands. The selection method, either single tree or in small groups, is used to produce stands containing trees of all ages or sizes.

The selection of a harvesting method to benefit wildlife is a matter of scale based on time, size of management unit, objectives of management unit, and size and intensity of the timber harvest. Time involves successional stages in which the plant density, life-form, and structure provide food and cover for various wildlife species. Food and cover resources together with wildlife species change over time on the same tract of land.

The selection of management unit size depends on the mobility of species native to the area. Large units are required for species such as deer and elk, but such units readily accommodate species with smaller home ranges. Animals of low mobility and specific habitat requirements—small birds and mammals—are adversely affected by large timber harvests because their entire home range is destroyed.

Domestic livestock production is in varying degrees of conflict with big game species—deer, elk, and bighorn sheep—due to direct competition for forage. In addition, competition exists between the wildlife species. A variety of grazing systems, including rotation, deferred, rest-rotation, and high intensity-short duration, have been developed to reduce soil degradation and reduce competition between domestic livestock and wildlife.

8.9 RECOMMENDED READING

Black, H. 1981. Effects of forest practices on fish and wildlife production. *Proc. of Tech. Session, Orlando, FL. 29 Sept. 1981*. Soc. of Amer. Foresters, Wash., DC.

Cahalane, V. H. 1939. Integration of wildlife management with forestry in the Central states. *J. For.* 37:162–166.

Davis, L. S., and K. N. Johnson. 1987. *Forest Management*. McGraw-Hill Book Co., New York.

Harris, L. D. 1980. Forest and wildlife dynamics in the Southeast. *Trans. N. Amer. Wildl. and Nat. Resour. Conf.* 45:307–322.

Leopold, A. 1930. Environmental controls for game through modified silviculture. *J. For.* 28:321–326.

Mitchell, G. E. 1950. Wildlife-forest relationships in the Pacific Northwest region. *J. For.* 48:26–30.

National Academy of Science. 1970. *Land Use and Wildlife Resources*. Natl. Res. Council, Div. of Biol. and Agric., Wash., DC.

Oliver, C. D. 1986. Silviculture: The past 30 years: The next 30 years. *J. For.* 84(4):32–42.

Smith, D. M. 1986. *The Practice of Silviculture*. John Wiley and Sons, New York, NY.

Society of American Foresters. 1980. *Choices in Silviculture for American Forests*. Soc. of Amer. Foresters in cooperation with Wildl. Soc., Wash., DC.

Stoddart, L. A., A. D. Smith, and T. W. Box. 1975. *Range Management*. McGraw-Hill, New York.

Van Lear, D. H. 1987. *Silviculture Effects on Wildlife Habitat in the South: An Annotated Bibliography, 1980–1985)*. Tech. Paper 17, Dept. of For., Clemson Univ., Clemson, SC.

Wenger, K. F., ed. 1984. *Forestry Handbook*. John Wiley and Sons, New York.

Chapter 9

HABITAT RELATIONSHIPS—
A SYSTEMS APPROACH TO
MANAGEMENT

A system is a set or arrangement of related or connected things forming a unity or whole. Biologists, for example, seek to understand how animal populations, as wholes, are related to the use of different plant life-forms for food and cover in a wildlife habitat system. A system is also a way of thinking about or executing a complex set of instructions in a systematic order to achieve a stated objective, as in a monitoring scheme.

Over the past 20 years many techniques have been developed to assist field biologists in managing wildlife populations and habitats. In many cases, as for basic habitat models or score cards, hand manipulation of data is adequate and still a common procedure. Some of the visual charts, graphs, and figures that have been developed over the years are very effective in presenting the flow of information in plant/animal management systems. These types of manual and visual tools are also preliminary forms suitable for integrating into sophisticated computer simulation models. Moreover, a point is reached when hand manipulation of data is no longer practical; managers must then necessarily turn to the computer to increase their efficiency and to manage the sheer complexity of the data that has been developed.

Computers have, for the first time, given mankind the capability to sort through and organize millions of pieces of data very quickly to provide comparisons, mathematical analyses, and statistical summaries. While the computer clearly has many advantages over hand manipulation of data, the user must always keep in mind that computers are controlled and programmed by humans who also enter the data, computer output is only as good as the quality of the stored information.

The rapid development of microcomputers has brought about a marked change in the way wildlife habitat and species information is used in decision making. The period from 1975 to the present has been one of explosive development not only in increasingly powerful computing devices but increasingly complex computer software to store and retrieve information as well. These improvements have led to better and more efficient ways of analyzing the data that has been accumulating in books, periodicals, and storage files for years. As a result resource managers are now developing larger databases and experimenting with increasingly sophisticated and realistic habitat simulation models to use in solving management problems.

Models reduce the ambiguity and complexity inherent in understanding the involved and intricate network of relationships found in every ecosystem. Both manual and computer techniques directed at understanding and producing management plans have a place in resource management. In this chapter examples of both kinds of techniques are presented. The purpose is to suggest ways in which various techniques can be used singly or, more often, combined into powerful habitat relationship systems to assist managers in decision making.

9.1 HABITAT MODELS

A model is an intellectual representation of a real-world situation (Fig. 9.1). In developing a model the modeler must identify a representation (abstraction) that accurately and comprehensively defines the problem. Once a model has been developed the results must constantly be interpreted and decisions must be made whether

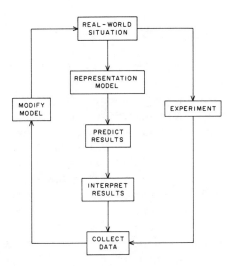

Figure 9.1. Models are a representation of a real-world situation.

more data is needed or a modification of the model is required.

Humans use models everyday to make decisions because they are constantly drawing on experience and learned patterns of using data stored in memory (brain). As experiences and patterns of thinking accumulate and are in turn used to make further decisions, we are testing or retesting a consciously or subconsciously formed model. If our decision is correct, our mental model is validated and will be used again in similar circumstances. If our decision proves unsound, we revise the model in light of this additional experience and new data: then we try again. In effect, we routinely build models in our mind and repeatedly test them by making decisions and assessing the outcomes of these decisions.

We build models because they help us categorize and organize data and thoughts and communicate ideas. Conceptual Equation 3 (Table 2.2) is a model categorizing and organizing all the direct factors affecting the ability of an animal to survive and reproduce. As a conceptual equation it is a useful tool to synthesize an immense body of data and to communicate ideas. Models can also be used to make predictions, and frequently they can also be used to compare management alternatives.

In wildlife and natural resource management, decisions must be made even though there is often a lack of good data on the use of habitat by a particular wildlife species. Unfortunately, this is the rule rather than an exception. But even when data is lacking and the biological processes are not well understood, techniques have been developed to improve decision making under conditions of uncertainty.

Factors to Consider

Models do not just happen. They have to be deliberately planned to solve a problem, be it simple or complex. Since models are a product of human thinking, they must be developed so that others can use and understand what the model does and what it tells them. Because a transfer of knowledge takes place between the model and user, the model must have a *structure* that facilitates the transfer. Models are made up of an input function, some form of manipulative function or algorithm, and an output function (Fig. 9.2A). Users need not understand the algorithm that manipulates the data, but they must be able to understand the structure of both the input and the output and how the latter can be used in decision making. Although the user may not understand the internal workings of a particular model, he or she cannot be relieved of the responsibility of assessing the validity of the results.

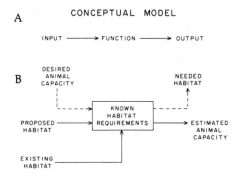

Figure 9.2. Models require an input, a manipulative function and an output.

Models use different levels of *resolution,* which refers to the programmed ability to separate components (Starfield and Bleloch 1986). For example, the results of one model may address total herbage production while those of another separate production into woody plants, forbs, and grasses. If resolution is too coarse, as may be the case with total production in the first example, the model is generally inadequate. If the resolution is too fine, too much time is spent entering or manipulating irrelevant detail. Resolution also has to take into account the level or scale of decision making. For example, models designed to provide answers on a regional scale are invariably too coarse for local use.

Models also vary in the degree of *detail* incorporated. In estimating herbage must the user record quantities to the nearest gram (ounce) or to the nearest kilogram (pound)? The question is: Will the decision to be made be affected by using the coarser herbage quantity? The impact of the effects of detail can often be determined by conducting a sensitivity analysis.

A *sensitivity analysis* is a means of systematically determining how changes in a variable or parameter affect the model's output. If a large change is needed in an input variable to effect a small change in the output, the variable being treated is not very sensitive. However, if a small variation in an input variable results in a large change in the output, the variable may be too sensitive for practical field use.

The modeler must constantly think of how the output is to be used in decision making and how much effort is required to collect data for input. The variables to be included or excluded define the *scope* of the model. If 3 variables will provide a 95% confidence interval while 2 variables will provide a 90% confidence interval in the output, is the inclusion of the third variable justified? Another very important factor in model development is *time.* Is the model to be used for present time (3–5 years), ecologic time dealing with plant succession, or instantaneous time such as in population equations? In model building most of the questions relating to scope, time, sensitivity, detail, and resolution are automatically taken into account by knowledgeable programmers.

Types of Models

Models are generally of two types: conceptual or empirical. *conceptual* models represent thoughts and ideas in a qualitative and deductive way rather than in terms of numbers. In Figure 9.2B, input, program function, and output have been formulated in a detailed and recognizable way to indicate how a conceptual model operates. While the arrows show a one-way flow of information, they can be reversed so that biologists seeking to maintain the population level of a certain species can determine the amount of habitat needed whether it currently exists or is proposed.

Equation 3 (Table 2.2) is also a conceptual model enumerating the factors influencing an animal's or population's chances of surviving and reproducing. The equation is used in this book to focus attention on the specific factors biologists must constantly consider when working with wildlife management problems. Another example of a conceptual model is Chapman's equation (Equation 2, Table 2.1) which uses biotic potential (BP) and environmental resistance (ER) to explain the factors affecting a population. The latter equation proved useful in communicating ideas between practicing field biologists for a number of years when very little scientific information was available for decision making. Conceptual models are the forerunners of the more sophisticated *empirical* models which utilize both qualitative and quantitative data.

Empirical models are based on the ideas and relationships contained in conceptual models but augmented by numeric data to provide a systematic means of making a decision. Quantitative empirical models are often *deterministic*. This means they predict a future event by using measurable sets of factors and relationships. For example, if the rate of increase (r) in a population of animals is known to be 0.02 and the base population is 1000, then the next year's increase is:

$$P_1 = r \times P_2 = .02 \times 1000 = 20$$

Present-day wildlife biologists can only hope that sometime in their career they will possess this type of data and that all their problems will be as simple to solve as the one in the example.

Increasingly models are being developed to deal with *stochastic* phenomenon. These types of models are designed to include the statistical probabilities that some special event will or will not happen. Stochastic models can operate at a level of uncertainty as, for example, required in pattern recognition models (Williams et al. 1977). Pattern recognition models were developed as a direct response to the needs of wildlife managers who must often make decisions in an environment of considerable uncertainty. Such methodologies are becoming increasingly useful as biologists become more familiar with this format and the ease with which field data can be collected and structured into a model (Grubb 1988).

The following example shows how the pattern recognition procedure is used. Consider a hypothetical turkey habitat with 10 units occupied and 10 units unoccupied. For the units to be occupied, some attributes necessary for perpetuating the population must exist. By selecting five attributes from the occupied sites that previously have been identified as contributing to turkey habitat, a model can be developed. These same attributes may or may not occur in 10 units of habitat that are unoccupied. Using presence-absence information, a set of probabilities are computed for each attribute (Table 9.1). For attribute 1, which occurs on 7 occupied sites, the probability is $7/10 = 0.7$, and for the unoccupied sites it is $3/10 = 0.3$. The same computation is used for the other attributes for occupied and unoccupied sites.

Table 9.1. Data from 20 sample units with turkey habitat attributes.

Habitat Attributes	Occupied (10 units)	Unoccupied (10 units)
1. Five openings from 10–20 acres	7 (0.7)	3 (0.3)
2. Two trees per acre >20 in dbh	6 (0.6)	2 (0.2)
3. Average trees >9 in dbh	8 (0.8)	4 (0.4)
4. Oak >10 in dbh present	9 (0.9)	5 (0.5)
5. Herbage production in openings >1,000 lbs/ac	4 (0.4)	7 (.07)

A new unit of habitat is to be evaluated for the probability of being occupied by turkeys. In the new unit the following information is obtained to be compared with the occupied and unoccupied information in Table 9.1:

Attribute	Present
1	yes (+)
2	yes (+)
3	no (−)
4	no (−)
5	yes (+)

A set of probabilities is computed using Bayes' theorem (Mason 1970). The computations for the habitat to be evaluated are as follows for the model of occupied sites:

$$P_O = (0.70)\ (0.60)\ (1–0.80)\ (1–0.90)\ (0.40) = 0.00336$$

Attribute numbers 1 and 2 are derived directly from Table 9.1, but attribute number 3 in the list above does not occur on the site to be evaluated (as indicated by the minus sign), so the computation is 1–0.80, the 0.80 coming from the frequency data for occupied habitat for attribute 3 in Table 9.1. The same procedure is followed for all the minus signs in occupied and unoccupied sites.

Next the model of unoccupied sites is calculated:

$$P_U = (0.30)\ (0.20)\ (1–0.40)\ (1–0.50)\ (0.70) = .0126$$

With these calculations in hand the probability of the unknown site being inhabited can be calculated by the following formula:

$$P_H = \frac{P_{PO} \times P_O}{(P_{PO} \times P_O) + (P_{PU} \times P_U)}$$

where

P_H = Probability of evaluated habitat being turkey habitat
P_O = Probability of attributes being in occupied habitat
P_U = Probability of attributes being in unoccupied habitat
P_{PO} = Prior probability of occupied habitat
P_{PU} = Prior probability of unoccupied habitat

Since the chances that the site to be evaluated is turkey habitat is unknown (P_H), a prior probability, (best guess) such as a fifty-fifty chance that it will be occupied (P_{PO}) or unoccupied (P_{PU}), is assigned, thus $P_o = 0.5$ and $P_u = 0.5$. With this estimate the calculated probability that the site is turkey habitat is:

$$P_H = \frac{0.5 \times 0.00336}{(0.5 \times 0.00336) + (0.5 \times 0.0126)} = 0.22$$

The conclusion is that given the set of inventory data for the five attributes for the site to be evaluated, the probability that it is occupied turkey habitat is 22%.

Pattern recognition is a quantitative model used for predicting whether a set of habitat factors will be used by an animal. Other quantitative models can be linear or take the form of curvilinear regressions as, for example, the uninhibited and inhibited growth models (Chapter 2). The habitat capability model for the Abert squirrel in ponderosa pine is curvilinear and associates squirrel density with habitat quality and with what a given quality is, on the average, capable of producing (Fig. 7.1). Quantitative models are difficult to develop because they require several years of data for replication of results.

Another approach to decision making (Chapter 10) is the use of decision tree and Delphi techniques (Crance 1987), or an expert system. Because of the recurrent need to make many decisions in short time frames with inadequate data, such as in national forest plans, a trend has developed to use techniques that produce results more quickly and less expensively than models which depend heavily on experimentation or the collection of field data.

The intent of a decision tree model is to chart out and document a course of management action based upon a logical assessment of a set of alternative rules. Pairs of empirical propositions are presented and one is then selected over another. This process of selection is repeated until a decision is reached; both the format used and the logical steps taken are analogous to those of a dichotomous key for keying out a plant or animal species. Decision trees are often depicted on a chart showing the pathways that can be taken to a decision. These visual displays help the user visualize the total set of alternatives and provide some confidence that what the users has selected is the right decision.

The Delphi technique has had considerable use in the social and behavioral sciences but in recent years has become more popular in solving natural resource problems. Delphi is a process similar to the expert opinion technique except that it is more detailed and time consuming. Members of the group doing a Delphi do not know each other; the process of focusing on the selection of alternatives and their outcomes are generally conducted through the mail. In addition, some statistical analysis can be done in the Delphi analytical process.

The expert system approach to models relies on the knowledge and experience accumulated by professional workers. The procedure is to select 4–5 practicing biologists or resource managers and convene them as a problem-solving group. The group is presented with a set of circumstances—for example, how will a proposed timber harvest affect deer, elk, and turkey habitat? Each member of the group independently quantifies his or her opinion using a preselected scale such as 1–5 or 1–10 and then an attempt is made to get the group to arrive at a consensus opinion. Despite the absence of scientifically validated data, in many instances subjective nonnumerical expressions, such as the use of the adjectives "poor" to "good," are adequate for arriving at meaningful conclusions.

The expert system approach can be very effective if the experts composing the group are chosen carefully and have many years of experience or background with the species they are considering. The knowledge and experience factors of the experts become evident if, for example, 5 members of a group each have 5 years of experience working with animals whose habitat they are evaluating, then the accumulated knowledge is 25 years. With this much experience, the choices made by each individual are very close to those of other members of the group. An example of how to develop and use an expert system is best presented by the process for developing a local habitat model.

Developing a Local Habitat Model

Decision making at the local level can often be enhanced by developing a model for a specific area or project to display alternative treatment effects. Models for decision making need not always be in a form suitable for computer use, and indeed paper models are often the precursor of more detailed computer models. The initial step in developing a habitat model involves defining the requirements by kinds, amounts, and distribution of food and cover for a species or group of species. The second step is to quantify these habitat components and group them into quality classes (poor to excellent) to rate a habitat for a given species or group of species.

One common approach is to rate habitat for individuals or groups of species on a scale from 1–5. The rating is based on the opinion of a group of experts or sometimes even on one person's opinion. The rating by the experts or manager is based on the knowledge of the food and cover requirements of the species involved and how these requirements are met by the existing habitat or will be met as the habitat is changed by a proposed timber treatment. Rating of habitat can be done on any scale (project area, planning unit, watershed, district) depending on need.

Table 9.2 presents a rating for an imaginary district with 4 planning units. The scores and ratings for the example district are based on 4 units for 6 species. Each unit is given a score for habitat quality by species where 0 = none or eliminated, 1 = poor, and 5 = optimum. Species rated for each unit vary due to differences in vegetation types

Table 9.2. Wildlife species ranked by quality of habitat on a 4-unit district.

Wildlife Species	Unit Score[1]				District Score[2]	Rating
	A	B	C	D		
Deer	3	3	2	1	9	45%
Elk	2	4	5	2	13	65%
Turkey	1	3	2	3	9	45%
Squirrel	5	4	1	4	14	70%
Cottontail	4	3	1	2	10	50%
Grouse	2	5	3	1	11	55%
Total	17	22	14	13	66	55%

[1]Scores are based on a scale of 1–5, 5 is optimum.
[2]The maximum rating possible is 120 (6 species × optimum score of 5 × 4 units).

and successional stages supporting differing amounts and types of food and cover.

In the example the total score is 66. Column totals give scores for each unit, while row totals show the score for each species. The base value of the district is then calculated in order to compare actual habitat value to an ideal or optimum habitat. If each habitat was given a rating of 5 (Excellent), then 6 species × 5 maximum ratings × 4 units equals 120. The district base is ideally 120 but the present quality is 66; this can be expressed as a rating of 55% (66/120). The assumption underlying the rating is that the lower the percentile the fewer the animals the unit on a district can support. Contrariwise, the more closely the rating approaches 5 the greater the number of animals a unit on the district can support.

Additional species can be added to the model to more adequately reflect the total wildlife population native to the district. By so doing the base (denominator) will be increased but the quality rating will always remain between 1–100%. A scale of this type easily lends itself to setting goals and levels of management intensity (Table 9.3). The district goal might be to raise deer habitat quality at 55%. Alternatively, improvement of habitat might be undertaken for all species having similar requirements or for only featured species to bring the habitat quality to any level consistent with the financial resources available.

It is wise to record on a form the information used to arrive at these manually maintained ratings (Fig. 9.3). References are critical if habitat ratings are to be included in management plans or environmental impact statements since natural resource management plans end up in litigation with increasing frequency. The basic assumption for the local habitat model is that habitat quality over the long run determines population density of the targeted species (Conceptual Equation 4, Table 4.6).

Table 9.3. Dollar values required to achieve a management goal for a particular species.

Species	Estimated Habitat Quality		Improvements	
	Present	Desired	Needed	Cost
Deer	45%	55%	1. 20 openings 10 ac ea	$ 40,000
			2. Water developments	10,000
			3. Aspen regeneration	35,000
			4. Seeding grasses and forbs	50,000
			Total	$135,000

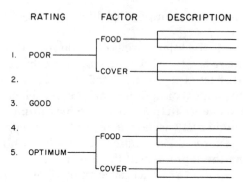

Figure 9.3. Document each rating by references or expert opinion, as though it would have to be defended in court.

Table 9.4. Matrix for wildlife stand structure model.

Wildlife Species:

Vegetation Type:

Single Storied Stand													
Use	:1	:2a	:2b	:2c	:3a	:3b	:3c	:4a	:4b	:4c	:5a	:5b	:5c
Food	:	:	:	:	:	:	:	:	:	:	:	:	:
Cover	:	:	:	:	:	:	:	:	:	:	:	:	:

Multistoried Stand													
Use	:1	:2a	:2b	:2c	:3a	:3b	:3c	:4a	:4b	:4c	:5a	:5b	:5c
Food	:	:	:	:	:	:	:	:	:	:	:	:	:
Cover	:	:	:	:	:	:	:	:	:	:	:	:	:

Code	Structural Stage	Size[1]
1	Grass-Forb	
2	Seedling-Sapling	(0–4.9 ft ht)
3	Pole	(5–11.9 in dbh)
4	Sawtimber	(12–16 in dbh)
5	Mature	(>16 in dbh)

Canopy Closure[1]

a	10–40%
b	41–70%
c	>70%

Habitat Rating Scale

(To be used to rate each cell in the above matrix)

5	Optimum	(1.0)
4		(0.8)
3	Moderate	(0.6)
2		(0.4)
1	Low	(0.2)
0	No value	(0.0)

[1]Size classes and canopy closure classes vary according to forest type.

Stand Structure Models

Associating wildlife species with forest stand structure in a matrix is an excellent way to present data in a clear and concise manner. Simple stand structure models are routinely organized and presented as manually constructed paper-based models, but the more comprehensive models dealing with large numbers of animals, food sources, cover conditions, complex energy flows, are subject to frequent revision and thus best maintained in a computer. In Table 9.4, the y axis of the matrix consists of two rows, one for food, the other for cover. The x axis contains spaces for "n" columns—in this case 13, one for each structural class.

The matrix is completed by entering into each cell the information for each species in each vegetation type. In some cases an association is at a crude level of "use" or "no use." If, however, sufficient information is available a quantitative rating scale, such as noted in Table 9.4, is used. The structure of such a model lends itself well to summarizing and communicating information on a management unit. It also serves well in cases where vegetation type and canopy closure percentages vary widely. Once a stand structure/wildlife species model has been developed, it can be augmented by linking it to tree growth and yield models suitable for evaluating both present and future stand conditions/habitat relationships.

Models and Decision Making

In the past several years there have been adequate warnings given to managers and developers about the proper use of models in making decisions (Verner et al. 1984) and there have been suggestions presented to improve the model development process (Jeffers 1980). Before undertaking a major project to develop a model for use by managers, it is incumbent upon the developer to seek help and advice from those with experience and background not just in the process itself but also in the subject area of the model. The best situation is to have several people with varied backgrounds working as a team (Bunnell 1989) as is often done in database development (Patton 1978) or in an expert system using a Delphi technique (Crance 1987, Mollohan 1988).

In developing models do not overlook biologists with many years of experience. These people may never have developed a model on paper, but every day they are in the field they build models in their mind and use them to make decisions. Trying to duplicate this process for a model requires that developers put themselves in the manager's position to understand the kinds of information he or she needs to make a decision.

Because of the large number of laws and regulations requiring many different types of data, managers now are in critical need of information to make decisions. If you do not want your information used, then it better not be written down. Once on paper it will get used even if it is just a WAG (Wild A** Guess). If you do write it down in a publication or use it in a presentation, then you are obligated to provide some quality assessments or consequences of its use.

9.2 HABITAT INDICES

The structure and life-form of vegetation are the most important factors in creating wildlife food sources and cover; therefore, describing forest wildlife habitat involves measuring the amounts and analyzing the relationships between various kinds of trees, shrubs, and herbaceous vegetation. In most instances biologists must use data developed by foresters who are oriented to forest management practices for habitat analysis. Traditional methods for inventorying timber stands use absolute measures of stand density, such as basal area. Basal area is the area of a horizontal cross section of a tree measured at a height of 1.4 m (4.5 ft) above the ground. All the trees on a plot are measured after which the areas of the individual trees are summed and the total stated on a unit area basis (hectare or acre). The resulting value is a measure of tree density.

Basal area is used in forestry because it is highly correlated with the growth and volume of forest stands. However, basal area does not adequately describe habitat because the same basal area for two different stands can result from different stand characteristics. For example, a basal area of 23 m^2/ha (100 ft^2/ac) could be made up of 450 trees/ha (181/ac) of 25 cm dbh (10 in) or 50 trees/ha (20 trees/ac) of 76 cm (30 in) (Table 9.5). Obviously these two stands have quite a different cover value for a given wildlife species. In pole-sized stands, a basal area of 13.8 m^2/ha (60 ft^2/ac) may provide sufficient cover for elk, whereas in a sawtimber-sized stand, 20.7 m^2/ha (90 ft^2/ac) of basal area are needed to provide cover (McTague and Patton 1989).

Another common forest measurement is canopy closure. Tree canopy closure is preferred over basal area by wildlife biologists in part because understory herbage production is directly related to the amount of light reaching the forest floor. In addition, tree canopy provides a measure of thermal cover for wildlife species such as deer and elk. However, canopy closure is generally of qualified utility because its use is largely restricted to mature stands of overstory trees.

Tree and shrub size and density are the two best measures of a stand for quantifying habitat for a particular wildlife species.

Table 9.5. Number of trees per unit area needed for a given basal area.

Dbh (cm)	BA m^2	No. Trees (per ha) Basal Area					
		17	23	29	34	40	46
25	0.051	333	450	568	666	784	901
38	0.114	149	201	254	298	351	403
51	0.202	84	114	144	168	198	228
64	0.316	54	73	92	108	127	146
76	0.456	37	50	64	75	88	101

Trees per ac = Trees per ha × 0.404
Basal area per ac = Basal area per ha × 4.356

Dbh (in)	BA ft^2	No. Trees (per ac) Basal Area					
		75	100	125	150	175	200
10	0.55	136	181	227	272	318	363
15	1.23	61	81	101	122	142	162
20	2.18	34	46	57	69	80	92
25	3.41	22	29	37	44	51	59
30	4.91	15	20	25	31	36	41

Trees per ha = Trees per ac × 2.471
Basal area per ha = Basal area per ac × 0.2296

However, other measures of relative size and density, such as the tree stand density index (SDI) developed by Reineke (1933) may prove beneficial. SDI was seldom used until the mid 1970s when its utility as a measure of tree density became more evident (Drew and Flewelling 1977). It provides a relationship between basal area, trees/ha (ac), average stand diameter, and stocking of a forested stand. It is a quantitative measure of forest stand characteristics developed from a hypothesis that all tree species in fully stocked stands follow an identical rate of self-thinning. In fully stocked stands with complete crown closure the following linear relationship exists between trees per ha (ac) and average stand diameter:

$$log\ N = k - 1.605\ log\ D$$
where
N = trees per unit of land area
D = quadratic mean diameter
k = a constant specific to tree species
log = logarithm to base 10

A reference diameter of 25.4 cm (10 in) is used to develop a set of curves so that the number of trees per unit of area (N) equals stand density index when the average stand diameter (D) equals 25.4 cm.

The equation is:

$$\log \text{SDI} = \log \text{N} + 1.605 \log(\text{D}/25.4)$$
$$\text{SDI} = 10^{(\log \text{SDI})}$$

For example, in a stand having an average diameter of 15 cm and 1500 trees per ha the computations are:

$$
\begin{aligned}
\log \text{SDI} &= 3.171 + 1.605 \times \log (15/24.5) \\
&= 3.171 + 1.605 \times (-0.2287) \\
&= 2.8090 \\
&= 10^{(2.8090)} \\
&= 644
\end{aligned}
$$

SDI is only now beginning to be used to describe wildlife cover (Smith and Long 1987). It may prove quite useful in quantifying wildlife habitat once it has been further developed and related to the habitat quality of a particular wildlife species. The value of a SDI lies in the insight it provides wildlife managers in trying to assess two stands with different average tree diameters and densities. SDI constructed for both stands may prove that they have the same wildlife habitat value (McTague and Patton 1989).

9.3 DATABASE MANAGEMENT

Before the introduction of computers on a broad scale natural resource agencies devoted considerable attention and resources to data storage and retrieval because it was a labor-intensive job and a highly useful undertaking if properly done. More commonly, an agency's natural resource data in the form of field reports was kept in files or boxes in some out-of-the-way corner. Once stored, managers found it almost impossible to locate data unless someone happened to remember where a box or file was last seen.

In the last 20 years much progress in data storage and retrieval has been made by converting the data contained in field reports first to key-sort punch cards and more recently to magnetic tape or disk. Concurrently, computer software to manipulate and retrieve stored data has been greatly improved. Data no longer needs to be held in an unorganized manner but can be entered into well-organized databases for later retrieval in a systematic way. Information in wildlife databases may be organized by guilds, life-forms, species habitat profiles, or as models of various kinds in habitat relationships systems.

Natural resource managers build databases because they must have ready access to the information necessary to make decisions. Databases store various types of information which when accumulated in sufficient breadth and depth can be used to develop models and build habitat relationship systems of sufficient complexity, sophistication, and utility to greatly assist managers in decision making.

A *database* is a collection of data organized in some rational and readily accessible fashion. In recent years the term "database" has become almost exclusively associated with computers. But databases are in truth a means of storing large amounts of information whether in a computer, a library card catalog, a dictionary, a subject file cabinet, or the most powerful database yet devised—the human brain. To be useful a database must be associated with a procedure for managing the data or information it contains.

Data for storage in a database must be structured and organized in a logical and consistent way. The fundamental structure is a record devoted to an identifiable entity (e.g., species). Within this record a varying number of fields are located, each dedicated to holding information or a specific characteristic associated with the entity. Similar types of data must be stored in similar locations in the database record. Thus, taxonomic data is not typically stored in a habitat database intended for management information. Irrelevant data takes up space and reduces efficiency. Every piece of information in a record needs to be relevant and concise.

One of the best ways to control the quality of data going into a computer database is to read it from a form whose format is easily followed by a data entry operator. The matrix in Table 9.4 is an example of a field data form which can be used to collect raw data in a format that can be readily entered in a relational database.

Database Management Systems

Broadly speaking database management systems can be grouped into three types: hierarchical, network, and relational. The network system is infrequently used in the natural resource field so is not discussed here. Until about 1980 almost all wildlife databases were organized in a hierarchical system. In Figure 9.4 the categories such as Animal Species and States are repeating groups so that there can be many animals listed under many states as in this example:

Repeating Group	States			
	AZ	NM	CO	UT
Animal species:	Mule deer	Mule deer	Mule deer	Mule deer
	Pronghorn	Pronghorn	Pronghorn	Pronghorn

Because humans quite typically think, organize, and model their world in a hierarchical manner, this type of database scheme is both a

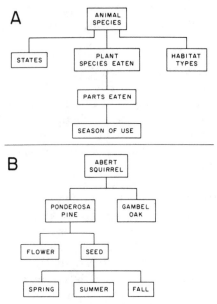

Figure 9.4. Flow of information in a hierarchical database. A definition tree showing the logical arrangement of data in repeating groups is <u>RUNWILD</u>
System 2000 (Lehmkuhl and Patton 1984).

comfortable and an intuitive approach to data management. But
hierarchical databases have major drawbacks; they are relatively
inflexible and hard to change. Furthermore, they usually provide
only one path to a particular data record due to the dendritic arrangement of the data.

The relational database management system arose out of a
concept developed by E. F. Codd (1970). The microcomputer had not
yet been developed so the relational approach was first conceived for
large mainframe computers. However, neither the type of computer
nor the available software packages had the capacity to link data to
show relationships in a fuller, more natural holistic way.

With rapid advances in miniaturization and increases in memory
and storage in the early 1980s, the power of microcomputers
increased so much that by the end of the decade they had evolved into
machines as powerful as the mainframe devices of 10 years before. As
a consequence, the larger and more complex systems developed for
mainframes were transformed and often augmented for use on a
microcomputer. Among the more practical systems of use to wildlife
biologists are the various relational database management systems.

A *relational database* is a collection of relations and attributes stored
together in a logical way. While the body of data is stored in a
hierarchical file system, a number of such files are linked in a variety

of ways not unlike ecological food chains and webs. The manner in which the data files are formatted and linked in a database are largely defined by the many ways in which the relational software packages have built upon Codd's initial formulations. Software packages to manage relational databases for microcomputers having up to 80 relations (tables), 800 attributes (columns), and tens of thousands of rows for data are now commercially available.

The term *relation* refers to an accumulation of facts (attributes) describing the same subject. An example of a subject might be students. The question is: What do students have in common? They have a first and last name, weight, date of birth, height, social security number, and so forth. All the characteristics pertaining only to students form a set of relations. The different characteristics just mentioned are attributes of each student. Therefore, an *attribute* is a fact about an inherent physical characteristic, property, or measurement of a particular subject. In a database scheme, a relation is a table and an attribute is a column in that table (Fig. 9.5). The relationships between animals and habitat can be described completely by using a system of attributes and relations in a relational database.

RELATION		
ATTRIBUTE	ATTRIBUTE	ATTRIBUTE
DATA CELL	DATA CELL	DATA CELL

Row →

TABLE		
COLUMN	COLUMN	COLUMN
DATA CELL	DATA CELL	DATA CELL

Figure 9.5. Habitat attributes and relations can be defined in tables and columns.

Intersecting rows and columns in a relational database create data cells for storing information describing the attribute characteristic. Some advantages of relational databases are:

1. They retain the simplicity and intellectually comfortable style associated with data in tabular form.

2. Two relations can be used to develop a third relation for either permanent or temporary use.

3. They have the ability to link and associate different relations.

4. New relations can be added or deleted at any time.

5. Some relational database software packages allow users to ask questions in a natural language structure.

In part, the utility of relational databases is found in the simplicity and in the control of data entered in the tables. If the tables are well organized in the system, any particular piece of data is entered only once. A relational database allows retrieval of information from linking tables so that the combined information provides answers to many questions.

9.4 RELATIONSHIPS SYSTEMS

A *relationships system* is information categorized and stored in a predesigned table-and-column format in a database. The use of relational databases implicitly imparts a very particular way of thinking and conceptualizing information. The procedures followed to link information forces the user to take into account the way in which the natural world is organized—not simply as hierarchical chains, but also as webs in a complex of relationships (e.g., an ecological system).

The easiest way to understand a relationships system is to go through the process of developing one for an actual field location with typical problems that might be found in other geographical areas. The example presented in the following sections is for a national forest in the Southwest.

Greg's Problem

Greg Goodwin is a wildlife staff biologist on the Coconino National Forest in Arizona who wants to store wildlife information in a microcomputer in an organized manner for easy retrieval for decision making. Greg has decided to use a relational database to store information because he knows that if he uses a good data model he will simultaneously generate a wildlife habitat relationships system. A data model is also a synonym for the term "database scheme" so the two can be used interchangeably.

Data models for developing habitat relationships are new to the wildlife profession because relational database software for storing and retrieving data conforming to relational theory has only recently become available for microcomputers. Moreover, Greg's problem has been considerably simplified because a habitat relationships data model has been developed for southwestern ponderosa pine, and the design for linking the relations can be used for other vegetation types with only slight modifications for local conditions (Patton and Severson 1989). Development of this data model, with its corresponding menu for each relation, has been in progress for ten years. The data model resulted from work on a database known as <u>RUNWILD</u> which provides information for national forests in Arizona and New Mexico to meet the requirements of the Forest and Rangeland Renewable Resources Planning Act of 1974 (Patton 1979b).

Even though a generic relationships data model is available for Greg's use, he still must go through the process of making a list containing information in each wildlife category that he routinely provides to decision makers such as the forest supervisor and district rangers. In addition, wildlife information for the forest also may be requested by members of a regional planning team or state and federal agencies such as the Game and Fish Department or the Fish and Wildlife Service.

The list that Greg produces contains four broad categories of information relating to (1) characteristics of wildlife species, (2) habitat characteristics used by wildlife, (3) wildlife associated with vegetation types, and (4) a local vegetation inventory. These categories are considered essential for sound wildlife management, monitoring, and reporting. Other categories such as taxonomic or biological information can be added to the list, but those Greg has identified are applicable and basic to forest areas across North America. Large amounts of information are typically available for each category, but the important items to consider are those that will be used in decision making or facilitate a reporting process. The categories of information contain subject matter as follows:

1. Characteristics of wildlife species
   a. Scientific name
   b. Common name
   c. Legal status
   d. Life history information
      Home range
      Breeding seasons
      Life-form

2. Habitat characteristics used by wildlife
   a. Physical description of feeding areas
   b. Physical description of cover areas

3. Wildlife associated with vegetation types
   a. What wildlife species are associated with what vegetation types
   b. Value of the vegetation to different species

4. Local vegetation inventory
   a. Management unit identification
   b. Stand locations
   c. Physical characteristics of stands

Once the list is complete, the next step is to develop the categories into tables containing columns of data that are useful and can be programmed into a relational database. The usefulness of information in a table is greatly enhanced if the idea of a central theme is followed so that columns in the table are directly related to the theme.

One central theme emerging from Greg's list is the designation of characteristics for wildlife species. The characteristics identified follow those listed in Table 9.6. The question is: What do all wildlife species have in common? The answer may contain several items, but those shown below are the most appropriate for the local situation and in fact for other countries as well:

1. Common name
2. Scientific name
3. Life-form
4. Breeding season
5. Home range
6. Legal status

These six items can be included in a table named SPECIES and patterned after the format conforming to relational theory and habitat relationships. By convention, table names most often contain from 1–8 capital letters, but more are possible depending on the software used. Column names generally have a maximum of 7 letters and the first letter is capitalized. Abbreviations are necessary and should be easily recognized by personnel using the database. A sample of entries in data cells in SPECIES shows the format and types of data in the table.

Table: SPECIES

Cname	Sname	Lform	Breed	Range	Legal
Mule deer	*Odocoileus hemionus*	Mammal	Fall	600 ha	Game animal
Abert squirrel	*Sciurus aberti*	Mammal	Spring	40 ha	Game animal
Wild turkey	*Meleagris gallopavo*	Bird	Spring	-0-	Game animal
Pygmy nuthatch	*Sitta pygmae*	Bird	Spring	-0-	Nongame

Data cells have a physical and a logical domain. Cells assigned a null value (-0-) indicate no data is available or that it was not entered. The *physical domain* defines the type (integer, real, date, text) of data contained in each data cell in a column and the extent of the size of the field (20 characters, etc.). For example, sufficient space must be allocated to the species field to accommodate the longest scientific name.

Table 9.6. Tables and columns used in <u>HABREL</u>. A null value (-0-) is used for any attribute not yet entered or is unknown. Number in parenthesis is the number of spaces allocated to each attribute. This number is always a default of 3 to allow for a null value.

TABLE	Column
1. SPECIES	1. Cname (30) (Common name)
	2. Sname (30) (Scientific name)
	3. Subsp (15) (Subspecies)
	4. Scode (9) (Scientific code)
	5. Lform (3) (Life-form)

	Am = Amphibian
	      Bi = Bird
	      Fi = Fish
	      Ma = Mammal
	      Re = Reptile
	      In = Invertebrate

	6. Legal (3) (Legal status)

	Fl = Federal listed
	      Sl = State listed
	      Ga = Game animal
	      Fb = Furbearer
	      Ng = Nongame

	7. Range (15) (Home range)
	User defined
	8. Breed (3) (Breeding season)

	Sp = Spring (Mar–May)
	      Su = Summer (Jun–Aug)
	      Fa = Fall (Sep–Nov)
	      Wi = Winter (Dec–Feb)

2. HABITAT	1. Cname (30) (Common name)
	2. Series (3) (Vegetation type)
	3. Reuse (3) (How physical resources are used)

	Fa = Feeding area
	      Ca = Cover area
	      Fo = Food item
	      Sh = Shelter
	      Sp = Special requirements
	      Ba = Breeding area

	4. Order1 (20) (1st level use)
	Feeding or Cover area

	Air
	      Opening
	      Topography
	      Tree stand
	      Water

Table 9.6. Continued.

TABLE	Column

Food item

Vertebrates
Tree-shrub parts
Herbaceous parts

Aquatic
Artifacts
Fungi
Arthropods

Shelter

Live tree
Snag
Cave
Log-stump
Ground
Understory
Woody debris
Litter
Rock fissures

Special requirements

User defined

5. Order2 (20) (2nd level use)

Air

Over water
Over vegetation
Over all terrain

Opening

Grass-forbs
Shrubs
Small sapling (<5'–1.9")
Single tree

Topography

Cliff-ledge
Talus slope
Canyon bottom
Rock outcrop
Landscape

Tree stand age

Young (5'–4.9")
Intermediate (5"–13.9")
Mature (14"–20"+)
Old growth (Old-growth conditions)

Water

Seeps-springs
Streams-rivers
Ponds-lakes

Table 9.6. Continued.

TABLE	Column
	Bank
	Marsh
	Rain pools
	Vertebrates
	Amphibians
	Birds
	Fish
	Small mammals
	Medium-large mammals
	Reptiles
	Tree-shrub parts
	Cones
	Twigs
	Buds
	Pollen
	Leaves
	Roots
	Needles
	Acorns
	Bark
	Soft fruits
	Nuts
	Seeds
	Sap
	Herbaceous
	Flowers
	Shoots
	Nectar
	Misc. vegetation
	Honey
	Grass
	Forbs
	Woody stems (browse)
	Aquatics
	Submergents
	Emergents
	Algae
	Plankton
	Aquatic insects
	Artifacts
	Carrion
	Eggs
	Bones
	Garbage
	Fungi
	Mistletoe
	Hypogeous fungi
	Shelf fungi

Table 9.6. Continued.

TABLE	Column

Arthropods

Flying insects
Insect larvae
Crawling insects
Crustaceans
Snails
Spiders
Worms

Shelter

Nest
Bed
Burrow
Roost
Cavity
Under bark
Bunchgrass
Brush

Special requirements

User defined

6. Order3 (15) (3rd level use)

Opening

Dry
Moist-wet

Water

Warm
Intermediate
Cold

Tree stand size

Yo	Ss = Dbh 5'–1.9"	(Small sapling)
	Ls = Dbh 2"–4.9"	(Large sapling)
In	St = Dbh 5"–7.9"	(Small tree)
	Sm = Dbh 8"–10.9"	(Small-medium tree)
	Me = Dbh 11"–13.9"	(Medium tree)
Ma	Ml = Dbh 14"–16.9"	(Medium-large tree)
	Lt = Dbh 17"–19.9"	(Large tree)
	Ma = Dbh 20"+	(Mature tree)
Og	Og = Old-growth conditions	

Special requirements

User defined

7. Order4 (15) (4th level use)

Opening or Shelter

Sandy soil
Rocky soil

Table 9.6. Continued.

TABLE	Column
	Tree stand density
	Thin (11–40% overstory) Moderate (41–70% overstory) Dense (>70% overstory)
	Water
	Milky-muddy
	Special requirements
	User defined
	8. Order5 (15) (5th level of use)
	Tree stand understory
	Little to none Grass-forbs Shrubs Small trees
	Special requirements
	User defined
	9. Order6 (15) (6th level of use)
	Tree stand overstory
	Single story Multistory
	Special requirements
	User defined
	10. Value1 (3) (Degree of preference when available)
	Used Little use Unknown
	11. Value2
	Used
	5 = High use 4 = Moderately high use 3 = Moderate use
	Little use
	2 = Moderately low use 1 = Low use
3. ECOS	1. Cname (30) (Common name)
	2. Form (3) (Vegetation formation)
	Fo = Forest Wo = Woodland De = Desert Sc = Scrubland Gr = Grassland

Table 9.6. Continued.

TABLE	Column
	3. Series (3) (Vegetation series, Southwestern U.S.).

Al = Alpine
Mc = Mixed conifer
Pp = Ponderosa pine
Pj = Pinyon-juniper
Rh = High-elevation riparian
Rl = Low-elevation riparian
Gh = High-elevation grassland
Gl = Low-elevation grassland
Ch = Chaparral

4. Assoc (3) (Association with Series)

Ha = Considered typical habitat
Fr = Occurs but is fringe habitat
Ra = Rarely occurs in type

5. Season (3) (Presence in Series)

Yl = Yearlong resident
Su = Summer
Wi = Winter
Sm = Seasonal movement (up-down slope)
Tr = Transient

TABLE	Column
4. AREA	1. Unit (30) Management unit name or number
	2. Sub (30) Subunit name or number
	3. Stand (3) Stand number
	4. Series (3) Vegetation series
	5. Area Number of hectares (acres) in stand
	6. Canopy Percent canopy closure of stand
	7. Trees Density of trees in stand
	8. Dbh Average diameter (cm, in) of trees in stand
	9. Vol Volume (cubic meters, board feet) of stand
	10. Ba Basal area (square meters, feet) of stand

DEFINITIONS

Opening: <10% canopy, or vegetation not in trees.
Small trees (5′ ht–1.9″ Dbh).

The *logical domain* includes all the meaningful information about the subject of the column. For example, the physical domain of both the scientific and common names is 30 characters but common name is not entered in the same column as scientific name because their meaning is different. Figure 9.6 is an example of data types and their physical and logical domain for a table named AREA that is one of the HABREL database tables.

The SPECIES table displays several characteristics common to tables in relational databases.

1. Every row has the same number of columns (Cname, Sname, etc.).

2. Each column contains the same kind of fact (logical domain) in each row. For example, the Sname for *Sciurus aberti* conveys the same fact about *Sciurus aberti* as Sname conveys about *Odocoileus hemionus*.

3. There is only one entry for each row for each fact. That is, there can be only one Cname for each species.

4. The order of the rows is unimportant. To get information about the Abert squirrel, it is not necessary to know the location of the row containing the name.

5. The order of the columns is not important. In the SPECIES table, Lform could just as well be placed before Cname.

6. No two rows are exactly the same.

TABLE : AREA

COLUMN NAME (LOGICAL DOMAIN)	ABBREV.	TYPE OF DATA	LENGTH (PHYS. DOMAIN)
Unit	Unit	Text	30 characters
Subunit	Sub	Text	30 characters
Stand	Stand	Text	3 characters
Vegetation series	Series	Text	3 characters
Area	Area	Integer	Open field
Percent canopy	Canopy	Text	3 characters
Number of trees / ha	Trees	Integer	Open field
Average tree diameter	Dbh	Real	Open field
Volume / hectare	Vol	Integer	Open field
Basal area / hectare	Ba	Integer	Open field

Figure 9.6. Physical and logical domain of data types.

After Greg has identified the categories of information for the forest database, he then determines the relationships between the information in the categories. The purpose of determining relationships is to create themes or subjects for each table and for linking tables with common columns. For example: How are different species of wildlife related to vegetation type? What vegetation successional stages provide food and cover for different wildlife species? How will this information be included in the tables? This review process is the basic building block for tables and columns included in a habitat relationships data model. It is important to consider all possible combinations of information in each category and between categories. Any one relationship may prove to be important for decision making.

The process of efficiently developing tables and determining relationships comes with experience, but fortunately the generic data model has for most uses solved this problem. There are ways to improve the design of the data model by combining and coding some attributes (columns) to gain efficiency, but they are beyond the scope of this book. The remaining job for Greg is to enter his data into the generic model in a relational database and make changes for local situations.

Design Criteria

The first attributes of habitat include how animals use a particular physical factor—whether for a feeding area, a cover area, a food item, for shelter (nest tree, cavity, burrow, etc.), or special requirements (aspen trees, soft soil, etc.). The feeding and cover area attributes are the most difficult to determine for a habitat relation (table). Of all the possible physical habitat combinations for feeding and cover areas for animals to use within a forest, the most descriptive are: air, openings, tree stands, and water. The food, shelter, and other category are less difficult to describe for what an animal needs to survive and reproduce.

As data accumulates in the HABITAT relation, the data model develops into a multispecies model corresponding to a three-dimensional matrix which is an extension of the matrix models now being used in several Forest Service regions.

The three categories of information along with individual items in each category, should be developed with management biologists to determine what subjects generally are used at the local management level for decision making. Each of the three broad categories can be used as the subject of a table to define the contents of columns containing more detailed information for a matrix. Discussions with forest biologists in the Southwest have lead to six design criteria that a habitat classification system must meet (Patton 1990).

1. Information in the system should generally be suitable for inclusion in a matrix table similar to the habitat models used by several Forest Service regions.

2. The system should be adaptable for storage and retrieval on a microcomputer using commercial relational database software such as Oracle, Rbase, and so forth.

3. Data in the system should be the type that could be used in decision making.

4. The system should provide the capability to answer nested questions such as "What wildlife species are found in ponderosa pine in the winter?"

5. The system should provide the capability to show what wildlife species use the same physical habitat factors. For example, it should be able to answer the question, "What species use old forest conditions in ponderosa pine for cover?"

6. Data should be capable of being displayed without a high degree of redundancy in rows and columns in a matrix table.

The above criteria could all be met if restrictions were enforced to describe habitat by relations and attributes conforming to relational theory.

Design Problems

Conforming to the six design criteria will create tradeoff problems in how data are stored and retrieved from tables. Tradeoffs can be made without deviating from restrictions, depending on how the data is displayed. For example, the following cases illustrate the different ways of storing or displaying information.

Case 1

Species	Vegetation type
Mule deer	Ponderosa pine
Mule deer	Mixed conifer
Mule deer	Pinyon juniper
Mule deer	Spruce-fir

Disadvantage: Mule deer must be entered more than once.

Case 2

Species	Vegetation type
Mule deer	Ponderosa pine, Mixed conifer, Pinyon juniper, Spruce-fir

    Disadvantage: More than one vegetation type is entered in one column.

Case 3

Species	Ponderosa pine	Mixed conifer	etc.
Mule deer	X	X	X

    Disadvantage: Each vegetation type requires a separate column.

Case 4

Species	Vegetation type
Mule deer	Ponderosa pine
	Mixed conifer
	Pinyon juniper
	Spruce-fir

    Disadvantage: Current techniques do not allow storing data in this structure in a table with blank spaces all relating to species (although it can be done by other means).

Case 1 provides the information required, but it has considerable redundancy because mule deer has to be entered for each vegetation type in which it occurs. In Case 2 mule deer is entered only once in the Species column but in return all vegetation types also have to be entered in one column. Case 3 is not efficient because it requires a column for every vegetation type. Case 4 is the preferred way to present information in reports and management plans.

The problem is one of storing data in tables, columns, and rows so that it can be printed on an 8- by 10-inch sheet of paper as depicted by Case 4. It can be done in Case 2 by selecting information with a condition clause and setting the width of the column in the statement to force the vegetation types into a vertical order; the Species column will contain only one name. Case 1 is a perfectly good table, but it can lead to visual clutter if there is a long list of vegetation types for each animal.

The Relationships Data Model

Four tables form the basic relationships data model for the Coconino National Forest. These are SPECIES, HABITAT, ECOS, and AREA (Fig. 9.7) in a model named HABREL—an acronym for *Hab*itat

*Rel*ationships. <u>HABREL</u> is not a software package. It is a file name for accessing a habitat relationships data model that can be stored in a microcomputer. The name of the model can be changed to describe any information contained in the tables and columns in the model. The categories of information developed by Greg conform quite clearly to the categories in the generic tables, columns, and menus (Table 9.6). The general model can be entered into any commercial database that uses relational commands. The reader should remember that the tables and columns listed in Table 9.6 comprise the relationships data model. The data model is the most important consideration, not the particular relational software selected to store the model.

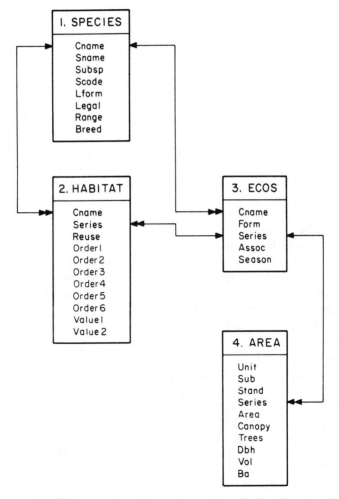

Figure 9.7. <u>HABREL</u> is a habitat relationships database for the Coconino National Forest, Arizona.

Data in the SPECIES table applicable to a particular animal is entered only once, so in relational theory this type of table is referred to as an entity table. The SPECIES table is also the *core table* for HABREL because data in the other tables ultimately is linked back to this table as it contains the scientific name and life-form of individual species. If a species common name is not entered in the proper column in each table and all the cells in each row filled, the core table will not be able to perform its function of logically organizing the information in the other relational tables. The following data, for example, associates animals with habitat:

Table: HABITAT

Cname	Series	Reuse	Order1	Order2	Order3	Order4	Order5	Order6	Value1	Value2
Mule deer	Pond. pine	Cover	Tr. stand	Inter.	Dbh 11–13.9	Mod.	Gr-forb	Mult.	Used	Mod.
Abert sq	Pond. pine	Cover	Tr. stand	Inter.	Dbh 11–13.9	Mod.	Gr-forb	Mult.	Used	High

It must now be noted that in completing the HABITAT table several columns are devoted to describing the nature of the physical characteristics of the areas in which the animals are found. Thus, columns are named by the word "order" followed by a rising sequence of numbers. These characteristics are preferably stated in quantitative terms, but if numerical data is lacking, qualitative terms may be used. Each order and its rising number provides increasing detail about the nature of the habitat generally characterized by the column "Order1". By separating data into finer levels of detail managers can best describe the habitat conditions most appropriate to the nature of their particular problem. Additional orders beyond 6 can be added in cases where more detailed data is available and may be useful for decision making. At some level of detail the complete niche for a particular species can be described. In many cases, however, detailed information beyond Order2 is not available so a general term such as "Intermediate" must be used (Table 9.6).

In addition to a hierarchical classification of habitats in the HABITAT table, the "use" status of an area is defined by the "Value" column followed by another series of rising numbers which describes these characteristics in increasing detail. Habitat use can often be described only in terms of used or little-no use, but if further detail can be added, it should be put in subsequent columns. Text in common language is generally used for the data entered in the columns, but codes, if they are well understood by the users, can be substituted: examples are Pp for ponderosa pine and Fo for forest. Entries or values are selected from a menu provided for each column.

A sample menu of terms used in the database is shown in Table 9.6. The entries in the following table use codes or abbreviations which correspond to the same data in the table above.

Table: HABITAT

Cname	Series	Reuse	Order1	Order2	Order3	Order4	Order5	Order6	Value1	Value2
Mule deer	Pp	Co	St	In	11–13.9	Mo	Gf	Mu	Us	Mo
Abert sq	Pp	Co	St	In	11–13.9	Mo	Gf	Mu	Us	Hi

The ECOS (ecosystem) table provides information on the adaptability of a species to various vegetation types. Since a given species is usually found in several vegetation types, ECOS table commonly contains the same data in several column cells but no two rows of data contain identical data—that is, one column cell in the table will always be different in each row. Data in the ECOS table below shows that mule deer and elk both occur in pinyon-juniper but at different seasons. This is not the case for the Abert squirrel for which pinyon-juniper is fringe habitat. When several hundred species have been entered into an ECOS table it becomes a powerful tool for associating species with vegetation types.

Table : ECOS

Cname	Form	Series	Assoc	Season
Mule deer	Forest	Ponderosa pine	Habitat	Yearlong
Abert squirrel	Forest	Ponderosa Pine	Habitat	Yearlong
Mule deer	Woodland	Pinyon-juniper	Habitat	Yearlong
Elk	Woodland	Pinyon-juniper	Habitat	Winter
Abert squirrel	Woodland	Pinyon-juniper	Fringe	Yearlong

Data associated with the nature of the stand is entered into the AREA table. Stand is defined as part of a subunit (Sub), while a subunit is defined as part of a unit. Because information in these hierarchical elements is in numerical form, totals can be computed for any management unit or subunit. However, text names can be used to describe unit, sub, and stand, and the names of the columns can be changed as well if local areas are better identified that way. The following data in the AREA table describes the characteristics of a stand. For reference its location is delineated on a map.

Table: AREA

Unit	Sub	Stand	Series	Area	Canopy	Trees	Dbh	Vol	Ba
100	20	33	Pp	35 ha	60%	333/ha	25cm	$131m^3$	$17m^2$
300	15	20	Pp	25 ha	30%	198/ha	51cm	$109m^3$	$14m^2$

The % sign and metric units are normally not included in a table since they are understood by the names of the column. They are simply provided here for clarity.

How The Relationships System Works

Each of the four types of tables contains key columns designed to retrieve data from more than one table at the same time. Thus data from one table can be logically linked to data in other tables by both direct and indirect routes. In Figure 9.8, SPECIES is directly linked to ECOS by Cname and ECOS is directly linked to AREA by Series. Therefore, SPECIES is indirectly linked to AREA through ECOS which contains both Cname and Series.

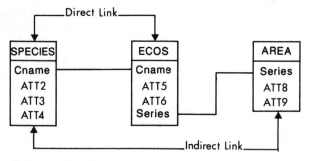

Figure 9.8. Direct and indirect routes link relations (tables) and attributes (columns).

In short, direct links between SPECIES, HABITAT, and ECOS are provided by Cname and between HABITAT, ECOS, and AREA by Series (Fig. 9.7). An example of a question that can be asked using the direct link is: What endangered species (Legal and Cname in SPECIES) are found in ponderosa pine yearlong (Cname, Series, and Assoc in ECOS)?

Since the Series attribute is described in HABITAT, ECOS, and AREA, and Cname is described in SPECIES, HABITAT, and ECOS, then SPECIES and AREA are indirectly linked through either HABITAT or ECOS. This indirect link provides the route for associating data in SPECIES with data in AREA. For example: What wildlife species (Cname in SPECIES) should be found in a management unit (Unit in AREA) containing ponderosa pine (Series in ECOS)?

Data can also be retrieved from one table. The statement that can be made from the one row of data for mule deer in the HABITAT table defined previously is:

Moderately dense, multiple-story ponderosa pine stands, ranging from 11–14 inches dbh, with a grass-forb understory have a moderate cover value for mule deer.

To link one table with another, both tables must have at least one common column that is unique. This column is called the *key column*. It can be a numeric or fact character(s); for example, a species or common name. A key column is any column or combination of columns identifying a row. No other row in the table can have the value of the key column(s). All the other tables in the relational database must contain the key column.

In wildlife databases the key column is usually the scientific name, common name, or a code developed for the scientific name. Such codes are commonly formed with a combination of letters from the genus name and species epithet. This code has not been standardized, except for plants, but many state and federal agencies use 2 letters from the genus and 2 from the species; however, this combination can lead to duplicates which must be avoided. To eliminate duplicates, 3 letters for species and 3 for genus, with an additional 3 spaces for subspecies, are adequate. If there are no subspecies of major concern, then the last three spaces can be used to number duplicates if they exist. For example, the scientific name for mule deer is *Odocoileus hemionus* with the code being Odohem. For ponderosa pine, *Pinus ponderosa*, the code is Pinpon. If plants and animals are both included in Sname, another column is needed to identify whether the Sname is a plant or an animal. In this case the Sname and a new column King (Kingdom) become the key columns.

Determine the Linking Columns

If the key columns have been carefully determined, the linking columns have ipso facto been determined (Fig. 9.7). The placement of the key column is important and depends upon the type of relationship between the tables. To determine the placement of the links, the type of relationship between tables and columns must be determined.

One-to-One Relationships

In this case one row in one table is linked only to one row in another table. The link between two tables is indicated by a solid line with arrows indicating the linking columns.

One-to-Many Relationships

One row in one table is linked to many rows in another table. The linking of SPECIES to ECOS in the HABREL system is a one-to-many type of relationship. For example, the Cname for mule deer is found only once in the SPECIES table, but it can be found many times in the ECOS table because mule deer occur in many vegetation types. To indicate the relationship in a diagram, one arrow identifies the one side of the relation and two arrows the many side.

Many-to-Many Relationships

This relationship can be identified by many rows in one table linking many rows in another table. Many-to-many relationships are not efficient and need to be redesigned as discussed below. Whatever the case, at least one key column must link all tables.

Rules for Placing the Links

In one-to-one relationships the link should be a comparatively stable attribute or it should be taken from the table which by definition contains the key column. In one-to-many relationships the linking column should be whatever column is defined as the key column in the first table; for example, the Cname in the SPECIES table. The entries in the key column from the SPECIES table are included in many rows in the ECOS table, which is the many side of the relationship. In many-to-many relationships the links are not placed in either table, so a new table must be created that contains attributes linking both tables.

Determine The Constraints

Most relational databases by definition require more than one table, but information from one table can be queried only if linking data has been entered in another table. For example, the Cname in the ECOS table must also appear in the SPECIES table before a common name can be located for wildlife that use ponderosa pine for food and cover. An input constraint must therefore be put in place and enforced that data cannot be entered into the ECOS table until the Cname has been entered into the SPECIES table.

One of the safest and most certain methods of constraining entry is to write data entry rules into the input software for each table. Thus, the software governing the HABITAT table input might contain a code which will elicit an error message if the operator enters data on a new plant or animal without first having entered the data in the SPECIES table. Software rules can also be written to prevent input of duplicate entries in all or only specified tables.

Entering, Editing, and Retrieving Data

Relational databases can only be used with relational database software. Information is entered, edited, manipulated, or retrieved in all such systems, including HABREL, by commands dictated by relational theory. These commands are specific to the four functional categories: entering, editing, manipulating, and retrieving data.

Data is entered, edited, and loaded with the appropriate commands from the keyboard which send it to a printer, another file, or the computer screen. In addition, *project, join,* and *select* commands are used to create new tables from existing ones or to select information from existing tables. Tables and columns can be deleted, added, and expanded and data elements and data types can be changed with easy-to-use commands.

The *project* command produces a new table from columns in a source table. It can not only retrieve information from a large database in a very short time, but it also is useful for building an ad hoc table needed for a specific report. Such ad hoc tables can be erased following use to reduce storage space since they can always be created again.

The *join* command produces a new table by taking information from two tables, but only if the two tables possess a common linking column (Fig. 9.8). In most cases the join command is used to create a temporary table that can be erased following use.

One important feature of relational databases is the ability to make computations. For example, acres is one of the columns within the hierarchy of management units in the AREA table. The data in these rows and columns can be added for any unit by using the appropriate command. In addition, averages, ranges, and so forth. can be computed using similar commands.

A detailed account of all the commands associated with editing, manipulating, and retrieving data in relational databases is inappropriate here, but one other command, *select,* must be described because its use is important in understanding the following section. The select command chooses information from one or more columns in a single table, the contents of which are restricted by a *"where"* clause (Fig. 9.9). The select command rapidly searches through thousands of data cells to provide only such data identified in the "where" clause.

Although the select command is simple, it is one of the most useful and commonly used commands for retrieving information. An example of the select command using the data in HABREL (Table 9.6) is as follows:

Select Cname from SPECIES where Lform equals Ma.

The output from the command is a list of common names of mammals contained in the database:

Cname	Lform
Mule deer	Ma
Elk	Ma
Mantled ground squirrel	Ma

As more information is added to the wildlife literature, and as management decisions become more complex, the use of increasingly powerful microcomputers using relational database software will make the biologist's job much easier. There is no type of suitable habitat information that cannot be entered into relational databases for developing useful wildlife habitat relationships.

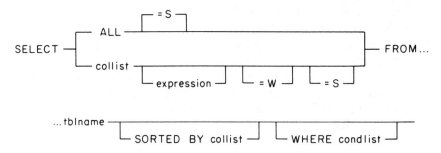

Figure 9.9. Structure of the Select command with a "where" clause.

Other Ways to Retrieve Data

In addition to the various means of manipulating and retrieving data reviewed above, pseudo tables, called *view* tables, can be created which display columns from up to five tables simultaneously. Obviously all tables that are to be included in a view table must have common linking columns (Fig. 9.7). Views are defined and named, and they are available any time the database is open. A view table is useful in helping a manager choose from a variety of ways in which data might be retrieved and printed out for decision making. View tables can be constructed faster than tables formed by join commands and require a minimum of storage because the view table is destroyed when the database is closed.

An example of a view table called NAMES, which contains all the columns from SPECIES and ECOS (Fig. 9.7), would, for example, provide the information needed to answer the following request:

VIEW NAMES WITH Cname Sname Series Season
FROM SPECIES ECOS

The retrieved information would appear on the screen or be printed as follows:

Cname	Sname	Series	Season
Mule deer	Odocoileus hemionus	Pp	Yearlong
Abert squirrel	Sciurus aberti	Pp	Yearlong

Several views can be defined so that with this single command many combinations of columns of data can be retrieved from a variety of tables. The view method of retrieving data provides much the same output as software using Boolean logic in statements cast in near natural language.

Boolean Logic and Near Natural Language

There is growing interest in creating software that can duplicate human intelligence so that computers can interpret verbs and nouns to answer a question in simple human language rather than in the abbreviated language of the computer. This area of software development is referred to as artificial intelligence. While artificial intelligence is not yet sufficiently well developed for common usages, there are other approaches that make it easier to ask questions of a computer. Questions are asked in a more natural way by using synonyms and Boolean operators (greater than, equal to, less than, etc.).

One way to use natural language is to use commands such as "select" to build a dictionary of synonyms. For example, the words "list", "give me", and "what are" can all be used in place of select, once they have been defined as replacements. When a question is asked about a database or a command entered, such as, "list the species of wildlife in ponderosa pine," the computer understands the word "list" as "select".

The natural language option is very useful in reducing the time to get an answer from a relational database. In many cases the join command is used to answer nested questions manually; that is, the command is used several times to create temporary tables to get answers. With natural language the join is done automatically. Most natural language options have the ability to answer questions from up to five tables at one time. The natural language synonyms can also have "where" clauses to further define and restrict a definition. For example, "creepy critters" can be defined as

Cname where Lform equals Re

In this case all the information comes from the SPECIES table (Fig. 9.6). On the other hand, if wildlife is defined as Cname from the SPECIES table, Series as vegetation type from the ECOS table, Pp in the column Series as ponderosa pine (Table 9.6), and Co in the Reuse column as cover, then the question can be asked: What wildlife select the vegetation type ponderosa pine for cover?

There are several ways to ask the same question: What wildlife use ponderosa pine for cover? The power of using natural language now starts to become apparent. There are only four tables in Greg's HABREL data model, but the types of questions that can be asked to retrieve information for decision making are numerous.

Evaluate and Test the Design

The last step in Greg's problem is to evaluate and test the design of the structure of his database. In this step he looks for design flaws that might lead to the incorrect or redundant relating of data. Every table in the database must be evaluated by asking the following questions:

1. Does the table have a central theme or subject?

2. Is each column a physical or logical fact about the subject?

3. Does the table have a key column?

And finally Greg needs to ask: Is the database easy to use? The goal of good database design is to include enough detail so that "[t]he only question which cannot be answered is the question that is never asked" (Kroenke and Nilson 1986). In the final analysis the only way to test the database is to ask questions for which the outcome is known. At first, load just enough data into the database to answer such test questions. Test data can, in fact, be "dummy" information to make it easy to recognize a problem in the database when the answer is presented.

Determining the proper links and relationships between tables is as much an art as it is a science. Intuition plays an important role in determining the links between tables and the dependencies that are created. Small test databases well "worked over" will in time identify the anomalies that exist and once known, a solution to problems can be found. The wildlife biologist should never develop a scheme and load an entire database without first making many test runs.

Tradeoffs must be made in database design. For example, one table may have to be split into two tables for logical reasons but doing so may then lead to redundancy of data and more work for the data entry operator. Tables with many columns, however, are generally not very efficient, except in the case of an entity table. The developer of the data model also has to recognize the capacity of a printer to print information on a given paper size. This restriction often force more use of codes in the attributes in order to print the appropriate relationships on 8.5- by 11-inch paper.

While the <u>HABREL</u> habitat relationships model provides only four relations, its utility is found in the combinations of characteristics associating an animal with physical habitat and the value of that association if it is known.

9.5 SUMMARY

A model is a representation of a real world situation. Humans routinely build useful models because they help organize thoughts and communicate ideas. Models have a structure that includes an input function and an output function. They can be tested by using a sensitivity analysis to determine how much quantification of an input is required to produce an output, thus permitting one to identify those inputs which produced major changes in the output. Obviously such inputs must be monitored and controlled.

Conceptual models represent thoughts and ideas in a qualitative and deductive way rather than in a quantitative way. Conceptual models precede empirical qualitative models. Stochastic models are developed to deal with situations in which probabilities can occur, while deterministic models deal with mechanical situations in which outcomes are rigidly determined. Models can take various forms, such as pattern recognition, decision trees, or expert opinion panels. The expert system model is frequently used in local situations when inadequate amounts of quantitative data are available.

Information used to build models can be stored in computers as databases to be used for decision making. Databases use a network, hierarchical, or relational formats. Recent advances in microcomputer storage and relational software development make relational database design, development, and use practical for managers. A relational database is a collection of data linked by relations. Wildlife habitat can be completely described by relations and attributes in a relational database.

A relation can be thought of as a table and an attribute as a column in such a table. Commercial database software is now available for microcomputers that can deal with 80 relations and 800 attributes.

Relational database software provides commands to join, select, and project various data elements to display relationships between animals and habitats. Relationship systems can be derived from relational databases that include information on habitat and habitat use by different wildlife species.

Once a relational model has been developed, it can be queried using the appropriate relational commands. These commands can be used directly or changed to near natural language which uses Boolean operators and synonyms to answer questions of data in the database. Near natural language facilitates the asking of nested questions from five relations at one time that otherwise would be produced from several join commands.

9.6 RECOMMENDED READING

Grubb, T. G. 1988. *Pattern Recognition: A Simple Model for Evaluating Wildlife Habitat*. Res. Note RM–487, USDA For. Serv., Rocky Mountain For. and Range Exp. Sta., Fort Collins, CO.

Jeffers, J. N. R. 1982. *Modelling*. Chapman and Hall, New York, NY.

Kroenke, D. M., and D. E. Nilson. 1986. *Database Processing for Microcomputers*. Science Research Associates, Chicago, IL.

Starfield, A. M., and A. L. Bleloch. 1986. *Building Models for Conservation and Wildlife Management*. Macmillan Publishing Co., New York.

Verner, J., M. L. Morrison, and C. J. Ralph. 1984. *Wildlife 2000: Modeling Habitat Relationships of Terrestrial Vertebrates*. Univ. of Wisc. Press, Madison.

Chapter 10

WILDLIFE IN RESOURCE PLANNING

A good plan today is better than a perfect plan tomorrow.
Gen. George S. Patton

Previous chapters have dealt with the centrum of an animal, management systems, habitat models, and habitat relationships as though they were independent entities and not part of an overall system or process. None of these topics is an effective management tool by itself but must be integrated into a scheme which gives all the topics direction and purpose. Today's effective resource manager must devote time and energy to devise efficient ways that will get inventories done and projects completed within the systematic context of the planning process. This process basically requires the proper allocation of resources—time, personnel, and physical elements.

The land base for resource management is diminishing due to the increasing number of cultural landscape features needed by an increasing human population. Present conflicts in allocating land to wildlife habitat can only intensify with population growth. The problems of allocating resources are compounded by the fact that different wildlife species have different food and cover requirements; this requires setting aside a variety of habitats to accommodate these species. The allocation of habitat to forest wildlife takes place in the planning process at the forest level and is based on the mandated principles of multiple use. A further complication in the allocation process results from legislation requiring the maintenance of viable populations of all native vertebrate species in the planning area. This legislative mandate is a consequence of the biological fact that virtually all animal species are included in and vital to the functioning of an ecosystem.

The balance between resource use, timber, water, livestock, wildlife and ecosystem functioning, energy flow, and nutrient cycling reflects the system of political and biological decisions that result in forest plans. The political decision determines how much acreage is designated for a given habitat and how many animal species will be maintained in the same habitat. The biological decision determines the effect of the political decision on animal and plant species in the management unit and provides alternatives whenever conflicts arise.

Aldo Leopold (1933) defined five biological factors involved in managing wildlife:

1. Regulation of animal numbers.
2. Predator control.
3. Refuges for protection.
4. Propagation of species.
5. Habitat manipulation.

These five items are all accounted for in Conceptual Equation 3 (Table 2.2) and have been successfully used by state and federal wildlife biologists for over 50 years. But now a sixth factor must be added to Leopold's original formulation:

6. Land management planning.

Land management planning clearly is not in the centrum of any animal species since this factor does not directly affect an animal's chance to survive and reproduce. But, land management planning does affect the six factors of Conceptual Equation 3 so it properly belongs in the *web* from a biological or ecological viewpoint. Yet, if land management planning has been made by virtue of legislation—a factor of such importance that it requires considerable time by managers—then it takes on the characteristics of a seventh factor in the centrum regardless of the dictates of logic.

10.1 LAWS AND POLICY IN THE UNITED STATES

Land use planning depends in the first instance on legislative authority empowering a federal or state agency to act and in the second instance on a policy formulated by that agency to use its authority. A long list of legislative statues provides the authority for various U.S. agencies—the Fish and Wildlife Service, the Bureau of Land Management, and the Forest Service to manage natural resources, including fish and wildlife on public lands, and to cooperate with state agencies in conducting these management activities. While federal agencies have been authorized in a few cases to manage populations (e.g., migratory species) on public lands, the generally accepted division of responsibility is that habitat management on federal land is the responsibility of federal agencies but population regulation is controlled and enforced by state agencies under state law.

Because this book is particularly concerned with wildlife management across the continent, it is appropriate to summarize a few of the major laws controlling federal land management agencies.

Fish and Wildlife Coordination Act (1934):

Provides that fish and wildlife resources receive equal consideration with other resources in water development projects. Authorizes and directs the U.S. Fish and Wildlife Service and the relevant state agencies to coordinate water development programs with the other land management agencies, and to encourage the installation of fish and wildlife measures to protect and/or enhance habitats.

Multiple-Use, Sustained-Yield Act (1960):

Recognizes and clarifies the U.S. Forest Service authority and responsibility to manage fish and wildlife. Defines multiple use as:
. . . the management of all the various renewable surface resources of the National Forest so that they are utilized in the combination that will best meet the needs of the American people; making the most judicious use of the land for some or all of these resources or related services over areas large enough to provide sufficient latitude for periodic adjustments in use to conform to changing needs and conditions; that some land will be used for less than all of the resources; and harmonious and coordinated management of the various resources, each with the other, without impairment of productivity of the land, with consideration being given to the relative values of the various resources, and not necessarily the combination of uses that will give the greatest dollar return or the greatest unit output.

Sikes Act (1960):

Directs the Secretary of Agriculture and the Secretary of the Interior, in cooperation with state agencies, to plan, develop, maintain, and coordinate programs for the conservation and rehabilitation of wildlife, fish, and game. Provides for fish and wildlife conservation programs on federal lands, including authority for cooperative state-federal plans and authority to enter into agreements with states to collect fees for funding programs identified in those plans.

National Environmental Policy Act (1970):

Requires that a systematic, interdisciplinary approach be employed in planning actions that may impact the human environment. Mandates the preparation of a detailed statement on the effects of legislative proposals or other major federal actions significantly affecting the quality of human environments. Wildlife, fish, and their habitats must be fully considered in project planning and decision making.

Endangered Species Act (1973):

Provides the means whereby the ecosystems upon which endangered and threatened species depend may be conserved. Provides a program for the conservation of such endangered and threatened species. Authorizes such steps as may be appropriate to achieve the purposes of the international treaties and conventions, CITES, etc. Establishes federal policy enjoining all federal departments and agencies to conserve endangered and threatened species.

Forest and Rangeland Renewable Resources Planning Act (1974):

Directs the Secretary of Agriculture and the U.S. Forest Service to make an inventory and assessment of the present and potential renewable resources of the United States every 10 years. Wildlife and fish, as part of the renewable resources of the U.S., are included in the inventory and assessment, as is an analysis of present and anticipated uses, and the demand for and supply of renewable resources.

Federal Land and Policy Management Act (1976):

An organic act for the Bureau of Land Management in the Department of Interior. Authorizes public lands to be managed to provide habitat for wildlife and domestic animals.

National Forest Management Act (1976):

Provides for balanced consideration of all resources in U.S. national forest land management planning. Directs the Secretary of Agriculture to protect and improve the future productivity of all renewable resources on forest land, including improvements of maintenance and construction, reforestation, and wildlife habitat management on timber sales.

Specifies that habitat diversity includes variety, distribution, and relative abundance of plant and animal communities in a multiple-use context.

The Resources Planning Act of 1974 (RPA) and the National Forest Management Act of 1976 (NFMA) have had a particularly profound effect on the management orientation and activities of the U.S. Forest Service. Both acts required major changes in the management direction of the agency and demanded a massive commitment of human and financial resources to complete the assessment on a national scale and then prepare accompanying management plans at a forest level in a timely way. The RPA and NFMA acts called for cooperation with all the federal land management agencies as well as those of the 50 states and Puerto Rico. The mandated changes have, in turn, resulted in marked changes in the way state natural resource agencies manage their activities. Such a complete assessment of natural resources had never before been undertaken. What was to be assessed, how was it to be done, and how it was to be presented required extraordinary consultation and input from a wide variety of sources—federal, state and private. The end product was *An Assessment of the Forest and Rangeland Situation in the United States* (U.S. Dept. of Agric. 1981). Another assessment is to be completed at the beginning of each decade. Two years after completing the *assessment* the U.S. Forest Service responded to the NFMA by preparing management plans for each national forest. During the process of forest plan preparation and review, states developed techniques to respond to the federal agencies in both adversarial and advocacy roles, leading to a synergistic process of considerable benefit to the public. Plans acceptable to the public are only now being approved.

Legislative authority to manage does not automatically result in a management strategy by a state or federal entity. The agency must develop a policy to guide management in the allocation of staff, finances, and other resources to achieve their goals. *Policy,* in short, is the setting of a course of direction. The first step in formulating policy is the preparation of a statement which summarizes the objectives to be realized and the process by which they will be reached. Policy statements must be clear and easily communicated through a chain of command. An example of a policy statement is found in the U.S. Department of Agriculture and Forest Service Memorandum (1982) which begins as follows:

The purpose of this memorandum is to state the policies of the U.S. Department of Agriculture with respect to management of fish and wildlife and their habitats and to prescribe specific actions to implement the policies.

The Department's prime responsibility is to help maintain sufficient and efficient production capability of farm,
forest, water, and rangeland resources for the public benefit,
now and in the future, and to encourage and support proper
use, management, and conservation of those natural
resources. Programs to meet this mission are carried out
through research, education, technical and financial assistance to landowners, managers, producers and consumers,
and through the management of public land for which the
Department is responsible, in cooperation with State and
local agencies.

. . . It is the policy of the Department to assure that the
values of fish and wildlife are recognized, and that their
habitats, both terrestrial and aquatic, including wetlands, are
recognized, and enhanced, where possible as the Department carries out its overall missions.

. . . A goal of the Department is to improve, where needed,
fish and wildlife habitats, and to ensure the presence of
diverse, native and desired non-native populations of wildlife, fish and plant species, while fully considering other
Department missions, resources, and services.

. . . On public lands administered by the Department,
habitats for all existing native and desired non-native plants,
fish, and wildlife species will be managed to maintain at least
viable populations of such species.

. . . This will be accomplished through the forest planning
process in response to targets identified in the Forest and
Rangeland Renewable Resources Planning Act program and
public issues and concerns brought up in the planning
process, consistent with available resources. Habitat goals
will be coordinated with State Comprehensive Plans
developed cooperatively under Sikes Act authority and
carried out in forest management plans with State
cooperators. Monitoring activities will be conducted to determine results in meeting population and habitat goals.

. . . The U.S. Department of Agriculture recognizes the
rights of the individual States to manage fish and wildlife
populations under their jurisdictions. Departmental agencies will utilize their respective authorities to manage habitat
on public lands, to assist landowners in managing habitat on
private lands, and to encourage and assist the States,
Territories, and other Federal agencies in conducting
resource inventories and evaluating the status and potential
of fish and wildlife habitat.

This policy statement continues with sections dealing with
threatened or endangered species, economic losses from plant and
animal pests, authorities to implement the policy, and responsibilities for implementation and coordination.

The Secretary of Agriculture is a Cabinet position responsible for several agencies: Forest Service, Agriculture Research Service, Soil Conservation Service, and so forth. Policies for each of these agencies must be interpreted from the Secretary's memorandum. The subsequent interpretation of the Secretary's policy at a lower level is found in the Forest Service statement (U.S. Dept. of Agric. 1984) on fish and wildlife which contains the following:

> Serve the American people by maintaining diverse and reproductive wildlife, fish, and sensitive plant habitats as an integral part of managing National Forest ecosystems. This includes recovery of threatened or endangered species, maintenance of viable populations of all vertebrates and plants, and production of featured species commensurate with public demand, multiple-use objectives and resource allocation determined through land management planning.
> . . . Recognize the State wildlife and fish agencies as responsible for the management of animals and the Forest Service as responsible for the management of habitat.
> . . . Specify quantitative wildlife, fish, and sensitive plant habitat objectives and standards in the RPA Program, Regional guides, Forest Plans and Sikes Act schedules.
> . . . Develop a balanced program that meets goals by investing directly in habitat improvements when necessary to meet public demand and coordinate management activities that produce other resources to restore or mitigate habitat losses or provide improved habitat.
> . . . Use Wildlife and Fish Habitat Relationships classifications, models, and procedures in quantitative habitat evaluations, planning for diversity, viable populations, and management indicator species habitat productivity, and to support monitoring of fish and wildlife resources.
> . . . Give coequal consideration to wildlife and fish habitat with other resources in Forest Service programs.

The U.S. Forest Service policy statement interprets and adds to the department's policy due to the fact that the Forest Service has authorities that other departmental agencies do not. The intent in policy statements is to clarify and strengthen objectives and goals at each lower level of administrative responsibility. The aim of well-written, succinct policy statements is to clearly communicate direction so that each employee has simple guidelines to the performance of his or her tasks and also can adequately represent the agency to the public.

10.2 WHY PLAN?

Planning at some level is required by law for most state and federal natural resource agencies. It is necessary not only to discharge the agency's responsibilities but to properly respond to a world of change and competition for financial resources and space as well. Change can be orderly to meet specific goals and objectives or it can be random and nondirected. Human-induced change, to be orderly, must be controlled in some fashion. The planning process establishes criteria, principles, and procedures for control and documents the sequence in which events should happen. The planning process and the resultant plan are means to an end or goal.

Wildlife managers must compete with the natural resource management goals of other agencies for resources to manage for wildlife food and cover, but because wildlife is part of an interdependent biological community, forest wildlife managers must also cooperate and, as Leopold (1933), has stated, "so that there is a place to compete for." Given all the new statutes and regulations governing the behavior and actions of managers Durward Allen's (1973) observations are of particular relevance today: "Long-range planning, in terms of the natural development of life communities, is one of the realistic and efficient ways to bring land into condition for maximum production of wildlife crops."

The planning process is the forum in which competing demands and objectives are finally integrated to marshall all the available resources into a practical working tool for management and staff. While the role of wild animals in the functioning of an ecosystem is not well documented or thoroughly understood, it is already clear in a general way that they are crucial in nutrient cycling and energy transfer. By virtue of the ecological role wild animals have as an ecosystem component, the wildlife management function in an agency should be placed on an funding level equal to that which other natural resource functions, such as timber, water, and livestock, have had for many years.

To adequately manage all plant and animal populations at any unit level requires a detailed plan of action. Employing the knowledge and insights of a variety of professional personnel—foresters, ecologists, wildlife biologists—in the planning process is a necessity. The trend away from single species or single resource management to ecosystem management is a positive approach to insuring that few if any animal species go below viable population levels as a result of human activities—unless done as a planned objective. One exception to multiresource management and planning is managing for an endangered species when single species management is biologically sound and legally justifiable.

To compete for dollars and space, to cooperate in the formulation and selection of alternatives, and to integrate wildlife objectives into ecosystem management plans, forest biologists must have access to the kind of information outlined in the preceding chapters. Now as never before biologists as planners must investigate and understand how wildlife interacts with and affects other ecosystem components. Ecosystem managers have to insure that all the resources are fully considered and provided for in the planning process.

10.3 FACTORS TO CONSIDER

The wide-ranging and extensive body of laws and regulations relating to natural resource management directly affect both the planning process and the plan itself. As in preparing inventories, plans must be programmed at different scales, depending upon whether they are for national, regional, or local use. The scale of the plan will dictate not only the kind and detail of the information included but the process and format as well. Plans at the national and regional level must be presented in a standardized format to facilitate understanding by management agencies, conservation organizations, and environmental groups because these plans relate to the same laws and sets of conditions.

Most natural resource agency plans are one of two types: strategic or operational. *Strategic plans* guide an agency or organization in meeting long-term goals and are stated in economic, technical, and political terms. Strategic plans are often referred to as *long-term plans*. *Operational plans* guide specific actions and allocation of resources on a shorter time scale, such as a single year, several months or weeks. An operational plan deals with specific management units or parts, specific ecosystems and particular organisms therein, the relationships between the environmental characteristics of the unit and their organisms, and the ways in which these relationships are to be maintained or allocated. Operational plans are known in natural resource professions as management plans. While national and regional plans are important, most field biologists are involved in preparing the local or management unit plan, so I focus on this kind of planning.

Management Unit Plans (Operational)

All wildlife management plans have certain elements in common which must be included if the plan is to be a useful working document. Two of these common elements, both of high priority in the process, are the setting of goals and objectives. Goals are broad statements of central themes which set management direction. They are seldom quantifiable and generally have no time limit. Objectives are very specific statements of activities to be undertaken to realize the goals set for a management unit. Objectives can be measured or accounted for through a given time. One approach, known as *Management by Objective* (MBO), fits well with a planning methodology widely adopted by many federal and state agencies. The technique depends upon adopting specific objectives and determining how they will be measured. Agency personnel are then expected to account for the completion of each objective in terms of the measurement stipulated.

In earlier chapters the element of time was discussed in relation to succession (Chapter 2) and forest rotation and cutting cycles (Chapter 8). All these are factors common to management plans and must be included, but another time factor, planning period, is important and must be part of the plan. When time checkpoints are routinely built into management plans at the end of a given planning period—for example, every five years—they provide a major checkpoint to evaluate accomplishments and set new directions if required. Managers clearly need to know the starting date on which a plan comes into operation, but equally important is the date when the plan is either completed or to be revised. An outdated plan is as useless as no plan. The planning period is often dictated by state or federal law or regulations, as is the ability to revise a plan when such a need arises. Revisions can result from advancing technology, the need to correct errors or incorporate new data, or the expressions of public concern about a management activity. While the life of a plan may be clearly stated, it should not be so rigid that revisions and updates cannot be made as conditions change.

Collecting Background Information

Management unit plans require considerable data, much of which may already be available but which should at least include the following:

Geographic data—Topographic maps are the basic document needed to locate boundaries and forest vegetation types.

Historical data—Fire history, farming, logging, mining activity, and so forth.

Wildlife census data—Information on populations both current and historical.

Game kill records—Often used to supplement historical census data.

Aerial photographs—Used to associate vegetation with landscape features when transferred to topographic maps.

Satellite photographs and multispectral images—Often useful in establishing land use patterns surrounding a management unit.

Range condition and trend surveys—Provides information on herbage production and utilization.

Soil surveys—Used to establish vegetation potential and productivity.

Allotment analysis reports—Establish past and present livestock use of an area.

Timber inventory records—Stocking and stand tables are needed to estimate current and future growth and yield.

Vegetation type maps—Keyed to identify food and cover areas.

Exclosure location and data—Helps to establish successional trend and use patterns.

Water inventory—The location of seeps, springs, streams, and small catchments helps determine the need for water development projects for habitat improvement.

Recreation facilities—To determine human impact on the management unit.

Roads and trails—Identify access routes that impact the management unit.

Much of the information listed will presumably have been collected by various agencies in the course of their management activities, so often this planning phase is simply a matter of bringing information together and checking for missing data. I cannot urge strongly enough that as this data is accumulated for inclusion in a new plan it be simultaneously entered into a database so that complete up-to-date information is always available for future use.

10.4 MANAGEMENT PRINCIPLES

Natural resource management must be based on the ecological principles and biological laws which have derived from the accumulated knowledge arising out of research and experience. These principles and laws are seldom explicitly stated or identified in support of actual project objectives. A clear statement of the ideas, principles, and laws, upon which decisions are made will strengthen the biologist's or natural resource manager's position in proposing specific goals and objectives in a plan. In addition, several useful observations regarding mankind's relation to the natural world have given managers experience with the public which also provides the grounds for setting or constraining plan goals and objectives. I have endeavored to generalize these principles, laws, and observations in the following list of propositions to clarify and support wildlife management goals and objectives. I urge their consideration for inclusion in plans as appropriate.

Proposition 1. Plant communities change through ecological time under the influence of parent material, soil, climate, topography, associated organisms, and time (Conceptual Equation 1, Table 1.1).

Proposition 2. Plant communities change in real time (here and now) as a result of human activities or natural catastrophes.

Proposition 3. All animal species are directly dependent on plants and plant communities for food and cover.

Proposition 4. Various animal species possess different tolerances (ecological amplitude) to environmental factors, the presence or absence of particular plant species, and the composition of plant communities.

Proposition 5. Because animals depend upon plants and plant communities, animal species composition will change as plant communities change.

Proposition 6. Animal species are integral parts of ecosystem functioning, providing pathways for nutrient cycling and energy flow through food chains and webs.

Proposition 7. Humans are ecological dominants with the ability to make environmental changes that speed up, slow down, or alter ecosystem succession.

Proposition 8. Not all plants or animals have equal value in the minds of humans, so some will be favored over others at one time and others at another time.

10.5 DECISION-SUPPORT SYSTEMS

Prior to the mid 1970s the planning goals and objectives of resource agencies were simple, requiring only minimal amounts, types, and complexity of data. Therefore, information generally could be analyzed and presented without the aid of complicated automated systems. This is not the case today because types and amounts of information needed to meet the requirements of the natural resource laws and agency regulations implementing these laws are sufficiently complex that new ways of analyzing data have had to be developed and incorporated into the planning process.

In response to the needs for a national and regional system of planning, the Forest Service adopted a linear programming model called FORPLAN (Mitchell and Kent 1987). FORPLAN (FORest PLANning) was designed to provide management alternatives by presenting various levels and combinations of goods and services which a forest can provide. FORPLAN is the primary tool used in developing the forest plans required under the NFMA. FORPLAN software is used with data supplied for each analysis area. The data must be developed for management prescriptions and estimated costs, together with the outputs, effects, and benefits for each prescription or combination of goods and services. This information is then used to generate matrices of managerial options and associated benefits, costs, and so forth. The end product is a listing of the optimum combinations of prescriptions and associated benefits. Needless to say, the final decision as to which combination of prescriptions is selected continues to be the responsibility of skilled natural resource managers.

While FORPLAN greatly assisted the Forest Service in meeting its legal responsibilities under the NFMA for planning at the national and regional level, it provided little planning assistance at local levels. But as pointed out earlier, resource management planning is usually done at the forest subunit level. To assist managers at this level *Decision Support Systems* (DSS) are now being developed to provide alternatives for the planning and decision-making process on small units of land. This type of analysis tool is appropriately titled, for decision support systems link together a variety of different modules to form a systematic planning structure. The modules can be linked as individual forest models, such as there described in the preceding chapter, as matrices, plant and animal databases, geographic information systems, or linear program optimization models.

One DSS already in use for systematic instruction, research, and management application purposes is TEAMS (Terrestrial Ecosystem Analysis and Modeling System) (Covington et al. 1988). TEAMS includes a geographic information system, a relational database, an artificial intelligence software package, a multiresource simulation model, algorithms for evaluating costs and outputs, an optimization model, a graphics package, and programs to control the flow of information between modules (Fig. 10.1).

TEAMS was developed for use on a microcomputer so that it can be readily used by resource managers at the project level. One use to which TEAMS has been put is to simulate wildlife food and cover areas over time given an initial stand structure (Fig. 10.2). In addition, this DSS can respond to adjacency constraints where stands having certain structural characteristics cannot be adjacent to other stands. Decision support systems are powerful tools to model wildlife habitat if the data provided are of a quality adequate to formulate habitat algorithms.

FORPLAN and TEAMS are not the only management planning systems available to natural resource, managers but they are representative of the types of tools that are becoming available for planning, decision making, and decision support.

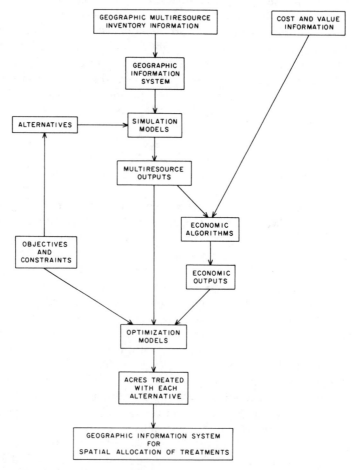

Figure 10.1. The flow of information between modules in TEAMS, a decision support system.

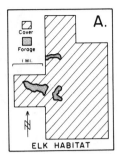 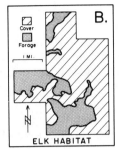 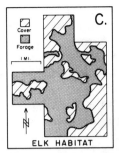

Figure 10.2. One output of TEAMS is simulation of food and cover areas for a given wildlife species.

10.6 INTEGRATED RESOURCE MANAGEMENT

As a result of the NFMA, some Forest Service regions developed a project implementation process (PIP) to insure that all projects and undertakings which generated human impacts—such as mining, range improvements, road construction, and timber sales—were carried out in accordance with previously developed plans for each forest (U.S. Dept. Agric. 1988). The PIP process grew out of and reflects the adoption of the policy to manage national forests on the basis of an integrated resource management (IRM) approach. IRM is based on the recognition that all natural resources are connected by an intricate web of interrelationships. As a result any activity undertaken with one resource affects other resources in the forest simultaneously. Because the interrelationships between natural resources are extraordinarily complex and not always well understood, the effects of a management activity on various elements of the forest ecosystem are difficult to predict.

The IRM process uses an interdisciplinary team for project design to predict the effects of a management activity. The team identifies all the resources involved, defines the interrelationships, and predicts the effects and impacts of the project on all the identified resources. To facilitate the work of the team in the project design, each project must have specific, quantifiable objectives. When the plan is in place with all impacts identified, the land manager can then implement the project and use the interrelationships between the resources to accomplish objectives.

The interdisciplinary team prepares a brief report to determine whether the proposed project is technically and economically feasible. Based on the report line officers decide whether to proceed with the project. The team must also develop a reasonable range of

alternatives so that responsible officials can select those to be implemented and then determine what National Environmental Policy Act
(NEPA) documentation is appropriate. Finally, the project is completed and evaluated for the success or failure of project design.

IRM is another powerful tool for planning and implementing the
objectives of a plan which can impact a variety of natural resources. It
offers the wildlife biologist, as part of an interdisciplinary team, an
unsurpassed opportunity to reduce the negative impacts on or to
improve the habitat of a specific species or forest wildlife in general.
IRM and its associated PIP, coupled with a decision support system
such as TEAMS, assures natural resource managers and the public
that the most modern and scientific methods are being used for planning and decision making.

10.7 SUMMARY

Leopold (1933) suggested five ways in which to manage wildlife:
regulation of animal numbers, predator control, provision of refuges,
modification of numbers by propagation, and habitat manipulation.
Thanks to a vastly improved understanding of the complexity of wildlife and ecosystems, and the inclusion of wildlife in legislation, it is
now appropriate to add land management planning as a sixth item
and an important factor in managing wildlife. The recognition of the
control role of land management in managing wildlife and all natural
resources led to the extensive legislation governing natural resource
utilization.

Legislation mandating natural resource planning in the United
States includes the Fish and Wildlife Coordination Act, the Multiple-
Use Sustained-Yield Act, the National Environmental Policy Act, the
Endangered Species Act, the Forest and Rangeland Renewable
Resources Planning Act, the Federal Land and Policy Management
Act, the Sikes Act, and the National Forest Management Act. This
legislation has required the appropriate agencies to develop national
policies aimed at managing natural resources in keeping with this
growing understanding at the national level. Similar legislation has
been put in place at the state level.

To respond effectively to federal legislation, agencies have had to
devote considerable attention to the planning function. The planning
process is the place where competing interests and alternatives are
resolved and integrated into a practical working tool. Plans are made
either to guide an agency in meeting long-term goals (strategic) or to
project specific actions on a short time scale (operational) to meet
particular objectives. The goals in either kind of plan set the direction
and determine objectives which are specific statements of activities

and how these are aimed at realizing the goals.

Management plans require considerable information including geographical data, historical data, census data, timber data, vegetation data, and related kinds of inventory data. Once this data is collected, it is best stored in a database to be used in the decision-making process. One of several useful systems for decision making is a decision support system (DSS) that links modules containing relational databases, linear program software, and geographic information systems together. DSS is particularly well suited to assisting natural resource managers in developing management plans for single forests or subunits.

The U.S. Forest Service has initiated a new approach to land management, Integrated Resource Management (IRM), which in the planning process takes into account all the highly complex and interrelated factors of forest ecosystems. IRM is an interdisciplinary approach to project design which depends upon assembling a team of specialists to identify the resources involved, define the interrelationships, and predict the effects of project implementation. IRM coupled with decision support systems puts to work some of the most powerful scientific and technical tools available to wisely manage forest ecosystems and wildlife simultaneously.

10.8 RECOMMENDED READING

Covington, W. W., D. B. Wood, D. L. Young, D. P. Dykstra, and L. D. Garrett. 1988. TEAMS: A decision support system for multiresource management. *J. For.* 86(6): 25–33.

Hoekstra, T. W., A. A. Dyer, D. C. Le Master, eds. 1987. *FORPLAN: An Evaluation of a Forest Planning Tool.* Gen. Tech. Rep. RM–140, USDA For. Serv., Rocky Mountain For. and Range Exp. Sta., Fort Collins, CO.

Smith, S. H., and A. H. Rosenthal, eds. 1978. *Concepts and Practices in Fish and Wildlife Administration.* Occ. Pap. No. 2, USDI Fish and Wildl. Serv., Wash., DC.

U.S. Dept. of Agric. 1980. *Land and Resource Management Planning.* Chap. 1920. For. Serv. Manual, Wash., DC.

Wildlife Management Institute. 1987. *Organization, Authority and Programs of State Fish and Wildlife Agencies.* Wildl. Manage. Inst., Wash., DC.

Chapter 11

THE WORLD ECOSYSTEM

The earth's land area of 149 million km² (57.5 million mi²) and water area of 361 million km² (139.4 million mi²) contains all the natural resources we will have for the foreseeable future, or perhaps for eternity, to maintain life for 1.5 million known plant and animal species. This number includes 4100 mammals, 8700 birds, 6300 reptiles, 3000 amphibians, 23,000 fishes, and 800,000 species of insects (Council on Environmental Quality 1981). However, these numbers are only for those species that have been identified and catalogued. New species are regularly being added as scientists discover them in the ocean depths, tropical forests, mountain tops, and arid lands throughout the world. The estimate for all the plant and animal species now in existence is 3–10 million.

11.1 HUMANS IN THE CENTRUM

When one thinks in terms of number of species and their populations, the reality of competition for space and natural resources becomes a worldwide issue, made especially critical by the fact that one species, *Homo sapiens,* is increasing at a rapid rate. In 7000 B.C. the earth was estimated to have between 5 and 10 million people; but by A.D. 1 the population had increased to 200–400 million (Council on Environmental Quality 1981). The time required for the world population to double has dropped from 173 years in 1800 to 41 years in 1979; this data indicates an estimated 5 billion plus people will occupy the earth by the year 2000 (U.S. Dept. of Commerce 1977). The rapid growth rate from 1800–1975 looks very much like an uninhibited growth curve (Figure 3.3).

Conceptual Equation 3 (Table 2.2), developed for a wild animal, is just as relevant to human populations. In some parts of the world even predators would have to be retained in the equation. The questions needing answers are: At what population level will the earth reach dynamic equilibrium with all the factors of Conceptual Equation 3? How can we use our natural resources, including wild animals, to keep a world ecosystem functioning to provide food and cover for all humans on earth?

One of the problems confronting mankind is that, unlike other animals, humans are ecological dominants. That is, we have the ability to drastically change our own habitat almost instantly with fire, mechanized equipment, or explosives. In a slightly longer time scale, habitat is changed by the axe and by overgrazing livestock, as is now the case in many underdeveloped countries. In this manner humans can continue to remove obstacles to food production to survive a little longer. Wild animals, however, cannot create new habitat for themselves—they can only create habitat for other species when they overuse the available resources. Yet, surviving a little longer is inextricably related to quality of life, which in turn is determined by the amount of food and habitat available together with cultural and social factors. Quality of life in each country is a reflection of that country's natural resource base, political system, cultural and religious heritage, and historical events. Durward Allen (1959) summarized survival in this fashion:

$$\frac{\text{Resources} \times \text{Culture}}{\text{Population}} = \text{Living Standard}$$

The idea expressed is simple, but it is neither well understood nor accepted by many governments.

International problems associated with wildlife management are not confined only to trade relations between producer and consumer nations, but also to production and consumption patterns within nations. In the former case developing countries in Africa and Latin American export natural resource products to developed countries to obtain revenue not only for development but to feed a growing population as well. Not uncommonly the most readily available and exportable products are raw natural resources, including living animals destined for pets and zoos, animal products such as skins, horns, and bones, and logs, plants, and so forth. International trade in wildlife and wildlife products is probably second only to habitat destruction as the major factor reducing wildlife populations in underdeveloped countries.

In recent years the distance between countries has, figuratively speaking, become shorter. But as noted, within country consumption patterns produce international imports. No country or region is any longer isolated—all are a part of the closed system so graphically depicted in photographs from the Apollo spacecraft (Figure 1.1). There is general agreement among scientists that air pollution is spreading, that acid rain affects vegetation growth on the land as well as organisms in aquatic environments, and that the destruction of tropical forests is impacting world climatology. For these and other reasons each country must come to understand that the ways in which it uses resources affects the rest of the world, not just economically but also environmentally.

Gifford Pinchot in 1947 wrote of natural resource use this way:

> Without natural resources life itself is impossible. From birth to death, natural resources are transformed for humans in food, clothes, shelter, and transport. Upon them we depend for every material necessity, comfort, convenience, and protection in our lives. Without abundant resources prosperity is out of reach. Therefore the conservation of natural resources is the fundamental material problem. It is the open door to economic and political process. The first duty of the human race on the material side is to control the use of the earth and all that there is in it. Conservation means the wise use of the earth and its resources for the lasting good of men. Conservation is the fore-sighted utilization, preservation, and/or renewal of forests, water, lands, and minerals, for the greatest good of the greatest number for the longest time. To the use of natural resources, renewable or nonrenewable, each generation has the first right. Nevertheless, no generation can be allowed needlessly to damage or reduce the future general wealth and welfare by the way it uses or misuses any natural resource. The rightful use and purpose of our natural resources is to make all the people strong and well, able and wise, well-taught, well-fed, well-clothed, well-housed, full of knowledge and initiative, with equal opportunity for all and special privilege for none.

While Pinchot clearly assumed a world order composed of only democratic, free-market nations, it is becoming clear that despite differing forms of political and economic organization, a growing sense of resource ethics is providing the grounds for a common unified action worldwide. When it is widely accepted that humans are control players in a functioning ecosystem and therefore part of the problem, *Homo sapiens* will have gone a long way in formulating solutions to present environmental problems because, as stated by Rene Dubos (1973), the earth will then be "humanized".

In his article Dr. Dubos argues that the human race has the ability to create a diverse environment by changing the distribution of living things, and adding human order and fantasy to the ecological determinism of nature. His thoughts on the general ecological law that "nature knows best" is provocative and should make all resource managers and resource users think about the role of humans in the world ecosystem.

11.2 ETHICS IN RESOURCE USE

Aldo Leopold (1949) probably formulated the most compelling rationale for the ethical use of natural resources in recent times when he stated:

[a] land ethic changes the role of *Homo sapiens* from conqueror of the land-community to plain member and citizen of it. It implies respect for his fellow-members, and also respect for the community as such.

... A land ethic, then reflects the existence of an ecological conscience, and this in turn reflects a conviction of individual responsibility for the health of the land. Health is the capacity of the land for self-renewal. Conservation is our effort to understand and preserve this capacity.

He further argued that every question of land use should be examined:

[I]n terms of what is ethically and aesthetically right, as well as what is economically expedient. A thing is right when it tends to preserve the integrity, stability, and beauty of the biotic community. It is wrong when it tends otherwise. . . . that economics determines all land use. . . . is simply not true. An innumerable host of actions and attitudes, comprising perhaps the bulk of all land relations hinges on investments of time, forethought, skill and faith rather than on investments of cash. As a land-user thinketh, so is he.

Leopold held that a conservation ethic arose out of and led to a state of harmony between humans, land, and animals. In his view, humans had necessarily to consider themselves a part of the ecosystem and integral to but not the dominating species in a community of plants and other animals. He urged that people worldwide develop a sense of the ethical value of land, comparable to the moral values assigned to the conduct between individuals.

In more recent times, noted conservationists have expanded upon and added to Leopold's ideas, but despite this augmentation the conclusions as practical implications remain virtually unchanged. For example, Stuart Udall in the *Quiet Crisis* (1963) writes:

Each generation has its own rendezvous with the land . . . By choice, or default, we will carve out a land legacy for our heirs. We can mis-use the land and diminish the usefulness of resources, or we can create a world in which physical affluence and affluence of the spirit go hand-in-hand . . . earlier civilizations have declined because they did not learn to live in harmony with the land . . . our triumph of technology holds a hidden danger: as modern man increasingly arrogates to himself over the physical environment, there is the risk that his false pride will cause him to take the resources of the earth for granted—and to lose all reverence for the land.

    . . . true conservation is ultimately something of the mind—an ideal of men to cherish their past and believe in their future. Our civilization will be measured by its fidelity to this ideal as surely as by its art and poetry and system of justice. In our perpetual search for abundance, beauty, and order we manifest both our love for the land and our sense of responsibility toward future generations.

    A land ethic for tomorrow . . . should stress the oneness of our resources and the live-and-help-live logic of the great chain of life.

While the central importance of a land ethic might be perfectly obvious and supportable in developed countries, urging a tribal Ethiopian, whose family is starving, to abide by such moral dictates is difficult to defend. Yet, if well-founded policies relative to conservation of resources and control of populations had been formulated and vigorously pursued, then these people would not be starving today. This is but another way of arguing that the ethical responsibility of every generation in every nation is to leave the country and the land it occupies in as good or better shape than it was when that generation inherited it. The Menominee Indians of Wisconsin have a saying that the land is not inherited from the ancestors but borrowed from the grandchildren. If this thought can be extended to all people, then there should be repayment to the land with compound interest.

11.3 INTERNATIONAL COOPERATION

Cooperation in wildlife management at the international level was inaugurated in 1916 between the United States and Canada with the signing of the Migratory Bird Treaty and between the United States and Mexico in 1936. These treaties were concluded to protect migratory birds and prohibit export of illegally taken birds and game animals. Prior to these treaties the Lacy Act of 1900 prohibited the transport of illegal game between states within the United States and between the United States and other countries. The most important international multicountry treaty relating to wildlife signed to date is the Convention on International Trade in Endangered Species (CITIES). This treaty restricts international commerce in species threatened with extinction. In addition, animal species can be exported only if they meet three criteria: evidence that the species can withstand exploitation, information on population trends, and information on trade and commerce.

Another major international wildlife preservation measure relates to whales. Because several species of whales are now close to extinction, the International Whaling Commission was established to set harvesting quotas. These quotas were so often violated by several countries, most notably Japan and USSR, that in 1982 the 38 member governments of the commission, under pressure from the world community, passed a resolution to eliminate all commercial whaling.

An international agency, the United Nations Educational, Scientific, and Cultural Organization (UNESCO), has encouraged the establishment of biosphere reserves in all 14 major biomes of the world (Udvardy 1975):

1. Tropical humid forest.
2. Subtropical and temperate rain forest and woodlands.
3. Temperate needle-leaf forest and woodlands.
4. Tropical dry or deciduous forests and woodlands.
5. Temperate broad-leaf forests and woodlands, and subpolar deciduous thickets.
6. Evergreen sclerophyllous forests, scrubs, and woodlands.
7. Warm deserts and semideserts.
8. Cold winter deserts and semideserts.
9. Tundra communities and barren arctic desert.
10. Tropical grasslands and savannas.
11. Temperate grasslands.
12. Mixed mountain and highland systems.
13. Mixed island systems.
14. Lake systems.

Biosphere reserves are selected not for their uniqueness, but to be representatives of the biomes in which they occur. One of their principal objectives is the preservation of genetic diversity. Each biosphere reserve is funded by the sponsoring country so as might be expected, developed countries such as the United States have established the greatest number of reserves. Another of the major objectives in establishing biosphere reserves is to provide research opportunities for visiting scientists investigating a variety of biological questions. The reserves also serve to identify similarities or differences between paired reserves in other regions.

Cooperation between national governments and international organizations is not the only source of protection for global natural resources. Private or semiprivate organizations such as the International Union for Conservation of Nature and Natural Resources and the World Wildlife Fund have worked steadily in the international preservation of specific plant and animal species as well as sensitive natural areas. They also undertake political lobbying and provide economic resources in support of natural resource preservation objectives.

11.4 ENDANGERED HABITAT

The world community has come to accept the view of endangered species, and a number of countries have enacted laws to protect indigenous plants and animals in danger of extinction. This same concept, however, has not been sanctioned for vegetation as habitat for the endangered animals. Governments are less likely to designate several thousand hectares of vegetation as endangered because the scale in hectares is less dramatic than the scale in several thousand animals.

Of the fourteen major world biomes, the tropical rain forest is the one being most rapidly devastated. The most productive biomes, measured in grams of carbon produced per square meter per year, are swamps, marshlands, and rain forests. Major rain forests occur in Central and South America, West Africa, and Southeast Asia between latitudes 10 degrees north and south of the equator. These areas are being reduced in size because of increasing populations which need food and living space. Each year millions of hectares are cleared to make room for cattle ranches and farming, with most of the cattle being exported to developed countries. Closed tropical rain forests are decreasing in size at an estimated rate of 10–20 million ha (25–49 million ac) (1–2%) per year. Diversity of plant and animal species increases from the poles to the equator. Approximately 60–80% of all species live in the tropics and about 40% live in the tropical rain forests (Council on Environmental Quality 1981). For example, as many species of plants are found in Panama as in all Europe; as many fish species occur in the Amazon basin as in the Atlantic Ocean.

Without doubt tropical rain forests contain a huge reservoir of plant and animal genetic material that is rapidly being depleted by human activities. An estimated 120 thousand to 1 million extinctions will occur in the tropical rain forests by the year 2000 (Council on Environmental Quality 1981). Most of these extinctions will result from destruction of habitat.

11.5 EPILOGUE

In early geologic time humans did not possess the ability to bring about change in their environment at a rate where in any sense they could be called *ecological dominants*. The lives of early people, by current standards, must have been very harsh and difficult. Improvement in survival techniques was slow but ultimately led to conditions that fostered increased natality with reduced mortality which started human populations on a slow road to exponential growth.

As information accumulated on better ways to kill animals for food and use their skins and bones for clothing, tools, and articles of convenience, the ability of humans to change their immediate surroundings slowly began to develop and accelerate. In this early period the human race had no need to be concerned about change because the world beyond was unknown, except for local areas that were included in a day's walk to hunt for the next meal.

Prior to several thousand years ago the changes brought about on earth were through the forces of nature: hurricanes, tornados, fire, volcanos, earthquakes, floods, glaciers, and so forth. Humans began to notice how these forces affected the landscape which contained the plants and animals that contributed food and clothing to their survival. Humans as yet were not able to drastically alter their environment to an extent that recovery was not possible, and the forces of nature by themselves were not changing world landscapes at a rapid rate.

In 1933 Aldo Leopold suggested that our ability to change the land was advancing at a rate where the land, through succession, could no longer heal itself. This rate of change is greater than that of evolutionary change which has permitted humans to adjust to their environments over thousands of years. If this is the case with a rapidly expanding world population, then is there any hope for future generations to have the quality of life that most people in developed countries enjoy today?

There seem to be two areas where progress could be made, but these cross religious and political boundaries that are now and will continue to be difficult to deal with because they affect individual and collective human values. The first area involves the slowing down of processes by curtailing some of the management practices and human activities that are responsible for the rapid changes taking place in the world ecosystem. These practices are the large-scale removal of forest lands to meet housing needs and the clearing of land worldwide for agriculture. To slow down these activities, humans will have to reduce the consumption of natural resources and this is not likely to happen in the near future. This brings up the second area which involves slowing down or keeping stable the world population of humans.

The philosophical reasons for controlling human populations are no different than those for controlling wildlife populations: to maintain a balance of resources needed for reproduction and survival with what can be produced by land on a sustained yield basis. In the early and mid 1970s there was much concern generated about world populations, but there does not seem to be the same level of concern today—at least it is not now a major news media item. So where does the wildlife biologist fit into the world ecosystem scheme?

First, biologists in the wildlife profession have experience with and knowledge of what goes on in nature at the local level when animals are first introduced into a new environment, and they know and understand the consequences of the effects on habitat of over-populations of animals. By analogy these same effects can be extended to human populations.

Field biologists, researchers, and wildlife academicians must broaden their horizons to look beyond local projects and become more involved at the national and international level. If the choice is not to get involved then the opportunities to work with and preserve wildlife at the local level may disappear as more harvesting is done to provide wood for more homes and the clearings used to produce more food for more people.

There is evidence that wildlife populations will crash when animal density exceeds the ability of the habitat to produce food and cover. Whether eventually this same ecological regulatory mechanism will be a factor in reducing human populations is certainly a matter of concern. But as humans mature from the individualistic concept of survival to a more caring view that we all share this planet together, then a more cooperative attitude of sharing and caring will probably develop. How long will it take for this new attitude to come about? Probably longer than the lifetime of anyone reading this paragraph. But the question remains, Can we wait this long before initiating some control action?

Slowing down resource consumption and human increases to reach a degree of dynamic equilibrium at the local, regional, and national level, and ultimately in the world ecosystem, is not unrealistic. However, biologists will have to be involved in the decision-making process by providing sound ecological and biological data to the decision makers. Getting involved is easily done in the planning process which was discussed in the previous chapter. The planning process is the place to start because of the opportunities to influence other people who make decisions. Other ways to get involved include working outside one's own "habitat" by going on educational trips to foreign countries, accepting an assignment to a foreign country to work on resource problems, or working with foreign students, managers, and visiting scientists in one's own agency.

The wildlife profession has progressed from game management to wildlife management, and the opportunity is now available to advance the ideas of ecosystem management to include both natural resources and humans—the user of resources. The wildlife profession has always been concerned for the future of wildlife and humans. With the dedicated and well-educated professionals now employed in state and federal land management agencies, the challenge of ecosystem management will be accepted with the great optimism that exists in the profession today.

11.6 SUMMARY

The world population is increasing at a rate so that by the year 2000 there will be over 5 billion inhabitants. Because humans are ecological dominants, they have the ability to drastically change their habitat in a short time by fire, mechanized equipment, or explosives. Using these techniques, humans can remove obstacles to food production, and the population will continue to increase. International problems in wildlife are related to producer and consumer nations. Developing countries exploit their natural resources for export to obtain revenue necessary to feed a growing population. When material is exported it is no longer available for local consumption.

Scientists generally agree that air pollution is spreading, acid rain is affecting vegetation, and the removal of tropical forests is widespread. Each country now needs to consider how its use of resources affects the rest of the world, not just economically but also environmentally. As seen from the Apollo spacecraft, we live in a closed system. Humans must consider themselves a part of the world community, a part of the ecosystem, and they must develop a worldwide conscience with regard to land that is comparable to their consciousness about relations with other people.

International cooperation involving wildlife started in the early 1900s with the signing of several migratory bird treaties. It has continued with agreements such as the Convention on International Trade in Endangered Species (CITIES). Another area of cooperation has been the International Whaling Commission, which in 1982 had 38 member countries. Programs sponsored by the United Nations, such as blosphere reserves, are encouraging countries to establish representative biomes to preserve genetic diversity. As might be suspected, the developed countries have established the most reserves.

The world community has accepted the idea of endangered animal and plant species, but this same concept has not been sanctioned for vegetation as habitat for animals that are endangered. Of the 14 major world biomes the tropical rain forest is the one being most rapidly devastated. These areas are being reduced at a rate of

10–20 million ha (25–49 million ac) per year because of increasing populations that need food and living space. Tropical rain forests contain a huge reservoir of plant and animal genetic material that is rapidly being depleted by human activities.

Humans now possess the ability to change the landscape at a rate that exceeds the capacity of the land to heal. This rate can be slowed down and brought back into balance by decreasing the rate of consumption of natural resources simultaneously with a reduction in the rate of increase in human populations. The philosophical reasons for controlling human populations are no different than those for controlling wildlife populations. Biologists with experience and knowledge of natural processes understand the consequences of the effects of an overpopulation of animals on habitat. By analogy these same effects could happen to human populations. Biologists must extend their horizons beyond the local level and become more involved in national and international affairs.

Slowing down resource consumption and human population growth means getting involved in the planning process where biologists can provide sound biological and ecological data to influence the decision makers. Other ways to be involved include working in foreign countries or visiting with foreign scientists working in our own agencies. The wildlife profession has dedicated and educated professionals who will take up the challenge of ecosystem management.

11.7 RECOMMENDED READING

Brady, N. C. 1988. International development and the protection of biological diversity. Pages 409–418 in E. O. Wilson, ed. *Biodiversity*. National Academy Press, Wash., DC.

De Vos, A., and T. Jones. 1967. Wildlife management and land use. Symposium proceedings, special issue. *E. African Agric. and For. J.*, Nairobi, Kenya.

Dubos, R. J. 1973. Humanizing the earth. *Amer. Assoc. Adv. Sc.*, Annual Meeting. B. Y. Morrison Memorial Lecture. USDA, Agric. Res. Serv., Wash., DC.

Lovelock, J. E. 1988. The earth as a living organism. Pages 486–489 in E. O. Wilson, ed. *Biodiversity*. National Academy Press, Wash., DC.

Owen-Smith, R. N. 1982. *Management of large mammals in African conservation areas. Symposium Proceedings.* Centre for Resource Ecology, Univ. of Witwatersrand, Johannesburg, S.A.

Watt, K. E. 1972. Man's efficient rush toward deadly dullness. *Nat. Hist. Magazine.* Amer. Museum of Nat. Hist., Wash., DC.

Appendix A

ESTIMATING WILDLIFE POPULATIONS

Using Pellet Counts and Lincoln Index for Populations Estimation

A. The following data illustrates the *pellet count technique* for mule deer:

	Daily Forage Rate (DFR)	Daily Deposition Rate (DDR)
Antelope	1.36 kg (3 lb)	13
White-tailed deer	1.36 kg (3 lb)	13
Mule deer	2.04 kg (4.5 lb)	13
Elk	4.98 kg (11 lb)	13
Cattle	9.07 kg (20 lb)	12

Number of sample plots = 400
Size of sample plots = 0.004 ha
Days of herd occupancy = 365
(Time between counts)
Pellet groups counted = 500
Pellet groups/plot = 500/400 = 1.25

1. Sample area = No. plots × size
= 400 × .004   = 1.6 ha

2. Groups/area/time = $\dfrac{\text{Groups counted}}{\text{Sample area}}$

Groups/ha/yr = $\dfrac{500 \text{ groups}}{1.6 \text{ ha}}$ = 312.5

3. Animals/area/time[1] = $\dfrac{\text{Groups/ha/yr}}{\text{DDR} \times \text{Time}^{1}}$

Deer/ha/yr = $\dfrac{312.5}{13 \times 365}$ = 0.0659

Deer/ac/yr = Deer/ha/yr × 0.4046
= 0.0659 × 0.4046   = 0.0266

Deer/km²/yr	$= \text{Animals/ha} \times 100 \text{ ha}$	
	$= 0.0659 \times 100$	$= 6.59$
Deer/mi²/yr	$= \text{Deer/km}^2 \times 2.5899$	
	$= 6.59 \times 2.5899$	$= 17.06$

or

$$= \text{Deer/ac/yr} \times 640$$
$$0.0266 \times 640 \qquad = 17.02$$

4. Forage consumed/time[1]

kg/ha/yr	$= \text{Deer/ha/yr} \times \text{DFR} \times \text{Time}$	
	$= 0.0659 \times 4.5 \text{ kg} \times 365$	$= 108.24$
lb/ac/yr	$= \text{kg/ha/yr} \times 0.8922$	$= 96.57$

[1]Time must be in days because DDR and DFR are daily rates.

B. The formula for the *Lincoln index* is:

$$\frac{M}{N} = \frac{m}{n}$$

where

M	$=$	number marked during 1st trapping period
m	$=$	number of marked animals recaptured in the 2nd trapping period
n	$=$	number of animals captured in 2nd trapping period
N	$=$	population estimate

For example, 35 squirrels were trapped in the 1st trapping period, and 20 marked and 5 unmarked (25 total) in the 2nd trapping period. The population estimate is:

$$\frac{35}{N} = \frac{20}{25}$$
$$20\,N = 25 \times 35$$
$$N = 44$$

Confidence limits can be calculated for N using the formula:

$$S_E = N \sqrt{\frac{(N - M)\,(N - n)}{Mn\,(N - 1)}}$$

where

S_E = standard error

For a 95% confidence limit, two standard errors are added and subtracted from the population such that:

$$S_E = 44 \sqrt{\frac{(44 - 35)\ (44 - 25)}{35(25)\ (44 - 1)}} = 2.96$$

Upper limit = 44 + 2(2.96) = 50
Lower limit = 44—2(2.96) = 38

Appendix B

DESCRIPTIONS OF
FOREST COVER TYPES

1. White-Red-Jack Pine

This type occurs on smooth to irregular plains and tablelands of the northern Lake states and provinces and in parts of New York, New England, Quebec, and the Maritime provinces. Eastern white pine, red pine, and jack pine are the major components. Associates include eastern hemlock, aspens, northern white cedar, maples, and birches. Precipitation averages 63–114 cm (25–45 in), distributed evenly throughout the year.

2. Spruce-Fir

Flat plains and tablelands in the Lake states and provinces, New England states, and the Maritime provinces at high elevations in the Appalachian Mountains are occupied by the Spruce-Fir type. Red spruce, white spruce, and balsam fir are associated with northern white cedar, eastern hemlock, eastern white pine, and tamarack. Hardwoods include maples, birches, and aspens. Normal precipitation is 76–100 cm (30–40 in).

3. Longleaf-Slash Pine

This type is restricted to the flat and irregular southern Gulf Coastal Plains of the United States. Local relief is less than 90 m (300 ft). Major species include longleaf and slash pine. Associated species include shortleaf, loblolly, and Virginia pine with a mixture of oaks and hickories. Annual precipitation ranges from 100–152 cm (40–60 in).

4. Loblolly-Shortleaf Pine

The loblolly-shortleaf type generally occurs on irregular Gulf Coastal Plains and the Piedmont where local relief is 90–305 m (300–1000 ft). This type is of the greatest extent in the South and Southeast. The major species are loblolly and shortleaf pine in association with oaks, hickories, maples. Annual precipitation is 100–152 cm (40–60 in), evenly distributed with a midsummer peak.

5. Oak-Pine

The Oak-Pine type occurs on diverse land forms from the southern-most ridges and valleys of the Appalachians westward across the Coastal Plains and north into the Ozark Plateaus. White oak, northern red oak, scarlet oak, black oak, chestnut oak, with loblolly pine, shortleaf, and Virginia pine are the major species. Precipitation is relatively high averaging 100–140 cm (40–55 in).

6. Oak-Hickory

Oak-Hickory is the most extensive type in the United States and occurs in the east in areas that are more mesophytic than the surrounding areas. White oak, northern red and black oak, with associates of pignut, shagbark, mockernut, and bitternut hickory are the major species in the type. The forest type is best developed in the Ozarks. Annual precipitation in the type averages between 89–114 cm (35–45 in).

7. Oak-Gum-Cypress

This type is characterized by the vegetation of the Mississippi Valley and other bottom lands in every southern state, the cypress savanna west of the Everglades in Florida, the mangrove swamps south of the Everglades, and the east coast of Florida, Georgia, and the Carolinas. Blackgum, sweetgum, baldcypress, and Nuttall, shumard, overcup, and cherrybark oak are the major species in the type. Other hardwoods such as willows, maples, sycamore, cottonwoods, and beech are present depending on location. Annual precipitation varies from 152 cm (60 in) in Florida to 89 cm (35 in) in the northern extremity of the type in Indiana.

8. Elm-Ash-Cottonwood

The Elm-Ash-Cottonwood type occurs in narrow belts along major streams or scattered areas of dry swamps largely on the lower terraces and flood plains of the Mississippi, Missouri, Platte, Kansas, and Ohio Rivers from the Dakotas, Minnesota, and Ohio south through Kansas and Missouri. Cottonwood is the dominate species in the type with associates of green ash, white ash, American elm, and black willow. Precipitation varies from 25 cm (10 in) near the foothills of the Rocky Mountains to 127 cm (50 in) in the southern and northeastern areas.

9. Maple-Beech-Birch

The Maple-Beech-Birch type is best developed in the New England states and Maritime provinces. Sugar maple, American beech, and yellow birch are the dominant tree species. Associates are red maple, eastern hemlock, white ash, black cherry, sweet birch, and northern red oak. Typically, the type occurs on open high hills and low mountains. Annual precipitation ranges from 101–122 cm (40–48 in).

10. Aspen-Birch

The Aspen-Birch type is found within the Great Lakes region. Quaking aspen, bigtooth aspen, balsam poplar, paper birch, and gray birch are the dominant species. Balsam fir and maples are also common. The type is closely associated with moraines and outwash plains of recent glacial origin. Annual precipitation for the type is between 76–89 cm (30–35 in).

11. Douglas-fir

Just south and west of the Hemlock-Spruce forest and inland from the coast of British Columbia, Washington and Oregon extending south to northern California is the Douglas-fir forest. This forest contains at least 50% Douglas-fir with varying amounts of western hemlock, western redcedar, and redwood. Pacific yew and bigleaf maple are common understory species. The type constitutes one of the largest blocks of timber in the West.

In the Rocky Mountains the type is developed best in Idaho, Montana, British Columbia, and Alberta but its range extends southward to Utah, Colorado, New Mexico, and Arizona. Associated species are lodgepole pine, western larch, grand fir, and quaking aspen. The Douglas-fir forest is the most widespread type in the northern Rocky Mountains, but southward it is scattered. Precipitation is 100–203 m (40–80 in) in the extreme west to 50–76 cm (20–30 in) in the interior.

12. Western Hemlock-Sitka Spruce

Southward along the coast of British Columbia, Washington, Oregon, and northern California, western hemlock and sitka spruce are the dominant species, but Douglas-fir and western redcedar are common associates. Red alder and black cottonwood are also present. Sitka spruce generally exceeds 60 m (200 ft) in height and can attain ages of more than 1000 years. Bigleaf maple is the only major hardwood component. Precipitation ranges from 152–292 cm (60–115 in) annually. The type probably has the most productive understory of any coniferous forest in the United States.

13. Redwood

The Redwood type occupies only the low coastal mountains of northern California and the southwestern portion of Oregon. Redwoods are the tallest trees in the world, often reaching 90 m (300 ft). Douglas-fir, western hemlock, and sitka spruce are common associates. Annual precipitation is approximately 152 cm (60 in). Hardwood species found in the Redwood type are bigleaf maple, red alder, and willows.

14. Ponderosa Pine

Extending south through central British Columbia, Washington, Oregon, and into California is the narrow ponderosa pine type. Associated with ponderosa pine in the northern areas is Douglas-fir, lodgepole pine, grand fir, and western larch. South into California common associates are white fir, Jeffrey pine, and incense cedar.

Ponderosa pine occupies warm dry sites in the Rockies with precipitation between 38–50 cm (15–20 in). Its range extends from eastern British Columbia, Washington, and Oregon, into Idaho, Montana, Wyoming, South Dakota, and south through Colorado, New Mexico, and Arizona. At higher elevations Douglas-fir is associated with ponderosa pine. In the northern part of its range lodgepole pine, blue spruce, and quaking aspen are common species. In the middle parts—Utah and Wyoming—ponderosa pine does not occur.

15. White Pine

The western white pine type occurs in the high mountains of the northern Rocky Mountains of western Alberta, Montana, and northern Idaho. It is also present in scattered areas in the Cascade Mountains of Oregon and Washington. White pine is the dominant species in combination with sugar pine. Western redcedar is an associated species. Annual rainfall in the type is from 51–76 cm (20–30 in).

16. Lodgepole Pine

East of the ponderosa pine type in California is the lodgepole pine type. It extends northward as a disjunct population primarily in Oregon. The major associated species are red fir and mountain hemlock. The type occurs in pure stands in British Columbia, Idaho, and Montana but it ranges from Oregon south into Wyoming and Colorado. Common associated species are Douglas-fir, Engelmann spruce, subalpine fir, and aspens. Lodgepole pine as a pioneer species gets established following fire but it can also form a climax forest.

17. Larch

Located in the high mountains of Alberta, eastern Oregon, northern Idaho, and western Montana is the larch forest type. Western larch is the dominant species in association with white pine. Also Douglas-fir and grand fir may be present. Annual precipitation is 50–76 cm (20–30 in) in eastern Oregon and 50–127 cm (20–50 in) in the northern Rockies.

18. Fir-Spruce

Adjacent to the ponderosa pine but at higher elevations is another long, narrow band of vegetation comprised of silver fir, subalpine fir, red fir, white fir, and Engelmann spruce, and sometimes in combination with mountain hemlock. In the Southwest the type is developed best in Colorado but it also occurs in northwest Wyoming and in disjunct areas in western Montana and eastern Idaho. Fir-Spruce occurs in small areas at high elevation in Arizona and New Mexico. Species in the type are subalpine and white fir with Engelmann spruce, blue spruce, and aspens. Annual precipitation ranges from 56 cm (22 in) in the Rocky Mountains to 127–190 cm (50–75 in) in the Sierras.

22. Pinyon-Juniper

The Pinyon-Juniper type occurs in the Basin and Range province of Utah, Nevada, southern Idaho, and southeast Oregon. Large areas of Pinyon-Juniper occur in Arizona and New Mexico. Singleleaf and Colorado pinyon with Utah, alligator, and oneseed juniper are the main species. The name "pygmy forest" characterizes the pinyon and juniper woodland. The trees occur as dense-to-open woodland and savannah woodland. Annual precipitation is about 25 cm (10 in).

23. Alaska Western Hemlock-Sitka Spruce

Along the Alaska and British Columbia coast is a narrow band of western hemlock and sitka spruce that extends southward along the coast through British Columbia to Washington and Oregon. Western hemlock is the dominant species with varying amounts of sitka spruce. Mountain hemlock and Alaska-cedar occur at higher elevations. Lush vegetation grows as an understory in the coastal forest where rain and fog provide abundant moisture. Annual rainfall ranges from 150 (60 in) to over 200 cm (79 in) annually. Black cottonwood, alders, and willows also occur in the type. Western redcedar is a minor component.

24. Alaska Spruce-Hardwoods

The Spruce-Hardwood forest of interior Alaska is the westward extension of the Boreal forest which traverses over 6,400 km (4,000 mi) from Newfoundland to Alaska. In interior Alaska the dominant species is white spruce but paper birch is also important. In addition, there are mixtures of dense stands of birches, aspens, and poplars that occupy large areas. Annual precipitation can be from 50 cm (20 in) to 100 cm (40 in).

Appendix C

SUMMARY LIFE HISTORY
FOR REPRESENTATIVE SPECIES

Amphibians

Northern leopard frog
    Scientific name: *Rana pipiens*
    Breeding season: March–June
    Forest regions: II, III, IV, V, VI
    General habitat: Meadows, ponds, lakes, streams
    Comments: Primarily nocturnal

Tiger salamander
    Scientific name: *Ambystoma tigrinum*
    Breeding season: March–June but variable
    Forest regions: III, IV, V, VI, VII
    General Habitat: Damp meadows and forests with soft soil
    Comments: Many subspecies

Western toad
    Scientific name: *Bufo boreas*
    Breeding season: January–September
    Forest regions: I, II, III, IV
    General habitat: Near meadows, springs, streams
    Comments: Active at dusk

Birds

American dipper
    Scientific name: *Cinclus mexicanus*
    Forest regions: I, II, III, IV
    General habitat: Mountain streams with fast flowing water
    Food habits: Insects and aquatic invertebrates
    Comments: Bobbing motion, walks underwater on stream bottom

American robin
    Scientific name: *Turdus migratorius*
    Forest regions: I, II, III, IV, V, VI, VII
    General habitat: Meadows, cut areas, open woods
    Food habits: Insects and plant material
    Comments: Best known of all U.S. birds

Bald eagle
    Scientific name: *Haliaeetus leucocephalus*
    Breeding season: December–January
    Mating behavior: Mates for life
    Young per season: 1–2
    Legal status: Federal endangered species
    Forest regions: I, II, III, IV, V, VI, VII
    General habitat: Near large lakes and rivers
    Food habits: Primarily fish, small rodents, carrion

Bewick's wren
    Scientific name: *Thryomanes bewickii*
    Forest regions: II, IV, V, VI, VII
    General habitat: On the ground in brush areas
    Food habits: Insects and spiders
    Comments: Cavity nester, uses a singing perch

Black-capped chickadee
    Scientific name: *Parus atricapillus*
    Forest regions: I, II, III, IV, V, VI
    General habitat: Mixed coniferous and deciduous forests
    Food habits: Insect and coniferous plant material
    Comments: Cavity nester

Black-chinned hummingbird
    Scientific name: *Archilochus alexandri*
    Forest regions: II, III, IV, VI
    General habitat: Mountain meadows and woodlands
    Food habits: Insects
    Comments: Male has a mating and feeding territory

Blue grouse
    Scientific name: *Dendragapus obscurus*
    Breeding season: April–May
    Mating behavior: Polygamous
    Young per season: 7–10
    Legal status: Game species
    Forest regions: I, II, III, IV
    General habitat: Conifer forest
    Food habits: Insects, buds, fruits
    Comments: Nests on ground

Blue-gray gnatcatcher
  Scientific name: *Polioptila caerulea*
  Forest regions: II, III, IV, V, VI, VII
  General habitat: Pinyon-juniper and deciduous woodland
  Food habits: Insects
  Comments: Flicks tail while feeding

Brown creeper
  Scientific name: *Certhia americana*
  Forest regions: I, II, III, IV, V, VI, VII
  General habitat: Coniferous and mixed forests
  Food habits: Bark insects, moves upward when feeding

Chipping sparrow
  Scientific name: *Spizella passerina*
  Forest regions: I, II, III, IV, V, VI, VII
  General habitat: Edges of coniferous and deciduous forests
  Food habits: Insects

Common raven
  Scientific name: *Corvus corax*
  Forest regions: I, II, III, IV, V, VI
  General habitat: All types of forest land
  Food habits: Generalist
  Comments: Intelligence matches that of a dog

Cooper's hawk
  Scientific name: *Accipiter cooperii*
  Forest regions: II, III, IV, V, VI, VII
  General habitat: Primarily deciduous forest types with meadows and clearings
  Food habits: Small birds and mammals

Dark-eyed junco
  Scientific name: *Junco hymenalis*
  Forest regions: I, II, III, IV, V, VI, VII
  General habitat: Coniferous and mixed forests
  Food habits: Ground feeder on seeds and fruits

Downey woodpecker
  Scientific name: *Picoides pubescens*
  Forest regions: I, II, III, IV, V, VI, VII
  General habitat: Deciduous woodlands
  Food habits: Insects and mast
  Comments: Cavity nester

Eastern screech-owl
    Scientific name: *Otus asio*
    Forest regions: II, III, IV, V, VI, VII
    General habitat: Deciduous and mixed forests
    Food habits: Rodents and insects
    Comments: Defends a nesting and feeding territory

Evening grosbeak
    Scientific name: *Coccothraustes vespertinus*
    Forest regions: II, III, IV, V, VI
    General habitat: Coniferous and mixed forests
    Food habits: Seeds and fruits

Fox sparrow
    Scientific name: *Passerella iliaca*
    Forest regions: I, II, III, IV, V, VI, VII
    General habitat: Undergrowth in coniferous forests
    Food habits: Insects and seeds
    Comments: Make considerable noise when scratching for
                              food under brush

Golden-crowned kinglet
    Scientific name: *Regulus satrapa*
    Forest regions: I, II, III, IV, V, VI, VII
    General habitat: Old coniferous and deciduous forests
    Food habits: Primarily insects

Great horned owl
    Scientific name: *Bubo virginianus*
    Forest regions: I, II, III, IV, V, VI, VII
    General habitat: All forest types
    Food habits: Birds, small mammals,
    Comments: Largest of the eared owls

Hairy woodpecker
    Scientific name: *Picoides villosus*
    Forest regions: I, II, III, IV, V, VI, VII
    General habitat: Deciduous forests
    Food habits: Insects and mast
    Comments: Cavity nester

Hermit thrust
    Scientific name: *Catharus guttatus*
    Forest regions: I, II, III, IV, V, VI, VII
    General habitat: Coniferous and mixed forests
    Food habits: Ground insect feeder

Lincoln's sparrow
  Scientific name: *Melospiza lincolnii*
  Forest regions: I, II, III, IV, V, VII
  General habitat: Deciduous forest with meadows and bogs
  Food habits: Seed

Long-eared owl
  Scientific name: Asio otus
  Forest regions: II, III, IV, V, VI, VII
  General habitat: Open deciduous and coniferous forests
  Food habits: Small mammals

Mountain bluebird
  Scientific name: *Sialia currucoides*
  Forest regions: I, II, III, IV
  General habitat: Open areas near forest edge
  Food habits: Insects
  Comments: Cavity nester

Mourning dove
  Scientific name: *Zenaida macroura*
  Breeding season: April–September
  Mating behavior: Monogamous
  Young per season: 1–3
  Legal status: Game species
  Forest regions: II, III, IV, V, VI, VII
  General habitat: Open forest areas
  Food habits: Seeds, fruits, mast

Northern cardinal
  Scientific name: *Cardinalis cardinalis*
  Forest regions: IV, V, VI, VII
  General habitat: Mixed woodland
  Food habits: Primarily seeds but some insects

Northern flicker
  Scientific name: *Colaptes auratus*
  Forest regions: I, II, III, IV, V, VI, VII
  General habitat: Open forested areas
  Food habits: Insects, especially fond of ants
  Comments: Cavity nester

Northern goshawk
  Scientific name: *Accipiter gentilis*
  Forest regions: I, II, III, IV, V, VI
  General habitat: Coniferous forests
  Food habits: Birds and small mammals

Northern oriole
 Scientific name: *Icterus galbula*
 Forest regions: II, III, IV, V, VI, VII
 General habitat: Deciduous forests
 Food habits: Insects, seeds, fruits

Northern saw-whet owl
 Scientific name: *Aegolius acadicus*
 Forest regions: I, II, III, IV, V, VI, VII
 General habitat: Coniferous and deciduous forests
 Food habits: Insects and small mammals
 Comments: Cavity nester

Northern shrike
 Scientific name: *Lanius excubitor*
 Forest regions: I, II, III, IV, V, VI
 General habitat: Coniferous woodlands
 Food habits: Insects, small rodents
 Comments: Kills prey with bill

Olive-sided flycatcher
 Scientific name: *Contopus borealis*
 Forest regions: I, II, III, IV, V, VI
 General habitat: Spruce-fir forests
 Food habits: Flying insects
 Comments: Perches on tops of tall trees

Pileated woodpecker
 Scientific name: *Dryocopus pileatus*
 Forest regions: II, III, V, VI, VII
 General habitat: Deciduous and mixed coniferous trees
 Food habits: Insects and mast

Pine grosbeak
 Scientific name: *Pinicola enucleator*
 Forest regions: I, III, IV, V
 General habitat: Coniferous forest and mixed woodlands
 Food habits: Seed and fruits
 Comments: Slow moving and tame

Pine siskin
 Scientific name: *Carduelis pinus*
 Forest regions: I, II, III, IV, V, VI, VII
 General habitat: Coniferous forests and mixed woodlands
 Food habits: Seeds of hemlock, alder, birch
 Comments: Likes salt

Purple finch
    Scientific name: *Carpodacus purpureus*
    Forest regions: II, III, V, VI, VII
    General habitat: Coniferous forests and woodlands
    Food habits: Seeds

Red crossbill
    Scientific name: *Loxia curvirostra*
    Forest regions: I, II, III, IV, V, VI
    General habitat: Coniferous forests
    Food habits: Conifer seeds

Red-breasted nuthatch
    Scientific name: *Sitta canadensis*
    Forest regions: I, II, III, IV, V, VI, VII
    General habitat: Coniferous forests
    Food habits: Insects
    Comments: Cavity nester

Red-eyed vireo
    Scientific name: *Vireo olivaceus*
    Forest regions: III, IV, V, VI, VII
    General habitat: Deciduous forests
    Food habits: Insects

Red-headed woodpecker
    Scientific name: *Melanerpes erythroceohalus*
    Forest regions: IV, V, VI, VII
    General habitat: Open parklike conditions
    Food habits: Insects and mast

Ruby-crowned kinglet
    Scientific name: *Regulus calendula*
    Forest regions: I, II, III, IV, V, VI, VII
    General habitat: Coniferous and deciduous forests
    Food habits: Insects
    Comments: Small bird, able to feed on tips of twigs

Ruffed grouse
    Scientific name: *Bonasa umbellus*
    Breeding season: April–May
    Mating behavior: Promiscuous
    Young per season: 10–12
    Legal status: Game species
    Forest regions: I, II, III, IV, V, VI
    General habitat: Coniferous and deciduous forests near edge
    Food habits: Buds, fruits, seeds

Rufus-sided towhee
Scientific name: *Piplo erythrophthalmus*
Forest regions: II, III, IV, V, VI, VII
General habitat: Brush around forest edges
Food habits: Insects and seeds
Comments: Ground feeder

Sharp-shinned hawk
Scientific name: *Accipiter striatus*
Forest regions: I, II, III, IV, V, VI, VII
General habitat: Dense coniferous forests
Food habits: Small birds, rodents, and insects

Solitary vireo
Scientific name: *Vireo solitarius*
Forest regions: II, III, IV, V, VI, VII
General habitat: Coniferous and deciduous forests
Food habits: Insects

Song sparrow
Scientific name: *Melospiza melodia*
Forest regions: I, II, III, IV, V, VI, VII
General habitat: Forest edges and streamside vegetation
Food habits: Insects and seeds

Spruce grouse
Scientific name: *Dendragapus canadensis*
Breeding season: April–May
Mating behavior: Monogamous
Young per season: 10–12
Legal status: Game species
Forest regions: I, II, III, V
General habitat: Coniferous forests
Food habits: Needles, buds, fruit

Steller's jay
Scientific name: *Cyanocitta stelleri*
Forest regions: I, II, III, IV
General habitat: Coniferous and deciduous mixed forests
Food habits: Insects and seeds

Three-toed woodpecker
Scientific name: *Picoides tridactylus*
Forest regions: I, II, III, IV, V
General habitat: Coniferous forests
Food habits: Insects
Comments: Important species in controlling insects; cavity
                        nester

Violet-green swallow
    Scientific name: *Tachycineta thalassina*
    Forest regions: I, II, III, IV
    General habitat: Coniferous and deciduous forests
    Food habits: Flying insects
    Comments: Ponderosa pine is major habitat
        Cavity nester

Western flycatcher
    Scientific name: *Empidonax difficilis*
    Forest regions: I, II, III, IV
    General habitat: Deciduous or coniferous forests near water
    Food habits: Flying insects
    Comments: Cavity nester

Whip-poor-will
    Scientific name: *Caprimulgus vociferus*
    Forest regions: II, IV, V, VI, VII
    General habitat: Deciduous forests
    Food habits: Insects
    Comments: Ground nester

White-crowned sparrow
    Scientific name: *Zonotrichia leucophrys*
    Forest regions: I, II, III, IV, V, VI, VII
    General habitat: Forest edges, meadows, clearings
    Food habits: Insects
    Comments: Nests on or near ground

Wild turkey
    Scientific name: *Meleagris gallopavo*
    Breeding season: April–May
    Mating behavior: Polygamous
    Young per season: 10–12
    Legal status: Game species
    Forest regions: II, IV, VI, VII
    General habitat: Deciduous or coniferous forests
    Food habits: Insects, mast, fruits, seeds

Wilson's warbler
    Scientific name: *Wilsonia pusilla*
    Forest regions: I, II, III, IV, V
    General habitat: Deciduous forests with shrubs
    Food habits: Insects
    Comments: Nest on or near ground

Winter wren
>   Scientific name: *Troglodytes troglodytes*
>   Forest regions: I, II, III, V, VI, VII
>   General habitat: Coniferous and deciduous forests
>   Food habits: Insects
>   Comments: Cavity nester

Wood duck
>   Scientific name: *Aix sponsa*
>   Legal status: Waterfowl
>   Forest regions: II, III, V, VI, VII
>   General habitat: Deciduous trees near water
>   Food habits: Acorns, fleshy fruit, vegetable matter
>   Comments: Cavity nester

Yellow-bellied sapsucker
>   Scientific name: *Sphyrapicus varius*
>   Forest regions: II, III, IV, V, VI, VII
>   General habitat: Deciduous and coniferous forests
>   Food habits: Tree sap and insects
>   Comments: Cavity nester

Fishes

Brown trout
>   Scientific name: *Salmo trutta*
>   Breeding season: October–January
>   Young per season: Up to 3000
>   Legal status: Game fish
>   Forest regions: II, III, IV, V, VI
>   General habitat: Fast-flowing, freshwater streams
>   Food habits: Young (aquatic insects), Adults (other fish)
>   Comments: Introduced from Europe and Asia

Brook trout
>   Scientific name: *Salvelinus fontinalis*
>   Breeding season: September–January
>   Mating behavior: Spawns in small, headwater streams
>   Young per season: Up to 5000
>   Legal status: Game species
>   Forest regions: II, III, IV, V, VI
>   General habitat: Clear, cool mountain streams
>   Food habits: Aquatic insects
>   Comments: Native to Canada and northeastern U.S.

Cutthroat trout

Scientific name:	*Salmo clarki*
Breeding season:	February–May
Mating behavior:	Needs stream gravel
Young per season:	Up to 4500
Legal status:	Game species
Forest regions:	I, II, III, IV
General habitat:	Alpine streams and lakes
Food habits:	Aquatic insects
Comments:	Many subspecies often called native trout

Largemouth bass

Scientific name:	*Micropterus salmoides*
Breeding season:	April–July
Mating behavior:	Needs aquatic vegetation and gravel
Young per season:	Up to 100,000
Legal status:	Game species
Forest regions:	V, VI, VII
General habitat:	Quiet, warm water
Food habits:	Smaller fish
Comments:	Most sought after warm water fish

Rainbow trout

Scientific name:	*Salmo gairdneri*
Breeding season:	April–June
Mating behavior:	Needs stream gravel
Young per season:	Up to 12,000
Legal status:	Game species
Forest regions:	I, II, III, IV, V, VI
General habitat:	Fresh and salt water
Food habits:	Insects and small fish
Comment:	Seagoing form is Steelhead trout

Smallmouth bass

Scientific name:	*Micropterus dolomieui*
Breeding season:	April–June
Mating behavior:	Bottoms of lakes and rivers
Young per season:	Up to 14,000
Legal status:	Game species
Forest regions:	V, VI, VII
General habitat:	Cool, clear water
Food habits:	Minnows and crayfish

Mammals

Abert squirrel
    Scientific name: *Sciurus aberti*
    Breeding season: February–March
    Mating behavior: Promiscuous, mating bouts
    Young per season: 2–4
    Legal status: Game species
    Forest regions: IV
    General habitat: Primarily ponderosa pine
    Food habits: Inner bark of ponderosa pine
    Comments: Several subspecies including the Kaibab squirrel

Beaver
    Scientific name: *Castor candensis*
    Breeding season: January–February
    Mating behavior: Monogamous
    Breeding age: 2
    Young per season: 4
    Legal status: Game species (fur)
    Forest regions: I, II, III, IV, V, VI, VII
    General habitat: Lakes or slow moving water
    Food habits: Deciduous trees of poplar, willow, aspen

Black bear
    Scientific name: *Ursus americanus*
    Breeding season: June–July
    Mating behavior: Probably monogamous
    Breeding age: 3
    Young per season: 2 every other year
    Legal status: Game animal, predator
    Forest regions: I, II, III, IV, V, VI, VII
    General habitat: Coniferous and deciduous forests
    Food habits: Omnivorous—fruits, berries, small mammals, and sometimes large mammals such as deer or cattle

Bobcat
    Scientific name: *Felis rufus*
    Breeding season: Year-long
    Mating behavior: Monogamous
    Breeding age: 1
    Young per season: 2–4
    Legal status: Game animal (fur)
    Forest regions: II, III, IV, V, VI, VII
    General habitat: Coniferous and deciduous forests
    Food habits: Small mammals

Coyote
Scientific name:	*Canis latrans*
Breeding season:	January–May
Mating behavior:	Monogamous
Breeding age:	1
Young per season:	5–7
Legal status:	Game animal (fur), predator
Forest regions:	I, II, III, IV, V, VI, VII
General habitat:	Open coniferous and deciduous forests
Food habits:	Small mammals, deer, fawns

Deer mouse
Scientific name:	*Peromyscus maniculatus*
Young per season:	4–5/litter and 3–4 litters
Forest regions:	I, II, IV, V, VI
General habitat:	Coniferous and deciduous forests
Food habits:	Insects, seeds, fruits, nuts
Comments:	Prey species for many birds and mammals

Eastern cottontail
Scientific name:	*Sylvilagus floridanus*
Breeding season:	February–September
Mating behavior:	Promiscuous
Young per season:	4–5/litter with 3–4 litters
Legal status:	Game species
Forest regions:	IV, V, VI, VII
General habitat:	Brush understory in forests and woodlands
Food habits:	Forbs, grasses, fruits, twigs

Elk
Scientific name:	*Cervus elaphus*
Breeding season:	September–October
Mating behavior:	Polygamous
Breeding age:	3
Young per season:	1
Legal status:	Game species
Forest regions:	II, III, IV
General habitat:	Coniferous forests, woodlands, meadows
Food habits:	Grass, forbs, browse

Gray fox
Scientific name:	*Urocyon cinereoargenteus*
Breeding season:	January–February
Mating behavior:	Monogamous
Breeding age:	1
Young per season:	3–5
Legal status:	Game species (fur), predator
Forest regions:	II, IV, V, VI, VII
General habitat:	Coniferous and deciduous forests
Food habits:	Small rodents, fruits, berries

Gray squirrel
    Scientific name: *Sciurus carolinensis*
    Breeding season: December–January
    Mating behavior: Promiscuous
    Breeding age: 10–11 months
    Young per season: 2–4, may have 2 litters
    Legal status: Game species
    Forest regions: V, VI, VII
    General habitat: Hardwood forests
    Food habits: Hard mast, fruits, fungi

Grizzly bear
    Scientific name: *Ursus arctos*
    Breeding season: June–July
    Mating behavior: Monogamous
    Breeding age: 3–5
    Young per season: 2
    Legal status: Federal endangered species
    Forest regions: I, III
    General habitat: Mountain, coniferous forests
    Food habits: Small and large mammals, fish

Hoary bat
    Scientific name: *Lasiurus cinereus*
    Forest regions: II, III, IV, V, VI, VII
    General habitat: Coniferous forests
    Food habits: Flying insects

Long-tailed vole
    Scientific name: *Microtus longicaudus*
    Forest regions: I, II, III, IV
    General habitat: Grassy areas away from water
    Food habits: Roots and bark

Long-tailed weasel
    Scientific name: *Mustela frenata*
    Breeding season: July–August, delayed implantation
    Mating behavior: Polygamous
    Breeding age: 1 for males, 3–4 months for females
    Young per season: 6–9
    Legal status: Game species (fur)
    Forest regions: II, III, IV, V, VI, VII
    General habitat: Varied forest conditions
    Food habits: Small mammals

Lynx

Scientific name:	*Felis lynx*
Breeding season:	January–February
Mating behavior:	Monogamous
Breeding age:	1
Young per season:	1–4
Legal status:	Game species (fur)
Forest regions:	I, III, IV
General habitat:	Dense coniferous forest
Food habits:	Small mammals, rabbits preferred

Marten

Scientific name:	*Martes americana*
Breeding season:	July–August
Mating behavior:	Promiscuous
Breeding age:	1
Young per season:	3–4
Legal status:	Game species (fur)
Forest regions:	I, II, III, IV, V
General habitat:	Dense coniferous forests
Food habits:	Small mammals especially red squirrels

Mink

Scientific name:	*Mustela vison*
Breeding season:	January–March
Mating behavior:	Polygamous
Breeding age:	1
Young per season:	5–6
Legal status:	Game species (fur)
Forest regions:	I, II, III, IV, V, VI, VII
General habitat:	Lakes and streams in coniferous and deciduous forests
Food habits:	Small mammals, fish

Moose

Scientific name:	*Alces alces*
Breeding season:	September–November
Mating behavior:	Polygamous
Breeding age:	2
Young per season:	1–2
Legal status:	Game species
Forest regions:	I, III, IV, V
General habitat:	Coniferous and deciduous forests with willow and aspen near water
Food habits:	Woody plants, aspen and willow

Mountain lion
  Scientific name: *Felis concolor*
  Breeding season: January–February
  Mating behavior: Monogamous
  Breeding age: 2–3
  Young per season: 1–4
  Legal status: Game species, predator
  Forest regions: II, III, IV, VII
  General habitat: Coniferous forests and woodlands
  Food habits: Small rodents, deer

Mule deer
  Scientific name: *Odocoileus hemionus*
  Breeding season: September–February
  Mating behavior: Polygamous
  Breeding age: 2
  Young per season: 1–2
  Legal status: Game species
  Forest regions: I, II, III, IV
  General habitat: Coniferous forests
  Food habits: Browse, forbs, grass

Northern flying squirrel
  Scientific name: *Glaucomys sabrinus*
  Breeding season: February–March
  Young per season: 3–6, may have 2 litters
  Forest regions: I, II, III, IV, V, VI
  General habitat: Dense coniferous and deciduous forest
  Food habits: Seeds, fruits, mast

Porcupine
  Scientific name: *Erethizon dorsatum*
  Breeding season: September–November
  Young per season: 1
  Legal status: Generally not protected
  Forest regions: I, II, III, IV, V, VI
  General habitat: Coniferous and hardwood forests
  Food habits: Bark and twigs

Pygmy shrew
  Scientific name: *Microsorex boyi*
  Forest regions: I, III, V, VI
  General habitat: Dense forests with moist areas
  Food habits: Insects
  Comments: Smallest mammal in North America, 2–4 grams

Raccoon
    Scientific name: *Procyon lotor*
    Breeding season: February
    Mating behavior: Monogamous
    Breeding age: 1
    Young per season: 3–6
    Legal status: Game species (fur)
    Forest regions: II, III, IV, V, VI. VII
    General habitat: Hardwood forests near water
    Food habits: Crayfish, fish, small rodents, eggs

Red fox
    Scientific name: *Vulpes vulpes*
    Breeding season: January–February
    Mating behavior: Monogamous
    Breeding age: 1
    Young per season: 4–10
    Legal status: Game species (fur)
    Forest regions: I, II, III, IV, V, VI, VII
    General habitat: Coniferous and deciduous forests with brushland
    Food habits: Small mammals, fruit, nuts

Red squirrel
    Scientific name: *Tamiasciurus hudsonicus*
    Breeding season: February–March
    Breeding age: 1
    Young per season: 4–5
    Legal status: Game species
    Forest regions: I, II, III, IV, V, VI
    General habitat: Coniferous forests
    Food habits: Seeds of conifers, mushrooms
    Comments: Stores cones in "middens"

River otter
    Scientific name: *Lutra canadensis*
    Breeding season: January–April
    Mating behavior: Monogamous
    Breeding age: 1
    Young per season: 2–3
    Legal status: Game species (fur)
    Forest regions: I, II, III, IV, V, VI, VII
    General habitat: Streams and lakes in coniferous and deciduous forests
    Food habits: Fish, small mammals, invertebrates

Snowshoe hare
    Scientific name: *Lupus americanus*
    Breeding season: February–August
    Breeding age: 1
    Young per season: 3–4 litter with 2–3 litters
    Legal status: Game species (fur)
    Forest regions: I, II, III, IV, V, VI
    General habitat: Coniferous and hardwood forests
    Food habits: Bark, buds, twigs
    Comments: Population rises and falls in a cycle

Striped skunk
    Scientific name: *Mephitis mephitis*
    Breeding season: February–March
    Mating behavior: Polygamous
    Breeding age: 1
    Young per season: 4–7
    Legal status: Game species (fur)
    Forest regions: II, III, IV, V, VI, VII
    General habitat: Coniferous and deciduous forests with brush
    Food habits: Small rodents, insects, fruits

Virginia opossum
    Scientific name: *Didelphis virginiana*
    Breeding season: Yearlong
    Mating behavior: Promiscuous
    Breeding age: 1
    Young per season: 5–7
    Legal status: Games species (fur)
    Forest regions: II, IV, V, VI, VII
    General habitat: Coniferous and deciduous forests near water
    Food habits: Will eat anything available (dead or alive)

White-footed mouse
    Scientific name: *Peromyscus leucopus*
    Young per season: 4–5 litter, 3–4 litters
    Forest regions: III, IV, V, VI, VII
    General habitat: Brushy areas in coniferous and deciduous forests
    Food habits: Seeds, nuts, insects

White-tailed deer
    Scientific name: *Odocoileus virginianus*
    Breeding season: September–February
    Mating behavior: Polygamous
    Breeding age: 1
    Young per season: 1–2
    Legal status: Game species
    Forest regions: II, III, IV, V, VI, VII

General habitat: Coniferous and deciduous forest with brush understory
Food habits: Browse, forbs, grass

Woodchuck
Scientific name: *Marmota monax*
Breeding season: March–April
Young per season: 4–5
Legal status: Game species
Forest regions: I, V, VI, VII
General habitat: Borders of deciduous forests and meadows
Food habits: Forbs, grasses

Reptiles

Coachwhip
Scientific name: *Masticophis flagellum*
Breeding season: April–May
Young per season: 4–16
Forest regions: II, IV, VI, VII
General habitat: Pine woodlands
Food habits: Insects, small mammals, snakes
Comments: A fast-moving snake

Common garter snake
Scientific name: *Thamnophis sirtalis*
Breeding season: March–May
Young per season: 7–85
Forest regions: II, III, V, VI, VII
General habitat: In forested areas with meadows and marshes
Food habits: Mice and small fish
Comments: Live-bearing, many subspecies

Common kingsnake
Scientific name: *Lampropeltis getulus*
Breeding season: March–June
Young per season: 3–24
Forest regions: II, IV, VI, VII
General habitat: Diverse forested areas
Food habits: Birds, small mammals, lizards, snakes
Comments: A constrictor

Gopher snake
Scientific name: *Pituophis melanoleucus*
Breeding season: March–May
Young per season: 3–24
Forest regions: II, III, IV, VI, VII
General habitat: Dry pine and oak woodlands
Food habits: Small rodents
Comments: Many subspecies

Milk snake
  Scientific name: *Lampropeltis triangulum*
  Breeding season: June–July
  Young per season: 2–17
  Forest regions: III, IV, V, VI, VII
  General habitat: Diverse forested areas
  Food habits: Small mammals, birds
  Comments: Many subspecies

Painted turtle
  Scientific name: *Chrysemys picta*
  Breeding season: May–July
  Young per season: 2–20, 2 times a year
  Forest regions: II, III, V, VI, VII
  General habitat: Slow moving streams
  Food habits: Aquatic insects, fish, frogs, plant material
  Comments: Most widespread turtle in North America

Racer
  Scientific name: *Coluber constrictor*
  Breeding season: April–May
  Young per season: 5–28
  Forest regions: II, III, IV, V, VI, VII
  General habitat: Diverse forested areas, meadows, grassland
  Food habits: Insects, small mammals, birds
  Comments: Not a constrictor

Ringneck snake
  Scientific name: *Diadophis punctatus*
  Breeding season: May–September
  Young per season: 1–10
  Forest regions: II, III, IV, V, VI, VII
  General habitat: Forested areas with moist areas
  Food habits: Earthworms, salamanders, young snakes
  Comments: Communal nest

Timber rattlesnake
  Scientific name: *Crotalus horridus*
  Breeding season: September–October
  Young per season: 5–17 young every two years
  Forest regions: V, VI, VII
  General habitat: Rocky situations and swamps in forested areas
  Food habits: Small mammals, birds
  Comments: May overwinter with other snakes

Western rattlesnake
<blockquote>

Scientific name:	*Crotalus viridis*
Breeding season:	March–May
Young per season:	4–21
Forest regions:	II, III, IV
General habitat:	Rocky outcrops in coniferous forests
Food habits:	Small mammals, lizards
Comments:	May overwinter in common den

</blockquote>

Appendix D

A COMPLETE LIFE HISTORY FOR ELK

SPECIES
   Common name: Elk
   Scientific name: *Cervus elaphus*
   Subspecies:
      C.e. nelsoni (Rocky mountain elk)
      C.e. manitobensis (Manitoba elk)
      C.e. roosevelti (Roosevelt elk)
      C.e. nannodes (Tule elk)
   Taxonomy:
      Order—Artiodactyla
      Family—Cervidae
   Weight: 227–363 kg (500–800 lb)
      Adult cows weigh about 272–295 kg (600–650 lb)
      Newborn calves weigh 14–16 kg (30–35 lb)
   Maximum ecological longevity: 20 years
   Young per year: Generally 1, twins are rare
   Gestation period: 210–225 days
   Breeding season: September–October, with several estrous cycles
   Mating: Polygamous
   Young born: May–June, usually in a secluded area. Cow-calf groups
      are formed and maintained through summer.
   Annual increase: 15–30%
   Antlers: Only males have antlers. Mature bulls have 6 points, male
      calves have buttons. Yearling bulls can have spikes without brow
      tines. Antlers are shed in March-April. Growth starts in May and
      continues until August when velvet is rubbed off. Weight of
      antlers is 11–14 kg (25–30 lb).
   Dentition: I 0/3, C 1/1, P 3/3, M 3/3 = 34 All permanent teeth are
      present at 36 months
   Major distribution: Arizona, New Mexico, Colorado, Utah, Nevada,
      California, Washington, Oregon, Idaho, Montana, Wyoming,
      British Columbia, and Alberta. Elk can live either in mountains or
      plains.
   Behavior: Gregarious. Bulls collect a harem of cows and calves.
      Young nonbreeding bulls are tolerated in harem. Combat
      between mature bulls for control of harem can result in death.
      Summer-winter migration or nonmigratory.

HAZARDS
Severe winters, drowning, rutting combat

PREDATORS
Mountain lions (mostly on young), wolves, coyotes (mostly on young), bears.

DISEASES
Anthrax, anaplasmosis, brucellosis, tick-born fever, foot rot, eperythrozoonosis

RESOURCES
Food:
Winter—Mostly grasses and shrubs
Summer—Transitions from grasses to forbs
Water: Free water is needed
Management Practices:
Food and cover requirements and management practices vary according to habitat conditions that the local population has adapted to. It is not wise to use data from another area far removed from the local management situation until there has been an effort to validate it. Some general guidelines follow that may be applicable for local populations. In general, elk should be free from human disturbance; some recommendations are:
1. 1.6 km (1 mi) of road/2.58 km^2 (1 mi^2) of habitat for primitive type roads
2. 0.8 km (0.5 mi) of road/2.58 km^2 (1 mi^2) of habitat for secondary roads
3. 0.4 km (.25 mi) of road/2.58 km^2 (1 mi^2) of habitat for primary roads
Approximately 40% of the occupied habitat should be in the following cover classes:
Hiding—20%
Thermal—20%
Hiding cover is any vegetation capable of hiding 90% of a standing adult elk at 60 m (200 ft).
Thermal cover is a forest stand at least 12 m (40 ft) in height with tree canopy cover of at least 70%. This is achieved in many closed sapling-pole stands and by all older stands. The other 60% of the habitat can consist of openings of 12–16 ha (30–40 ac) or distances across an opening of 365 m (1200 ft).
Water sources need to be no more than 1.6–2.4 km (1–1.5 mi) apart for maximum habitat use.
Space—In general, depending on habitat quality, a small herd (30–50) of elk requires approximately 400 ha (1,000 ac) each of winter or summer habitat.

HUMANS

Disturbance by humans is a major management problem in some areas.

MAJOR REFERENCES

Severson, Kieth E. and Alvin L. Medina. 1983. Deer and elk management in the southwest. *J. Range Manage.*, Monogr. No. 2., Soc. for Range Manage., Denver, CO.

Thomas, J. W., and D. E. Toweill (eds.). 1982. *Elk of North America: Ecology and Management*. Wild. Manage. Inst. Stackpole Books, Harrisburg, PA.

Appendix E

COMMON AND SCIENTIFIC NAMES
OF ANIMALS AND PLANTS

Includes names not mentioned in text which are provided for reference purposes since many of them occur in the seven forest regions.
* Indicates species described by a summary life history in Appendix C.

Common Name	Scientific Name
Amphibians	
American toad	*Bufo americanus*
Appalachian woodland salamander	*Plethodon jordani*
Arboreal salamander	*Aneides lugubris*
California newt	*Taricha torosa*
Carpenter frog	*Rana virgatipes*
Cascades frog	*Rana cascadae*
Clouded salamander	*Aneides ferreus*
Common gray treefrog	*Hyla versicolor*
Eastern newt	*Notophthalmus viridescens*
Foothill yellow-legged frog	*Rana boylei*
Green treefrog	*Hyla cinerea*
Jemez Mountains salamander	*Plethodon neomexicanus*
Long-toed salamander	*Ambystoma macrodactylum*
Marbled salamander	*Ambystoma opacum*
Mountain dusky salamander	*Desmognathus ochrophaeus*
Mountain yellow-legged frog	*Rana muscosa*
*Northern leopard frog	*Rana pipiens*
Olympic salamander	*Rhyacotriton olympicus*
Oregon ensatina	*Ensatina eschscholtzi*
Pacific giant salamander	*Dicamptodon ensatus*
Pacific treefrog	*Hyla regilla*
Pickerel frog	*Rana palustris*
Pine woods treefrog	*Hyla femoralis*
Red-legged frog	*Rana aurora*
Rough-skinned newt	*Taricha granulosa*
Southern chorous frog	*Pseudoacris nigrita*
Southern leopard frog	*Rana sphenocephala*

Common Name	Scientific Name
Southern toad	*Bufo terrestris*
Spotted frog	*Rana pretiosa*
Spring peeper	*Hyla crucifer*
Tailed frog	*Ascaphus truei*
*Tiger salamander	*Ambystoma tigrinum*
*Western toad	*Bufo boreas*
Wood frog	*Rana sylvatica*

Birds

Acadian flycatcher	*Empidonax virescens*
Acorn woodpecker	*Melanerpes formicivorus*
American crow	*Corvus brachyrhynchos*
*American dipper	*Cinclus mexicanus*
American redstart	*Setophaga ruticilla*
*American robin	*Turdus migratorius*
American swallow-tailed kite	*Elanoides forficatus*
American woodcock	*Scolopax minor*
Anhinga	*Anhinga anhinga*
Anna's hummingbird	*Calypte anna*
Ash-throated flycatcher	*Myiarchus cinerascens*
*Bald eagle	*Haliaeetus leucocephalus*
Band-tailed pigeon	*Columba fasciata*
Barn swallow	*Hirundo rustica*
Barred owl	*Strix varia*
Bay-breasted warbler	*Dendroica castanea*
*Bewick's wren	*Thryomanes bewickii*
Black-and-white warbler	*Mniotilta varia*
Black-backed woodpecker	*Picoides arcticus*
Black-billed cuckoo	*Coccyzus erythropthalmus*
*Black-capped chickadee	*Parus atricapillus*
*Black-chinned hummingbird	*Archilochus alexandri*
Black-headed grosbeak	*Pheucticus melanocephalus*
Black-throated blue warbler	*Dendroica caerulescens*
Black-throated gray warbler	*Dendroica nigrescens*
Black-throated green warbler	*Dendroica virens*
Blackburnian warbler	*Dendroica fusca*
Blackpoll warbler	*Dendroica striata*
*Blue grouse	*Dendragapus obscurus*
Blue jay	*Cyanocitta cristata*
*Blue-gray gnatcatcher	*Polioptila caerulea*
Boreal chickadee	*Parus hudsonicus*
Bridled titmouse	*Parus wollweberi*
Broad-tailed hummingbird	*Selasphorus platycercus*
Broad-winged hawk	*Buteo platypterus*
*Brown creeper	*Certhia americana*

Common Name	Scientific Name
Brown-crested flycatcher	*Myiarchus tyrannulus*
Brown thrasher	*Toxostoma rufum*
Brown towhee	*Pipilo fuscus*
Brown-headed nuthatch	*Sitta pusilla*
Bushtit	*Psaltriparus minimus*
California quail	*Callipepla californica*
Calliope hummingbird	*Stellula calliope*
Canada warbler	*Wilsonia canadensis*
Cape May warbler	*Dendroica tigrina*
Carolina chickadee	*Parus carolinensis*
Carolina wren	*Thryothorus ludovicianus*
Cassins's finch	*Carpodacus cassinii*
Cedar waxwing	*Bombycilla cedrorum*
Chestnut-backed chickadee	*Parus rufescens*
Chestnut-sided warbler	*Dendroica pensylvanica*
Chimney swift	*Chaetura pelagica*
*Chipping sparrow	*Spizella passerina*
Chuck-will's-widow	*Caprimulgus carolinensis*
Clark's nutcracker	*Nucifraga columbiana*
Common barn-owl	*Tyto alba*
Common loon	*Gavia immer*
*Common raven	*Corvus corax*
*Cooper's hawk	*Accipiter cooperii*
*Dark-eyed junco	*Junco hyemalis*
*Downy woodpecker	*Picoides pubescens*
Dusky-capped flycatcher	*Myiarchus tuberculifer*
Eastern screech-owl	*Otus asio*
Eastern wood pewee	*Contopus virens*
Elegant trogon	*Trogon elegans*
*Evening grosbeak	*Coccothraustes vespertinus*
Flammulated owl	*Otus flammeolus*
*Fox sparrow	*Passerella iliaca*
Gila woodpecker	*Melanerpes uropygialis*
*Golden-crowned kinglet	*Regulus satrapa*
Golden-fronted woodpecker	*Melanerpes aurifrons*
Grace's warbler	*Dendroica graciae*
Gray jay	*Perisoreus canadensis*
Great crested flycatcher	*Myiarchus crinitus*
Great gray owl	*Strix nebulosa*
*Great horned owl	*Bubo virginianus*
Green-tailed towhee	*Pipilo chlorurus*
*Hairy woodpecker	*Picoides villosus*
*Hermit thrush	*Catharus guttatus*
Hermit warbler	*Dendroica occidentalis*
Hooded warbler	*Wilsonia citrina*
House wren	*Troglodytes aedon*
Kentucky warbler	*Oporornis formosus*

Common Name	Scientific Name
Kirtland's warbler	*Dendroica kirtlandii*
Ladder-backed woodpecker	*Picoides scalaris*
Lazuli bunting	*Passerina amoena*
Lewis' woodpecker	*Melanerpes lewis*
*Lincoln's sparrow	*Melospiza lincolnii*
*Long-eared owl	*Asio otus*
MacGillivray's warbler	*Oporornis tolmiei*
Magnolia warbler	*Dendroica magnolia*
Merlin	*Falco columbarius*
Mississippi kite	*Ictinia mississippiensis*
*Mountain bluebird	*Sialia currucoides*
Mountain chickadee	*Parus gambeli*
Mountain quail	*Oreortyx pictus*
*Mourning dove	*Zenaida macroura*
Nashville warbler	*Vermivora ruficapilla*
*Northern cardinal	*Cardinalis cardinalis*
*Northern flicker	*Colaptes auratus*
*Northern goshawk	*Accipiter gentilis*
*Northern oriole	*Icterus galbula*
Northern parula	*Parula americana*
Northern pygmy owl	*Glaucidium gnoma*
*Northern saw-whet owl	*Aegolius acadicus*
*Northern shrike	*Lanius excubitor*
Nuttal's woodpecker	*Picoides nuttallii*
*Olive-sided flycatcher	*Contopus borealis*
Orange-crowned warbler	*Vermivora celata*
Ovenbird	*Seiurus aurocapillus*
Painted bunting	*Passerina ciris*
*Pileated woodpecker	*Dryocopus pileatus*
*Pine grossbeak	*Pinicola enucleator*
*Pine siskin	*Carduelis pinus*
Pine warbler	*Dendroica pinus*
Pinyon jay	*Gymnorhinus cyanocephalus*
Plain titmouse	*Parus inornatus*
Prairie warbler	*Dendroica discolor*
Prothonotary warbler	*Protonotaria citrea*
*Purple finch	*Carpodacus purpureus*
Pygmy nuthatch	*Sitta pygmaea*
*Red crossbill	*Loxia curvirostra*
Red-bellied woodpecker	*Melanerpes carolinus*
*Red-breasted nuthatch	*Sitta canadensis*
Red-cockaded woodpecker	*Picoides borealis*
*Red-eyed vireo	*Vireo olivaceus*
Red-faced warbler	*Cardellina rubrifrons*
*Red-headed woodpecker	*Melanerpes erythroceohalus*
Red-shouldered hawk	*Buteo lineatus*
Red-tailed hawk	*Buteo jamaicensis*

Common Name	Scientific Name
Rose-breasted grosbeak	*Pheucticus ludovicianus*
*Ruby-crowned kinglet	*Regulus calendula*
Ruby-throated hummingbird	*Archilochus colubris*
*Ruffed grouse	*Bonasa umbellus*
Rufous hummingbird	*Selasphorus rufus*
*Rufous-sided towhee	*Pipilo erythrophthalmus*
Scarlet tanager	*Piranga olivacea*
Scrub jay	*Aphelocoma coerulescens*
Siberian tit	*Parus cinctus*
*Sharp-shinned hawk	*Accipiter striatus*
*Solitary vireo	*Vireo solitarius*
*Song sparrow	*Melospiza melodia*
Spotted owl	*Strix occidentalis*
*Spruce grouse	*Dendragapus canadensis*
*Steller's jay	*Cyanocitta stelleri*
Strickland's woodpecker	*Picoides stricklandi*
Summer tanager	*Piranga rubra*
Swainson's thrush	*Catharus ustulatus*
Tennessee warbler	*Vermivora peregrina*
*Three-toed woodpecker	*Picoides tridactylus*
Townsend's solitaire	*Myadestes townsendi*
Townsend's warbler	*Dendroica townsendi*
Tree swallow	*Tachycineta bicolor*
Tufted titmouse	*Parus bicolor*
Varied thrush	*Ixoreus naevius*
Vaux's swift	*Chaetura vauxi*
Veery	*Catharus fuscescens*
*Violet-green swallow	*Tachycineta thalassina*
Virginia's warbler	*Vermivora virginiae*
Warbling vireo	*Vireo gilvus*
Western bluebird	*Sialia mexicana*
*Western flycatcher	*Empidonax difficilis*
Western screech-owl	*Otus kennicottii*
Western tanager	*Piranga ludoviciana*
Western wood pewee	*Contopus sordidulus*
*Whip-poor-will	*Caprimulgus vociferus*
White-breasted nuthatch	*Sitta carolinensis*
*White-crowned sparrow	*Zonotrichia leucophrys*
White-eyed vireo	*Vireo griseus*
White-headed woodpecker	*Picoides albolarvatus*
White-throated sparrow	*Zonotrichia albicollis*
White-winged crossbill	*Loxia leucoptera*
*Wild turkey	*Meleagris gallopavo*
Williamson's sapsucker	*Sphyrapicus thyroideus*
*Wilson's warbler	*Wilsonia pusilla*
*Winter wren	*Troglodytes troglodytes*
*Wood duck	*Aix sponsa*

Common Name	Scientific Name
Wood thrush	*Hylocichla mustelina*
Worm-eating warbler	*Helmitheros vermivorus*
*Yellow-bellied sapsucker	*Sphyrapicus varius*
Yellow-breasted chat	*Icteria virens*
Yellow-rumped warbler	*Dendroica coronata*
Yellow-throated vireo	*Vireo flavifrons*
Yellow-throated warbler	*Dendroica dominica*

Fishes

*Brook trout	*Salvelinus fontinalis*
*Brown trout	*Salmo trutta*
*Cutthroat trout	*Salmo clarki*
*Largemouth bass	*Micropterus salmoides*
*Rainbow trout	*Salmo gairdneri*
*Smallmouth bass	*Micropterus dolomieui*

Mammals

*Abert's squirrel	*Sciurus aberti*
*Beaver	*Castor canadensis*
Bighorn sheep	*Ovis canadensis*
*Black bear	*Ursus americanus*
*Bobcat	*Felis rufus*
Brush rabbit	*Sylvilagus bachmani*
Bushy-tailed woodrat	*Neotoma cinerea*
Coati	*Nasua nasua*
Collared peccary	*Dicotyles tajacu*
Cotton mouse	*Peromyscus gossypinus*
*Coyote	*Canis latrans*
*Deer mouse	*Peromyscus maniculatus*
Douglas' squirrel	*Tamiasciurus douglasii*
Eastern chipmunk	*Tamias striatus*
*Eastern cottontail	*Sylvilagus floridanus*
Eastern spotted skunk	*Spilogale putorius*
Eastern woodrat	*Neotoma floridana*
*Elk	*Cervus elaphus*
Ermine	*Mustela erminea*
Fisher	*Martes pennanti*
Fox squirrel	*Sciurus niger*
*Gray fox	*Urocyon cinereoargenteus*
*Gray squirrel	*Sciurus carolinensis*
Gray wolf	*Canis lupus*
Gray-collared chipmunk	*Eutamias cinereicollis*
*Grizzly bear	*Ursus arctos*

Common Name	Scientific Name
*Hoary bat	*Lasiurus cinereus*
Hoary marmot	*Marmota caligata*
Least chipmunk	*Eutamias minimus*
*Long-tailed vole	*Microtus longicaudus*
*Long-tailed weasel	*Mustela frenata*
*Lynx	*Felis lynx*
Mantled ground squirrel	*Spermophilus lateralis*
*Marten	*Martes americana*
Mexican woodrat	*Neotoma mexicana*
*Mink	*Mustela vison*
*Moose	*Alces alces*
Mountain beaver	*Aplodontia rufa*
*Mountain lion	*Felis concolor*
*Mule deer	*Odocoileus hemionus*
New England cottontail	*Sylvilagus transitionalis*
Nine-banded armadillo	*Dasypus novemcinctus*
*Northern flying squirrel	*Glaucomys sabrinus*
Pika	*Ochotona princeps*
*Porcupine	*Erethizon dorsatum*
*Pygmy shrew	*Microsorex boyi*
*Raccoon	*Procyon lotor*
*Red fox	*Vulpes vulpes*
*Red squirrel	*Tamiasciurus hudsonicus*
Red tree vole	*Phenacomys longicaudus*
Ringtail	*Bassariscus astutus*
*River otter	*Lutra canadensis*
Short-tailed shrew	*Blarina brevicauda*
*Snowshoe hare	*Lupus americanus*
Southern flying squirrel	*Glaucomys volans*
*Striped skunk	*Mephitis mephitis*
Townsend's chipmunk	*Eutamias townsendii*
Uinta chipmunk	*Eutamias umbrinus*
Vagrant shrew	*Sorex vagrans*
*Virginia opossum	*Didelphis virginiana*
Western gray squirrel	*Sciurus griseus*
Western jumping mouse	*Zapus princeps*
White-footed mouse	*Peromyscus leucopus*
*White-tailed deer	*Odocoileus virginianus*
Wild boar	*Sus scrofa*
Wolverine	*Gulo gulo*
*Woodchuck	*Marmota monax*
Woodland jumping mouse	*Napaeozapus insignis*
Woodland vole	*Microtus pinetorum*
Yellow-bellied marmot	*Marmota flaviventris*
Yellow-pine chipmunk	*Eutamias amoenus*

Common Name	Scientific Name

Reptiles

Common Name	Scientific Name
Broadheaded skink	*Eumeces laticeps*
California mountain kingsnake	*Lampropeltis zonata*
*Coachwhip	*Masticophis flagellum*
Coal skink	*Eumeces anthracinus*
*Common garter snake	*Thamnophis sirtalis*
*Common kingsnake	*Lampropeltis getulus*
Copperhead	*Agkistrodon contortrix*
Corn snake	*Elaphe guttata*
Eastern box turtle	*Terrapene carolina*
Eastern coral snake	*Micrurus fulvius*
Eastern diamondback rattlesnake	*Crotalus adamanteus*
Five-lined skink	*Eumeces fasciatus*
*Gopher snake	*Pituophis melanoleucus*
Green anole	*Anolis carolinensis*
Ground skink	*Scincella lateralis*
*Milk snake	*Lampropeltis triangulum*
Mole skink	*Eumeces egregius*
Northern alligator lizard	*Gerrhonotus coeruleus*
*Painted turtle	*Chrysemys picta*
Pigmy rattlesnake	*Sistrurus miliarius*
Pine woods snake	*Rhadinaea flavilata*
Plateau striped whiptail	*Cnemidophorus velox*
*Racer	*Coluber constrictor*
Rat snake	*Elaphe obsoleta*
*Ringneck snake	*Diadophis punctatus*
Rough green snake	*Opheodrys aestivus*
Rubber boa	*Charina bottae*
Sharp-tailed snake	*Contia tenuis*
Short-horned lizard	*Phrynosoma douglassii*
Sonoran mountain kingsnake	*Lampropeltis pyromelana*
Striped racer	*Masticophis lateralis*
Striped whipsnake	*Masticophis taeniatus*
*Timber rattlesnake	*Crotalus horridus*
Western diamondback	*Crotalus atrox*
Western fence lizard	*Sceloporus occidentalis*
*Western rattlesnake	*Crotalus viridis*
Western skink	*Eumeces skiltonianus*
Western terrestrial garter snake	*Thamnophis elegans*
Wood turtle	*Clemmys insculpta*

Common Name	Scientific Name
Trees	
Alaska-cedar	*Chamaecyparis nootkatensis*
Alder	*Alnus* spp.
Alligator juniper	*Juniperus deppeana*
American beech	*Fagus grandifolia*
American elm	*Ulmus americana*
Ash	*Fraxinus* spp.
Aspen	*Populus* spp.
Baldcypress	*Taxodium distichum*
Balsam fir	*Abies balsamea*
Balsam poplar	*Populus balsamifera*
Beech	*Fagus* spp.
Bigleaf maple	*Acer macrophyllum*
Bigtooth aspen	*Populus grandidentata*
Birch	*Betula* spp.
Bitternut hickory	*Carya cordiformis*
Black cherry	*Prunus serotina*
Black cottonwood	*Populus trichocarpa*
Black oak	*Quercus velutina*
Black willow	*Salix nigra*
Blackgum	*Nyssa sylvatica*
Blue spruce	*Picea pungens*
California red fir	*Abies magnifica*
Chestnut oak	*Quercus prinus*
Cottonwood	*Populus* spp.
Douglas-fir	*Pseudotsuga menziesii*
Eastern hemlock	*Tsuga canadensis*
Eastern white pine	*Pinus strobus*
Elm	*Ulmus* spp.
Engelmann spruce	*Picea engelmannii*
Grand fir	*Abies grandis*
Gray birch	*Betua populifolia*
Green ash	*Fraxinus pennsylvanica*
Hickory	*Carya* spp.
Incense-cedar	*Libocedrus decurrens*
Jack pine	*Pinus banksiana*
Jeffrey pine	*Pinus jeffreyi*
Loblolly pine	*Pinus taeda*
Lodgepole pine	*Pinus contorta*
Longleaf pine	*Pinus palustris*
Maple	*Acer* spp.
Mockernut hickory	*Carya tomentosa*
Mountain hemlock	*Tsuga mertensiana*
Northern red oak	*Quercus rubra*
Northern white-cedar	*Thuja occidentalis*
Nutall oak	*Quercus nuttallii*

Common Name	Scientific Name
Oak	*Quercus* spp.
Oneseed juniper	*Juniperus monosperma*
Overcup oak	*Quercus lyrata*
Pacific yew	*Taxus brevifolia*
Paper birch	*Betula papyrifera*
Pignut hickory	*Carya glabra*
Pinyon pine	*Pinus edulis*
Ponderosa pine	*Pinus ponderosa*
Poplar	*Populus* spp.
Quaking aspen	*Populus tremuloides*
Red alder	*Alnus rubra*
Red maple	*Acer rubrum*
Red pine	*Pinus resinosa*
Red spruce	*Picea rubens*
Redwood	*Sequoia sempervirens*
Scarlet oak	*Quercus coccinea*
Shagbark hickory	*Carya ovata*
Shortleaf pine	*Pinus echinata*
Shumard oak	*Quercus schumardii*
Silver fir	*Abies amabalis*
Singleleaf pinyon	*Pinus monophylla*
Sitka spruce	*Picea sitchensis*
Slash pine	*Pinus elliottii*
Southern red oak	*Quercus falcata*
Subalpine fir	*Abies lasiocarpa*
Sugar maple	*Acer saccharum*
Sugar pine	*Pinus labertiana*
Sweet birch	*Betual lenta*
Sweetgum	*Liquidambar styraciflua*
Sycamore	*Platanus* spp.
Tamarack	*Larix laricina*
Utah juniper	*Juniperus osteosperma*
Virginia pine	*Pinus virginiana*
Western hemlock	*Tsuga heterophylla*
Western larch	*Larix occidentalis*
Western redcedar	*Thuja plicata*
Western white pine	*Pinus monticola*
White ash	*Fraxinus americana*
White fir	*Abies concolor*
White oak	*Quercus alba*
White spruce	*Picea glauca*
Willow	*Salix* spp.
Yellow birch	*Betula alleghaniensis*

Common Name	Scientific Name
Shrubs	
Apache-plume	*Fallugia*
Azalea	*Rhododendron*
Barberry	*Berberis*
Bearmat	*Chamaebatia*
Blueberry	*Vaccinium*
Buckthorn	*Rhamnus*
Bush-honeysuckle	*Diervilla*
Buttonbrush	*Cephalanthus*
Calliandra	*Calliandra*
Ceanothus	*Ceanothus*
Cherry	*Prunus*
Chokecherry	*Aronia*
Cinquefoil	*Potentilla*
Cliffrose	*Cowania*
Condalia	*Condalia*
Cyrilla	*Cyrilla*
Dalea	*Dalea*
Elder	*Sambucus*
Euonymus	*Euonymus*
Gooseberry	*Ribes*
Hackberry	*Celtis*
Hawthorn	*Crataegus*
Hazelnut	*Corylus*
Holly	*Ilex*
Huckelberry	*Gaylussacia*
Juniper	*Juniperus*
Kalmia	*Kalmia*
Manzanita	*Arctostaphylos*
Menziesia	*Menziesia*
Mountainmahogany	*Cercocarpus*
Oak	*Quercus*
Palmetto	*Sabal*
Paperflower	*Psilostrophe*
Porliera	*Porliera*
Rabbitbrush	*Chrysothamnus*
Raspberry	*Rubus*
Rose	*Rosa*
Sagebrush	*Artemesia*
Saw-palmetto	*Serenoa*
Serviceberry	*Amelanchier*
Silktassel	*Garrya*
Snowberry	*Symphoricarpos*
Sumac	*Rhus*
Viburnum	*Viburnum*
Waxmyrtle	*Myrica*

Common Name	Scientific Name
Willow	*Salix*
Winterfat	*Eurotia*
Witchhazel	*Hamamelis*
Wolfberry	*Lycium*

Forbs

Agoseris	*Agoseris*
Alumroot	*Heuchera*
Arnica	*Arnica*
Balsamroot	*Balsamorhiza*
Brodiaea	*Dichelostemma*
Bundleflower	*Desmanthus*
Buttercup	*Ranunculus*
Cinquefoil	*Potentilla*
Clover	*Trifolium*
Croton	*Croton*
Dandelion	*Taraxacum*
Deervetch	*Lotus*
Dock	*Rumex*
Eriophyllum	*Eriophyllum*
Eupatorium	*Eupatorium*
Filaree	*Erodium*
Fleabane	*Erigeron*
Galax	*Galax*
Goldenrod	*Solidago*
Groundcherry	*Physalis*
Groundsmoke	*Gayophytum*
Hawkweed	*Hieracium*
Hollyfern	*Polystichum*
Knotweed	*Polygonum*
Lespedeza	*Lespedeza*
Lettuce	*Lactuca*
Lupine	*Lupinus*
Marshpurslane	*Ludwigia*
Medic	*Medicago*
Oxalis	*Oxalis*
Parthenium	*Parthenium*
Partridgeberry	*Mitchella*
Peavine	*Lathyrus*
Penstemon	*Penstemon*
Plantain	*Plantago*
Pussytoes	*Antennaria*
Ragweed	*Ambrosia*
Royal fern	*Osmunda*
Senna	*Cassia*

Common Name	Scientific Name
Snapweed	*Impatiens*
Strawberry	*Fragaria*
Sweetclover	*Melilotus*
Tansymustard	*Descurainia*
Thistle	*Cirsium*
Tickclover	*Desmodium*
Trailing-arbutus	*Epigaea*
Turkey-mullein	*Eremocarpus*
Vetch	*Vicia*
Violet	*Viola*
Waterlily	*Nymphaea*
Wild-buckwheat	*Eriogonum*
Willow-herb	*Epilobium*
Wintergreen	*Gaultheria*
Wyethia	*Wyethia*
Yarrow	*Achillea*

Grasses

Bluegrass	*Poa*
Bristlegrass	*Setaria*
Brome	*Bromus*
Cane	*Arundinaria*
Cupscale	*Sacciolepis*
Cutgrass	*Leersia*
Fescue	*Festuca*
Junegrass	*Koleria*
Muhly	*Muhlenbergia*
Needlegrass	*Stipa*
Panicum	*Panicum*
Paspalum	*Paspalum*
Saltgrass	*Distichlis*
Squirreltail	*Sitanion*
Wheatgrass	*Agropyron*
Wildrye	*Elymus*

Grasslike

Bulrush	*Scirpus*
Rush	*Juncus*
Sedge	*Carex*
Spikerush	*Eleocharis*

GLOSSARY

Note: Some of the definitions in this glossary are in the formulation stage because the ideas and concepts are new and still developing. These definitions will become more exact as experience accumulates with their application.

Age specific life table. Also called dynamic life table. A life table that follows a cohort of animals through survival and mortality until the last member has died.

Allogenic. Describes changes within an ecosystem caused from external forces such as fire, wind, drought, and so forth.

Allotment. A bounded unit of land on which livestock grazing is permitted.

Alpha diversity. See species richness.

Alpha species. A habitat specialist. A species that is generally restricted to an identifiable vegetation type, is small in size with a small home range, and a high reproductive rate.

Animal unit. The weight (1000 lbs, 455 kg) of one mature cow with a calf.

Animal unit month. The amount of forage required for one animal unit for one month (30 days).

Area regulation. A method for regulating a forest in which trees in an area of designated size are cut each year or other time period, but the volume is variable between areas.

Assessment. A determination and evaluation of the facts about a process or system.

Attribute. A fact about an inherent physical characteristic, property, measurement, or quality of an animal, plant, or habitat.

Autogenic. Describes changes within an ecosystem caused by the interations of plants and animals.

Bag limit. The number of animals that can legally be harvested by hunting in one day, month, or year.

Basal area. The cross-sectional area of a tree measured at breast height. Expressed in cubic meters per hectare or square feet per acre.

Beta diversity. Diversity between habitats or communities.

Biological diversity. The distribution and abundance of different plant and animal species and communities within a given area.

Bio-politics. The mixing of biology and politics in the decision-making process.

Biosphere reserve. A system of worldwide reserves set aside to preserve biomes on different continents.

Biotic potential. The theoretical number of young a female can produce under optimal conditions.

Boolean logic. Refers to logical operators of equal to, greater than, less than, and so forth.

Breeding season. The time of year when animals mate.

Browse. Woody plants eaten by ungulate species (e.g., deer, elk, cattle, and so forth).

Carnivore. An animals that eats meat.

Carrying capacity (K). The maximum number of a given species of animal that a habitat can support without damage to soil and vegetation resources.

Census. A complete count of a species in a given area.

Center of activity. An area within a home range in which an animal spends the majority of its time.

Centrum. The six factors—hazards, predators, diseases, genetics, resources, and humans—that directly affect an animal's chance to survive and reproduce.

Cienega. A meadow where moisture is constantly at or near the soil surface.

CITIES. Convention on International Trade In Endangered Species.

Clearcutting. The harvesting of all standing trees in a given area at the same time.

Climax. The last seral stage in succession.

Cohort. A group of individuals born into a population at the same time.

Commercial forest. A forest capable of producing at least 20 cubic feet per year (0.56 cubic meters per ha) per acre.

Commercial thinning. Harvesting selected usable trees before the final or regeneration cut.

Community. Two or more populations of plants or animals interacting in the same habitat.

Compensatory mortality. Another factor (hunting) is a replacement for the natural mortality caused by predators, disease, and so forth.

Competitive exclusion. No two animals can occupy the exact same niche at the same time.

Composite life table. A life table developed from mark-recapture data for two or more periods of time.

Concept. The expression of an idea in an abstract form.

Conceptual model. A model that expresses ideas in a qualitative manner.

Condition. An expression of the "health" of vegetation, particularly on rangelands used by grazing animals.

Constant. Describes an ecosystem that over time maintains the same density of plants or animals within a species.

Contagious disease. A disease that can be transmitted from one animal to another either by direct or indirect means.

Continuous grazing. The year-long use of an area by livestock in consecutive years.

Coordination measure. A wildlife habitat improvement practice accomplished by combining the work with other resource functions of range, water, timber, and so forth.

Core table. A table in a relational database that is linked to all the other tables. Data in the core table, except the linking column, is not replicated in any of the other tables.

Cover. The physical habitat component or landscape feature that provides protection from hazards, predators, and humans.

Crude birth rate. The number of individuals born per unit of population (100, 1000).

Crude density. The number of animals per unit of area regardless of what cultural features, such as roads or buildings, the area might contain that are not suitable habitat for the species.

Cruising radius. The distance an animal travels in one day in search of food and cover.

Cutting cycle. The interval of time between the harvesting of trees in an uneven-aged stand.

Cycle. A type of fluctuation in animal numbers with a predictable pattern of peaks and troughs separated by several years.

Database. A collection of organized data.

Database management system (DBMS). A system for organizing data in a database. Can be a hierarchical, network, or relational system.

Decimating factor. A factor in the centrum of an aniaml, that can kill directly, such as a predator, disease, or hazard.

Decision support system (DDS). A computerized analytical tool linking a variety of different resource modules to provide management alternatives in a systematic planning structure.

Decision tree. A type of flowchart model with statements leading to a decision.

Decreaser plant. A plant that decreases in number as a result of grazing pressure.

Deferred grazing. The delaying of grazing for a period of time during seasons when herbaceous growth is highest.

Delta notation. The use of a delta symbol to indicate change in number.

Demographic vigor. Relates to population vital statistics and dynamics expressed as a factor of population growth.

Density. In population, the number of organisms per unit area. Can also refer to the compactness of (weight) of water.

Density dependent factor. A factor that affects a population's ability to increase and is itself a result of an increasing density of animals.

Density independent factor. A factor that affects a population's ability to increase regardless of the density of the population.

Dependency (database). A set of data that depends on another set to fully describe its characteristics.

Detail. The degree of exactness needed by a habitat model for measurement of input factors.

Deterministic model. A quantitative empirical model in which relationships are fixed and there can be only one outcome.

Disease. An impairment of the normal state of a living animal or plant that affects the performance of the vital functions.

Distribution (wildlife). The spatial pattern of animals in an area (e.g., homogeneous, random, clumped, and so forth).

Double clutching. The removal of a clutch of eggs from a nest to fool the female into laying another clutch.

Drift fence. A fence constructed so as to collect snow behind the fence.

Dynamic equilibrium. The point at which a balance between plant and animal density is reached over time. Can also refer to carrying capacity, river continuum, or other processes involving a feedback mechanism.

Ecological amplitude. The amount of tolerance an organism has to environmental conditions.

Ecological carrying capacity. The equilibrium reached between plant growth and animal density in the absence of hunting.

Ecological density. The number of animals per unit of area, where the area is considered as suitable habitat for a particular species.

Ecological dominant. An organism which by its presence can alter the environment of other species.

Ecological longevity. The actual or realized life span of an organism in the wild.

Ecological niche. The way an animal uses its habitat and the special adaptations, either behavioral or physical, that facilitate this use.

Ecological time. The time commonly used in describing plant community replacement as in succession.

Economic carrying capacity. The equilibrium reached between plant growth and animal density when hunting is a factor. It is below ecological carrying capacity. Carrying capacity is determined by quantifying the density of animals desired to meet certain objectives.

Ecosystem. A defined unit of orgnisms interacting with the physical environment so that a flow of energy leads to a clearly defined trophic structure and the cycling of nutrients.

Ecosystem functioning. The process of energy flow and nutrient cycling in a ecosystemm.

Ecosystem structure. The physical characteristics of an ecosystem, such as soil, topography, or plants.

Edge effect. The boundary between two distinct vegetation life-forms or structures within life-forms.

Emigration. The act of animals leaving one population to join another.

Empirical model. A working model which uses either qualitative or quantitative data.

Environment. All the factors that affect an animal's chance to survive and reproduce.

Environmental dendrogram (species dendrogram). A systematic listing of the factors in the centrum and web of an animal ordered by their level of influence in the web.

Environmental influences. A factor, such as weather or succession, that alters wildlife welfare and/or decimating factors.

Environmental resistance. The factor or factors that limit biotic potential.

Epilimnion. The surface layer of a stratified lake. This layer is warmed by sunlight, has plenty of food and oxygen, and circulates but does not mix with lower layers.

Eutrophic. Describes a lake that has high levels of nutrients.

Eutrophy. Enrichment with nutrients in a lake.

Even-aged management. A management system designed to perpetuate even-aged stands.

Even-aged stand. A stand of trees in which all are in the same age class.

Exotic. Not native to an area.

Expert system model. A model based on the individual or collective experience of knowledgeable people.

Exponential growth model. See uninhibited growth model.

Extirpated. Driven out or eliminated from an area.

Fall overturn. A mixing of the layers in a stratified lake as a result of wave action and a change in water temperature in the fall.

Featured species. The one species selected for an area and for which management efforts are concentrated to fulfill its needs through coordination of timber and wildlife habitat activities.

Fecundity. Fertility. The number of offspring produced per female per year.

Field data recorder. An electronic device or small computer for storing field data.

Finite rate of increase. The rate of change (lambda) of population size which is the ratio of the total number of animals in one year to the total humber of animals the next year.

First law of thermodynamics. When energy is changed from one form to another, no energy is created or destroyed.

Fluctuate. To change in quantity over a short time.

Food availability. The food that is available during the various seasons. Also relates to the physical location of food.

Food chain. A path through which energy flows from plants to animals.

Food palatability. The acceptance of food by an animal. Relates to nutritive content and digestibility.

Food preference. The hierarchical ordering of food by frequency of use.

Food web. A series of interconnecting food chains.

Forage. Vegetative material that is eaten by a specific animal.

Forest. A vegetation community made up of more or less dense and extensive tree cover. An area whose dominant life-form is trees.

Forest commons. An area in a forest with certain characteristics that is used by many wildlife species.

Forest cover type. An ecological classification for vegetation expressed by a dominant life-form such as species of trees.

Forest region. A classification that groups forest cover types into broad geographical areas.

Forest regulation. The process of bringing a forest production and harvesting into balance to provide a sustained yield of wood volume.

Forest wildlife. All the animal species that depend on forests and trees for food and cover.

Forestry. The management of all resources (range, water, wildlife, timber, recreation, and so forth) on lands classified as forests.

Forb. A low herbaceous lant with broad leaves.

FORPLAN. An acronym for forest planning.

Foster parent. A substitute parent.

Function. A statement of dependency.

Game ranching. An area where wild animals are raised in a given area primarily for food.

Gamma diversity. Diversity across landscapes separated by geographical areas.

Gamma species. A habitat generalist. A species that can use many vegetation types, is generally large in size, with a large home range, and a low reproductive rate.

Geographic information system (GIS). A computerized means of making base maps with resource overlays.

Geographical range. The total area in which a species exists, including occupied and unoccupied habitats.

Geological time. The tens and hundreds of thousands of years required to effect regional and global change in topography and climate.

Gestation period. The length of time a female is pregnant.

Goal. A broad statement of central themes setting management direction.

Grazing capacity. The number of livestock that a given area can support.

Guild. A group of species exploiting the same class of resources in a similar way.

Habitat. The environment of and the specific place where an organism lives.

Habitat evaluation. An assessment of the "health" of habitat factors as they affect a single species or a group of species using the same resources.

Habitat evaluation procedure (HEP). The specific procedures used in habitat evaluation.

Habitat extent (size). The size of the combined area needed to maintain viable populations of native species when managed as a group.

Habitat index. A quantitative value to express habit evaluation. A type of habitat model.

Habitat model. A model based on habitat factors and how they relate to population density.

Habitat suitability index (HSI). A quantitative value that expresses how suitable a habitat is for a certain species or group of species. A type of habitat model.

Habitat unit. An abstract unit containing food and cover for a particular species.

Harvesting. The removal of animals from a population by hunting.

Hazard. Anything, other than disease or a man-made object, that is capable of causing injury or death to an animal.

Hedging. A "club" growth form from the effect of overuse of woody plants.

Herbage. All species of browse, forbs, and grasses produced in an area.

Herbicide. A chemical compound specifically formulated to kill plants.

Herbivore. An animal that eats plant material only.

Hierarchical dominance. A "pecking order" within a social group of animals.

Hierarchy. An ordering of factors according to some criteria.

Home range. The area in which an animal normally confines itself to obtain food and cover.

Horizontal diversity. Diversity that results from the arrangement of plant life-forms in different successional stages.

Hunting preserve. An area set aside on private land solely for sport hunting for a fee.

Hypolimnion. The lowest and coldest layer of a stratified lake.

Immigration. The act of animals joining a population from another population.

Increaser plant. A plant that increases in number with increased grazing pressure because other more desirable plants are consumed first.

Indicator (ecological). A plant or animal so closely associated with a particular environmental factor that the presence of the species is clearly indicative of the presence or absence of the factor.

Induced edge. An edge resulting from management practices such as fire, timber harvesting, grazing, seeding, or planting.

Infectious disease. A disease that is capable of invading body tissue.

Inflection point. The point on an inhibited growth curve where the population rate of increase starts to decrease.

Inherent edge. An edge resulting from the natural outcome of plant succession.

Inhibited growth model. The reduction of the uninhibited growth rate due to the directly acting factors of the centrum that result in an S-shaped curve.

Insecticide. A chemical compound specifically formulated to kill insects.

Instantaneous time. The time dealing with short rates of change, such as in population dynamics.

Integrated resource management. A management system that recognizes that all natural resources are connected by an intricate web of relationships and that the effects of a management activity on one resource will have some effect on the others.

Intermediate. Describes a tree that is between tolerant and intolerant to shade.

Intersect. A relational command used to create a new table from two other tables in a relational database.

Interspersion. The distribution of habitat units for a species over the landscape.

Intolerante. Describes a tree that cannot grow in shade.

Invader plant. A plant that originally was not part of the vegetation complex but becomes established as a result of grazing pressure.

Irruption. A type of fluctuation in animal numbers that is marked by a sharp increase in a short time followed by a population crash.

Isogram. A line connecting points of equal value.

Join. A relational command that combines two tables in a relational database. Similar to intersect but provides different combinations of data.

Juxtaposition. The location of food and cover adjacent to or in close proximity to one another.

Key column. A linking column in a relational database.

K-selection. Refers to the ability to maintain a population that fluctuates around a carrying capacity (K).

Law of dispersion. The potential density of game of low radius requiring two or more types is, within ordinary limits, proportional to the sum of type peripheries.

Law of the minimum. The resource factor in least supply is the factor limiting wildlife population growth.

Law of tolerance. Any factor that exceeds the limits of tolerance for a particular species becomes a limiting factor for that species.

Lentic. A type of freshwater habitat characterized by calm water.

Life equation. All the events that happen in a population over a one-year period.

Life-form. In management systems, a single category for a group of species that synthesizes information on where that group feeds or breeds. In taxonomy, life-form refers to the grouping of organisms by their common characteristics such as trees, shrubs, grass, fish, and birds.

Life history. All the events that happen from birth to death of an individual.

Life table. A compilation of demographic facts for a cohort of animals from birth to death.

Limiting factor. Any environmental factor in the centrum of an animal that limits the growth of a wildlife population (e.g., food, water, cover, etc.).

Limnetic zone. The zone of light penetration in a lake.

Lincoln index. A population estimation technique based on a ratio of the number of marked to unmarked animals in an area. Also known as Peterson index.

Linking column. A column with the same name and data type in two different tables in a relational database.

Littoral zone. The zone of a lake containing rooted plants. It is part of the limnetic zone.

Local population. A population of animals generally adapted to a particular area.

Logical domain. The meaningful information about the subject of an attribute.

Lotic. A type of freshwater habitat characterized by running water (e.g., springs, streams, rivers).

Management by objective (MBO). The setting of a course of action to achieve a stated management objective.

Management indicator species (MIS). A species selected to set management direction.

Man-made hazard. Anything made by humans (e.g., fences, roads, buildings) that is capable of causing injury or death to an animal.

Map scale. The ratio of actual distance on the ground to the same distance on a map.

Maximum sustained yield (MSY). Reducing a population to the inflection point where the maximum growth rate is achieved.

Migration. The seasonal movement of animals over a long distance.

Mobility. Refers to the potential of a species for travel, generally determined by the animal's home range and cruising radius.

Model. An abstraction of a real world situation.

Monitoring. The process of checking, observing, or measuring the outcome of a process or the changes deviating from some specified quantity.

Monogamous. Refers to a species in which a single male and a single female pair for the breeding season.

Mortality. The number of individuals in a population dying during a given period of time, usually a year.

Multiresource inventory. An inventory that includes many resources at the same time (e.g., vegetation measurements that pertain to timber, range, and wildilfe).

Multispecies management. The management of more than one species at the same time on the same area.

Natality. The number of new individuals added per year to a population.

Natural hazard. Anything not of human origin that is capable of causing injury or death to an animal, including wind, water, temperature, and topographic features.

Natural population. The set of all local populations that are not separated by a geographical or ecological barrier that would prevent breeding.

Near natural language. The means by which computers discriminate between words by using synonyms.

Nested plots. The placing of smaller plots inside larger plots to sample different plant and animal attributes.

Net reproduction rate (R). The average number of young born per female in each age group.

Nonadjacency constraint (NAC). A restriction that adjacent stands must be of different age or size classes. Used to regulate harvesting.

Objective. A specific statement of an activity to be accomplished to realize the goals of a management plant.

Old-growth forest. A forest containing large, mature trees mostly exceeding rotation age, with some dead trees and snags, and an understory containing decomposing logs and litter.

Obligatory relationship. A dependency of one organism on another organism or on a set of environmental conditions.

Oligotropic. Describes a lake that is low in nutrients.

Operational plan. A plan that guides specific actions and allocation of resources in a short time frme, such as one year or less.

Pathogen. A disease-producing agent.

Pattern recognition model (PATREC). A model based on the probabilities of occupied and unoccupied habitat according to predefined characteristics.

Persistent. Describes an ecosystem that maintains the same number of species over time.

Pesticide. A chemical compound used to control plants and animals classified as pests.

Peterson index. See Lincoln index.

Physical domain. The type of data (e.g., text, real, data) describing an attribute in a database.

Physiological longevity. The maximum life span of an organism under optimum living conditions.

Pioneering population. An initial population of animals introduced into unoccupied habitat.

Plantation effect. The result of evenly spaced trees in a forest.

Poaching. The illegal killing of wildlife.

Policy. The setting of a course of action.

Polygamous. Refers to a species in which one male or female breeds with more than one female or male.

Population. A number of animals of the same species inhabiting the same area.

Predator. An animal that kills and feeds on other animals.

Present time. The "here" and "now" time frame used by managers in the short run of 1–5 years.

Primary habitat. All the environmental factors in the centrum of an animal that are necessary to support a viable population.

Primary succession. Succession starting from bare soil.

Principle of complementarity. Information obtained under one set of experimental conditions might well not be the same as or even consistent with information obtained in a different set of procedures. In such cases the information obtained from the first experiment must be viewed as complementary to the information obtained in the second.

Profundal zone. The zone of a lake in which light does not penetrate.

Promiscuity. A type of polygamy in whicch one male or one female breeds with many of the opposite sex.

r selection. Refers to a high intrinsic rate of increase.

Refuge. An area established for the protection of specific wildlife species.

Relation. A collection of facts (attributes) describing the same subject.

Relational database. A database containing linking relations (table).

Relational system. Information categorized and stored in a predesigned table-and-column format in a relational database.

Resilience. Describes an ecosystem that has the power when under stress to return to its original state when the stress is removed.

Resolution. The programmed ability of a habitat model to separate components.

Resources. A compenent of the Centrum of an animal containing food, water, and cover.

Rest-rotation grazing. The deferment of grazing pastures in an allotment for one year on a rotating basis.

Restriction. A component of a "where" clause in a relational command that restricts the selection of data to that listed in the clause.

Riparian. Describes vegetation immediately adjacent to streams or along the edges of lakes and ponds that is different from vegetation of the surrounding area.

Risk analysis. An evaluation of the uncertainty in decision making.

Rotation. In forest management, the time interval between one regeneration cut and the next on the same area of land to perpetuate an even-aged stand.

Rotation-deferment grazing. A system of grazing which combines deferred grazing and rotation grazing. Grazing is delayed in each pasture, the deferment being rotated on a planned schedule between pastures.

Rotation grazing. Rotating livestock between different subdivisions of a range.

Salvage cut. The removal of trees from an area affected by a natural catastrophe such as fire or blow-down.

Scope. The number of variables that are included in a habitat model to make it usable at a given level of resolution.

Score card. A form of habitat model that uses a point system to arrive at total score to compare with a predetermined scale.

Secondary habitat. An area in which an animal may spend part of its time but which does not meet all its life requirements.

Secondary succession. Succession that starts when the climax stage has been distrubed.

Seed tree cut. A type of harvesting in which a cut removes almost all the trees in an area, but leaves a few scattered mature trees of good genetic stock to produce seed to regenerate the stand.

Select (database). A command to retrieve data from a relational database.

Selection cut. The selective removal of single trees or small groups of trees in a system of uneven-aged managment.

Sensitivity analysis. A means of systematically determining how changes in a variable or parameter affect a habitat model's output.

Seral stage. One stage of a particular life-form in the successional process.

Sere. All the stages in a successional process.

Shelter. A subset of cover associated with the rearing of young, such as a nest or burrow.

Shelterwood. A type of harvesting in which trees on a site are removed in a series of cuts over time to create an even-aged stand. Mature trees are left following an initial cut to provide both seed and shelter for seeds to germinate and for seedlings to get established.

Shrub. A woody plant of low growth stature with many branches.

Sign. Physical evidence that an animal was present, for example, tracks, feathers, scat, or bones.

Silviculture. The process of reproducing, growing, and improving stands of trees in a forest.

Single species management. The mangement of only one species at a time to meet a specific objective.

Snag. A dead or dying tree.

Species habitat profile. A hierarchical system for associating physical habitat factors with wildlife spcies at different levels of detail.

Species richness. The number of species occurring in a defined area. Often referred to as alpha diversity.

Specific birth rate. The number of individuals born per unit of population (100, 1000) by a specific age class of females.

Spring overturn. The mixing of water in the spring in a lake as a result of a change in temperature and increasing wind action.

Stability. Describes an ecosystem where the chance of a species going to extinction is low.

Stand. A group of trees that are sufficiently uniform in composition, age, structure, and spatial arrangement to be distinguishable from adjacent communities.

Stand density index. A relative measure of the growing space of a stree stand based on tree size and density.

Stand structure model. A matrix format table containing columns labeled with stand structure classes with rows under the columns containing a quantitative value for food and cover.

Stochastic model. A model designed to include probabilities that some special event will or will not happen.

Strategic plan. A plan stated in economic, technical, or political terms that is used to meet long-term goals.

Stratification. The separating or grouping of characteristics for sampling purposes.

Structure. The input and output processes of a habitat model that facilitate the transfer of knowledge from the model to its user. Used to describe vegetation, structure refers to the different plant life-forms and stages within life-forms particularly trees.

Succession. The replacement of one plant community with another over time. Can also refer to the replacment of one animal community with another over time as a result of changes in habitat brought about by plant succession.

Sustained yield. The number of animals that can be removed from a population over a period of years leaving not only a breeding population sufficient to maintain both the equilibrium number of animals but also a harvestable surplus.

Tension zone. The zone of constant change along the boundary of an inherent edge.

Territory. A defended area in the home range of an animal, particularly during the breeding season.

Thermocline. The thin, middle layer of a stratified lake in which temperature drops rather rapidly with depth.

Thinning. The remval of trees in a stand to increase the growth of remaining trees.

Time specific life table. A life table developed from animals captured in one period of time to determine their distribution of ages.

Tolerant. Describes a tree that can grow in shade.

Transparency. A property of water that allows light to penetrate it.

Tree. A woody plant with a single stem greater than 10.2 cm (4 in) measured at 1.37 m (4.5 ft).

Trend. An expression (up, down, increasing, decreasing) of the direction of change in the condition of vegetation compared to some base measurement.

Turbidity. The degree of clearness of water.

Umwelt. The totality of an animal's environment, of which some factors may not be obvious to humans.

UNESCO. United Nations Education, Scientific, and Cultural Organization.

Uneven-aged management. A system designed to perpetuate uneven-aged stands.

Uneven-aged stand. A stand containing trees of all age classes.

Uninhibited growth model. A population increasing at a rate which when plotted forms a characteristic J-shaped curve resulting from exponential growth.

Vector. An organism that is the carrier of a disease.

Vertical diversity. Diversity in height between two vegetation types of life-forms which portrays the effect of layering. Also referred to as contrast.

Viable population. The number of individuals sharing a common gene pool and able to reproduce and maintain the population from one generation to the next as in a local population.

View (database). A command used with a relational database to obtain information from more than one relation (table) at a time.

Virulent disease. A disease that multiplies rapidly to overcome an animal's immunological system.

Viscosity. The resistance to objects moving through water.

Volume. The amount of standing wood or wood removed from an area measured in board feet or cubic meters.

Volume regulation. A method for regulating a forest in which a given volume of wood is removed from an area over a specific period of time regardless of the size of area.

Watershed. A natural drainage area.

Web. The indirectly acting biotic and abiotic factors that affect the centrum of an animal.

Welfare factor. A factor in the centrum of an animal that indirectly reduces wildlife productivity when it is in short supply, such as food, water, or cover.

Where clause. A restrictive clause used with the select command in a relational databse.

Wild. An animal that is unrestrained or free roaming and not domesticated.

Wildlife. All species of amphibians, birds, fishes, mammals, and reptiles occurring in the wild.

Wildlife habitat relationships (WHR). A systems approach to comprehend all the associations with and uses of habitat by wildlife. It is a way of thinking about the complexities of plant-animal interactions.

Wildlife management. The art and science of manipulating the centrum of wild animal populations to meet specific objectives.

Wind break. A barrier that provides shelter from the wind on the downwind side.

Wind rowing. A method for controlling snow movement in open areas that involves the placement of logging debris perpendicular to the wind direction.

Yield. The surplus of game animals killed each year that otherwise would be lost to winter mortality. Is the result of seasonal fluctuations of population.

REFERENCES

Allan, P. F., L. G. Garland, and R. F. Dugan. 1963. Rating northeastern soils for their suitability for wildlife habitat. *Trans. N. Amer. Wildl. and Nat. Resour. Conf.* 28:247–261.

Allen, D. L. 1959. Resources, people and space. *Trans. N. Amer. Wildl. Conf.* 24:531–538.

_____, chair. 1973. *North American Wildlife Policy.* Wildl. Manage. Inst., Wash., DC.

_____. 1974. *Our Wildlife Legacy.* Funk and Wagnalls, New York, NY.

_____. 1979. *Wolves of Minong: Their Vital Role in a Wild Community.* Houghton Mifflin, Boston, MA.

Anderson, R. C. 1972. The ecological relationships of meningeal worm and captive cervids in North America. *J. Wildl. Dis.* 8:304–310.

Anderson, R. C., and A. K. Priestwood. 1981. Lungworms. Pages 266–317 in W. R. Davidson, ed. *Diseases of White-tailed Deer.* Misc. Pub. 7, Tall Timbers Res. Sta., Tallahassee, FL.

Andrewartha, H. G., and L. C. Birch. 1954. *The Distribution and Abundance of Animals.* Univ. of Chicago Press, Chicago, IL.

_____. 1984. *The Ecological Web: More on the Distribution and Abundance of Animals.* Univ. of Chicago Press, Chicago, IL.

Arnold, D. A. 1979. Deer on the highway. *Traffic Safety* 79:8–10.

Audubon Society. 1988. *The Audubon Society Field Guide to North American Fishes, Whales, and Dolphins.* Alfred A. Knopf, New York, NY.

Avery, T. E. 1975. *Natural Resources Measurements.* McGraw-Hill, New York, NY.

Baker, F. S. 1949. A revised tolerance table. *J. For.* 47: 179–181.

Bendell, J. F. 1974. Effects of fire on birds and mammals. Pages 73–138 in T. T. Kozlowski and C. E. Ahlgren, eds. *Fire and Ecosystems.* Academic Press, New York, NY.

Bennett, G. W. 1986. *Management of Lakes and Ponds.* Robert E. Krieger Publishing Co., Malabar, FL.

Bennett, H. 1950. Soil conservation and wildlife. Paper presented at 13th Annual Convention of Mich. United Cons. Clubs, 17 June, Ludington, MI.

Bennitt, R., and W. O. Nagel. 1937. A survey of the resident game and furbearers of Missouri. *Univ. of Missouri Studies* 12:77–85.

Bissell, H. D., B. Harris, H. Strong, and F. James. 1955. The digestibility of certain natural and artificial foods eaten by deer in California. *Calif. Fish and Game* 41:57–78.

Black, H. 1981. Effects of forest practices on fish and wildlife production. *Proc. of Tech. Session, Orlando, FL, 29 Sept. 1981.* Soc. of Amer. Foresters, Wash., DC.

Black, J. D. 1968. *The Management and Conservation of Biological Resources.* F. A. Davis Co., Philadelphia, PA.

Blockstein, D. E. 1990. Toward a federal plan for biological diversity. *J. For.* 88:15–19.

Bock, C. E., and J. F. Lynch. 1970. Breeding bird populations of burned and unburned conifer forest in the Sierra Nevada. *Condor* 72:182–189.

Boyce, S. G., and N. D. Cost. 1978. *Forest Diversity: New Concepts and Applications.* Res. Paper SE-194, USDA For. Serv., Southeastern For. and Range Exp. Sta., Asheville, NC.

Brady, N. C. 1988. International development and the protection of biological diversity. Pages 409–418 in E. O. Wilson, ed. *Biodiversity.* National Academy Press, Wash., DC.

Brown, E. R., ed. 1985. *Management of Wildlife and Fish Habitats in Forests of Western Oregon and Washington.* Part I. USDA For. Serv., Pacific Northwest Region in cooperation with USDI Bureau of Land Manage., Portland, OR.

Brown, J. L. 1969. Territorial behavior and population regulation in birds: A review and re-evaluation. *Wilson Bul.* 81:293–329.

Buech, R. R., K. Siderits, R. E. Radtke, H. L. Sheldon, and D. Elsing. 1977. *Small Mammal Populations After a Wildfire in Northeast Minnesota.* Res. Paper NC-151, USDA For. Serv., North Central For. Exp. Sta., St. Paul, MN.

Bunnell, F. L. 1989. *Alchemy and Uncertainty: What Good Are Models?* Gen. Tech. Rep. PNW-GTR-232, USDA For. Serv., Pacific Northwest Res. Sta., Portland, OR.

Bull, J., and J. Farrand, Jr. 1984. *The Audubon Society Field Guide to North American Birds: Eastern Region.* Alfred A. Knopf, New York, NY.

Cahalane, V. H. 1939. Integration of wildlife management with forestry in the Central states. *J. For.* 37:162–166.

Calhoun, J. B. 1952. The social aspects of population dynamics. *J. Mammal.* 33:139–159.

Call, M. W. 1982. *Terrestrial Wildlife Inventories.* Tech. Note 349, USDI Bureau of Land Manage., Denver, CO.

Carson, R. 1962. *Silent Spring.* Fawcett Publications, Greenwich, CT.

Caughley, G. 1966. Mortality patterns in mammals. *Ecology* 47:906–918.

———. 1970. Eruption of ungulate populations with emphasis on Himalayan Thar in New Zealand. *Ecology* 51:53–72.

———. 1977. *Analysis of Vertebrate Populations.* John Wiley and Sons, New York, NY.

———. 1979. What is this thing called carrying capacity? Pages 2–8 in M. S. Boyce and L. D. Hayden-Wing, eds. *North American Elk: Ecology, Behavior, and Management.* Univ. of Wyoming, Laramie, WY.

Caughley, G., and L. C. Birch. 1971. Rate of increase. *J. Wildl. Manage.* 35:568–663.

Chapman, H. H. 1936. Forestry and wildlife. *J. For.* 34: 104–106.

Chapman, J. A., and G. A. Feldhamer, eds. 1982. *Wild Mammals of North America.* John Hopkins Univ. Press, Baltimore, MD.

Chapman, R. N. 1928. The quantitative analysis of environmental factors. *Ecology* 9:111–122.

———. 1931. *Animal Ecology.* McGraw-Hill, New York, NY.

Cheatum, E. L., and C. W. Severinghaus. 1950. Variations in fertility of white-tailed deer related to range conditions. *Trans. N. Amer. Wildl. Conf.* 15:170–189.

Chew, R. M., B. B. Butterworth, and R. Grechman. 1958. The effects of fire on the small mammal populations of chaparral. *J. Mammal.* 40:253.

Christian, J. J. 1950. The adreno-pituitary system and population cycles in mammals. *J. Mammal.* 31:246–260.

Clements, F. E. 1916. *Plant Succession: An Analysis of the Development of Vegetation.* Carnegie Inst. Pub. 242, Wash., DC.

———. 1949. *Dynamics of Vegetation.* H. W. Wilson Co., New York, NY.

Codd, E. F. 1970. A relational model of data for large shared databanks. *Communications of the ACM* (June) 13(6).

Cody, M. L, and J. M. Diamond, eds. 1975. *Ecology and Evolution of Communities.* Harvard Univ. Press, Cambridge, MA.

Cole, G. F. 1958. *Range Survey Guide.* Federal Aid Project W-37-R, Montana Dept. of Game and Fish, Missouli, MT.

Cole, L. C. 1954. Some features of random population cycles. *J. Wildl. Manage.* 18:2–24.

Colinvaux, P. 1986. *Ecology.* John Wiley and Sons, New York, NY.

Conant, R. 1958. *A Field Guide to Reptiles and Amphibians.* Houghton Mifflin Co., Boston, MA.

Conner, R. N., and C. S. Adkisson. 1975. Effects of clearcutting on the diversity of breeding birds. *J. For.* 73:781–785.

Cook, D. B., and W. J. Hamilton, Jr. 1944. The ecological relationship of red fox food in eastern New York. *Ecology* 24:94–104.

Cook, S. F. 1959. The effects of fire on a population of small rodents. *Ecology* 40:102–108.

Cooperrider, A. Y., R. J. Boyd, and H. R. Stuart. 1986. *Inventory and Monitoring of Wildlife Habitat.* USDI Bureau of Land Manage. Serv. Center, Denver, CO.

Council on Environmental Quality. 1981. *Environmental Trends.* Exec. Office of the President, Wash., DC.

Covington, W. W., D. B. Wood, D. L. Young, D. P. Dykstra, and L. D. Garrett. 1988. TEAMS: A decision support system for multiresource management. *J. For.* 86(6):25–33.

Crance, J. H. 1987. *Guidelines for Using the Delphi Technique to Develop Habitat Suitability Index Curves.* Biol. Rep. 82, USDI Fish and Wildl. Serv., National Ecology Center, Fort Collins, CO.

Crawford, H. S., and R. W. Titterington. 1979. Effects of silvicultural practices on bird communities in upland spruce-fir stands. Pages 110–119 in R. M. DeGraff and K. E. Evans, eds. *Management of Northcentral and Northeastern Forest for Nongame Birds.* Gen. Tech. Rep. NC-51, USDA For. Serv., Northcentral For. Exp. Sta., St. Paul, MN.

Crawford, W. T. 1950. Some specific relationships between soils and wildlife. *J. Wildl. Manage.* 30:396–398.

Crouch, G. L. 1961. *Wildlife Populations and Habitat Conditions on Grazed and Ungrazed Bottomlands in Logan County, CO*. M.S. Thesis, Colorado State Univ., Fort Collins.

Cummins, K. W. 1973. Trophic relations of aquatic insects. *Ann. Rev. Entomel.* 18:183–206.

Curtis, J. P. 1959. *The Vegetation of Wisconsin*. Univ. of Wisconsin Press, Madison.

Dambach, C. A. 1944. A ten-year ecological study of adjoining grazed and ungrazed woodlands in northwestern Ohio. *Ecol. Monogr.* 14:256–270.

Dasmann, R. F. 1981. *Wildlife Biology*. John Wiley and Sons, New York, NY.

Dasmann, W. P. 1948. A critical review of range survey methods and their application to deer range management. *Calif. Fish and Game* 34:189–207.

———. 1951. Some deer range survey methods. *Calif. Fish and Game* 37:43–52.

Daubenmire, R. 1968. Ecology of fire in grasslands. Pages 209–273 in *Advances in Ecological Research*. Vol 5. Academic Press, New York, NY.

Davenport, L. A. 1940. Timber vs. wildlife. *J. For.* 38:661–666.

Davidson, W. R., F. A. Hayes, V. F. Nettles, and F. E. Kellog, eds. 1981. *Diseases and Parasites of White-tailed Deer*. Tall Timbers Res. Sta., Univ. of Georgia, Athens.

Davidson, W. R., and V. F. Nettles. 1988. *Field Manual of Wildlife Diseases in the Southeastern United States*. Southeastern Coop. Disease Study, Univ. of Georgia, Athens.

Davies, N. B. 1978. Ecological questions about territorial behavior. Pages 317–335 in J. R. Krebs and N. B. Davies, eds. *Behavioral Ecology: An Evolutionary Approach*. Sinauer Assoc., Sunderland, MA.

Davis, D. E. 1986. *Handbook of Census Methods for Terrestrial Vertebrates*. CRC Press, Boca Raton, FL.

Davis, L. S., and K. N. Johnson. 1987. *Forest Management*. McGraw-Hill, New York, NY.

Dayton, W. A. 1931. *Important Western Browse Plants*. Misc. Pub. 101, U.S. Dept. Agric., Wash., DC.

DeByle, N. V. 1981. *Songbird Populations and Clearcut Harvesting of Aspen in Northern Utah*. Res. Note INT-32, USDA For. Serv., Intermountain For. and Range Exp. Sta., Ogden, UT.

Decker, D. J., T. L. Brown, and W. Sarbello. 1981. Attitudes of residents in the peripheral Adirondacks toward illegally killing deer. *N. Y. Fish and Game J.* 28:73–80.

Denney, A. H. 1944. Wildlife relationships to soil types. *Trans. N. Amer. Wildl. Conf.* 9:316–223.

De Vos, A., and T. Jones. 1967. Wildlife management and land use. Symposium Proceedings, Special Issue. *E. African Agric. and For. J.*, Nairobi, Kenya.

Dice, L. R. 1952. *Natural Communities*. Univ. of Mich. Press, Ann Arbor.

Drew, T. J., and J. W. Flewelling. 1977. Some recent theories in yield-density relationships and their application to Monterey pine plantations. *For. Sc.* 23:517–534.

Driscoll, R. S., D. L. Merkel, D. L. Radloff, D. E. Snyder, and J. S. Hagihara. 1984. *An Ecological Land Classification Framework for the United States.* Misc. Pub. No. 1439, USDA For. Serv., Wash., DC.

Drury, W. H., and I. C. T. Nisbet. 1973. Succession. *J. Arnold Arboretum* 54:331–367.

Dubos, R. J. 1973. Humanizing the earth. *Amer. Assoc. Adv. Sc.,* Annual Meeting. B. Y. Morrison Memorial Lecture. U.S. Dept. Agric., Agric. Res. Serv., Wash., DC.

Duff, D. A. 1979. Riparian habitat recovery on Big Creek, Rich County, Utah. Pages 91–92 in O. B. Cope, ed. *Forum on Grazing and Riparian Stream Ecosystems.* Trout Unlimited, Denver, CO.

Eckholm, E. 1978. *Disappearing Species: The Social Challenge.* Paper 22, Worldwatch Inst., Wash., DC.

Edgerton, P. J. 1972. Big game use and habitat changes in a recently logged mixed conifer forest in northeastern Oregon. Pages 239–242 in *53rd Annual Conf. of the West. Assn. of State Game and Fish Comm.,* Portland, OR.

Edminister, F. C. 1954. *American Game Birds of Field and Forest.* Charles Scribner's Sons, New York, NY.

Edwards, R. Y., and C. D. Fowle. 1955. The concept of carrying capacity. *Trans. N. Amer. Wildl. Conf.* 20:589–602.

Einarsen, A. S. 1945. Some factors affecting ring-neck pheasant population density. *Murrelet* 26:2–9, 39–44.

_____. 1948. *The Pronghorn Antelope and Its Management.* Wildl. Manage. Inst., Wash., DC.

Elton, C. S. 1942. *Voles, Mice and Lemmings.* Clarendon Press, London.

_____. 1956. *Animal Ecology.* Sidgwick and Jackson, London.

_____. 1958. *The Ecology of Invasions by Animals and Plants.* Methuen and Co., New York, NY.

_____. 1966. *The Pattern of Animal Communities.* Methuen, New York, NY.

Elton, C. S., and M. Nicholson. 1942. The ten-year cycle in numbers of lynx in Canada. *J. Animal Ecol.* 11:215–244.

Emlen, J. M. 1973. *Ecology: An Evolutionary Approach.* Addison Wesley, Reading, MA.

Errington, P. L. 1945. Some contributions of a 15-year study of the northern bobwhite to a knowledge of a population phenomena. *Ecol. Monogr.* 15:1–34.

_____. 1967. *Of Predation and Life.* Iowa State Univ. Press, Ames. Everhart, W. H., and W. D. Youngs. 1981. *Principles of Fishery Science.* Cornell Univ. Press, Ithaca, NY.

Everhart, W. H., and W. D. Youngs. 1981. *Principles of Fishery Science.* Cornell Univ. Press, Ithaca, NY.

Eyre, F. H., ed. 1980. *Forest Cover Types of the United States and Canada.* Soc. of Amer. Foresters, Wash., DC.

Farb, P. 1967. *The Land, Wildlife, and Peoples of the Bible.* Harper and Row, New York, NY.

Flood, B. M., M. Sangster, R. Sparrowe, and T. Baskett. 1977. *A Handbook for Habitat Evaluation Procedures.* Resource Pub. 132, USDI Fish and Wildl. Serv., Wash., DC.

Forbs, R. D., ed. 1961. Forest wildlife management. Sec. 9:38–39 in *Forestry Handbook*. Ronald Press Co., New York, NY.

Ford-Robertson, F. C., ed. 1971. *Terminology of Forest Science, Technology Practice and Products*. Multilingual For. Term. Series No. 1. Joint FAO/IUFRO Comm., Soc. of Amer. Foresters, Wash., DC.

Fox, M. W. 1980. *The Soul of the Wolf*. Little, Brown and Co., Boston, MA.

Fox, R., and B. Rhea. 1989. Spotted owls in the Rockies. *J. For.* 87:41–45.

French, C. E., L. C. McEwen, N. D. MacGruder, R. W. Ingram, and H. W. Swift. 1956. Nutrient requirements for growth and antler development in the white-tailed deer. *J. Wildl. Manage.* 20:221–232.

Gabrielson, I. N. 1936. The correlation of forestry and wildlife management. *J. For.* 34:98–103.

Galli, A. E., C. F. Leck, and R. T. T. Forman. 1976. Avian distribution patterns in forest islands of different sizes in New Jersey. *Auk* 93:356–364.

Gallizioli, S. 1965. *Quail Research in Arizona*. Ariz. Game and Fish Dept., Phoenix.

Gashwiler, J. S. 1970. Plant and mammal changes on a clearcut in west-central Oregon. *Ecology* 51:1018–1026.

Gates, D. M. 1965. Radiant energy, its receipt and disposal. *Metero. Monogr.* 6:1–26.

Giles, R. H., Jr., and N. Snyder. 1970. Simulation techniques in wildlife habitat management. Pages 23–49 in D. Jameson, ed. *Modeling and Systems Analysis in Range Science*. Range Science Dept. Series No. 5., Colorado State Univ., Fort Collins.

Giles, R. H., Jr., and L. A. Nielson. 1990. A new focus for wildlife resource managers. *J. For.* 88:21–26.

Gleason, H. A. 1926. The individualistic concept of the plant association. *Bul. Torrey Bot. Club* 53:7–26.

_____. 1939. The individualistic concept of the plant association. *Amer. Mid. Nat.* 21:92–110.

Good, E. E., and C. A. Dambach. 1943. Effect of land use practices on breeding bird populations in Ohio. *J. Wildl. Manage.* 7:291–297.

Goodman, D. 1975. The theory of diversity-stability relationships in ecology. *Quart. Rev. of Biol.* 50:237–266.

Grinnell, J. 1943. *Philosophy of Nature*. Univ. of Calif. Press, Berkeley.

Gross, T. E., and D. P. Dykstra. 1989. Harvest scheduling allowing well-defined violations to age class nonadjacency constraints. Pages 165–175 in A. Techle, W. W. Covington, and R. H. Hamre, eds. *Multiresource Management of Ponderosa Pine Forests*. Gen. Tech. Rep. RM-185, USDA For. Serv., Rocky Mountain For. and Range Exp. Sta., Fort Collins, CO.

Grubb, T. G. 1988. *Pattern Recognition: A Simple Model for Evaluating Wildlife Habitat*. Res. Note RM-487, USDA For. Serv., Rocky Mountain For. and Range Exp. Sta., Fort Collins, CO.

Gullion, G. W. 1960. The ecology of Gambel's quail in Nevada and the arid Southwest. *Ecology* 14:518–536.

Gysel, L., and J. Lyon. 1980. Habitat analysis and evaluation. Pages 305–327 in S. D. Schemnitz, ed. *Wildlife Management Techniques Manual*. Wildl. Soc., Wash., DC.

Haapanen, A. 1965. Bird fauna of the Finnish forests in relation to forest succession. I. Ann. Zoo. Fenn. 2: 153–196.

Hagar, D. C. 1960. The interrelationships of logging, birds, and timber regeneration in the Douglas-fir region of northwestern California. Ecology 41:116–125.

Hakala, J. B., R. K. Seemel, R. A. Richey, and J. E. Kurtz. 1971. Fires effects and rehabilitation methods: Swanson-Russian River fires. Pages 87–89 in C. W. Slaughter, R. J. Burns, and G. M. Hansen, eds. Fire in the Northern Environment. Symposium Proceedings, USDA For. Serv., Pacific Northwest For. and Range Exp. Sta., Portland, OR.

Hall, F. C., and J. W. Thomas. 1979. Silvicultural options. Pages 128–147 in J. W. Thomas, ed. Wildlife Habitats in Managed Forests: The Blue Mountains of Oregon and Washington. Hdbk. 553, U.S. Dept. Agric., Wash., DC.

Halls, L. k. 1970. Nutrient requirements of livestock and game. Pages 10–24 in H. A. Paulsen, Jr., and E. H. Reid, eds. Range and Wildlife Habitat Evaluation. Misc. Pub. No. 1147, USDA For. Serv., Wash., DC.

Hansen, C. S. 1983. Costs of deer-vehicle accidents in Michigan. Wildl. Soc. Bul. 11:161–164.

Harlow, R. F., and R. L. Downing. 1969. The effects of size and intensity of cuts on production and utilization of some deer foods in southern Appalachians. Trans. Northeast Sec. Wildl. Soc. 26:45–55.

Harris, L. D. 1980. Forest and wildlife dynamics in the Southeast. Trans. N. Amer. Wildl. and Nat. Resour. Conf. 45:307–322.

_____. 1984. The Fragmented Forest: Island Biogeography Theory and the Preservation of Biological Diversity. Univ. of Chicago Press, Chicago, IL.

_____. 1988a. Edge effects and conservation of biotic diversity. Conservation Biology. 2:330–332.

_____. 1988b. Landscape Linkages. A Florida Films Production. Video tape. 25 minutes. Tele. 904–377–3456.

_____. 1988c. Reconsideration of the habitat concept. Trans. N. Amer. Wildl. and Nat. Resour. Conf. 53:137–144.

Harris, L. D., and W. H. Smith. 1978. Relations of forest practices to non-timber resources and adjacent ecosystems. Pages 28–53 in T. Tappen. Productivity on Prepared Sites. USDA For. Serv., New Orleans, LA.

Heinrichs, J. 1983. Old growth comes of age. J. For. 81:776–779.

Hewett, O., ed. 1954. A symposium on cycles in animal populations. J. Wildl. Manage. 18:1–112.

Hickman, C. P. 1966. Integrated Principles of Zoology. C. V. Mosby Co., St. Louis, MO.

Hill, E. P. 1972. The Cottontail in Alabama. Bul. No. 440, Alabama Agric. Exp. Sta., Auburn Univ.

Hoekstra, T. W., A. A. Dyer, D. C. LeMaster, eds. 1987. FORPLAN: An Evaluation of a Forest Planning Tool. Gen. Tech. Rep. RM-140, USDA For. Serv., Rocky Mountain For. and Range Exp. Sta., Fort Collins, CO.

Holbrook, H. L. 1974. A system for wildlife habitat management on southern national forests. Wildl. Soc. Bul. 2:119–123.

Holechek, J. L. 1982. Managing rangelands for mule deer. Rangelands 4:25–28.

_____ . 1983. Considerations concerning grazing systems. *Rangelands* 5:208–211.

_____ . 1988. An approach for setting the stocking rate. *Rangelands* 10:10–14.

Holling, C. S. 1973. Resilience and stability of ecological systems. *Ann. Rev. Ecol. Syst.* 4:1–23.

Holt, S. J., and L. M. Talbott. 1978. *New Principles for the Conservation of Wild Living Resources*. Wildl. Monogr. Wildl. Soc., Wash., DC.

Hooper, R. G. 1967. *The Influence of Habitat Disturbance on Bird Populations*. M.S. Thesis, Virginia Polytechnic Inst. Blacksburg.

Hooven, E. F. 1973. A wildlife brief for the clearcut logging of Douglas-fir. *J. For.* 71:210–214.

Hoover, R. L., and D. L. Wills, eds. 1984. *Managing Forested Lands for Wildlife*. Colorado Div. of Wildl. in cooperation with USDA For. Serv., Rocky Mountain Region, Denver, CO.

Horn, H. S. 1976. Succession. Pages 187–204 in R. M. May, ed. *Theoretical Ecology: Principles and Applications*. Blackwell Scientific Publications, Oxford.

Hornocker, M. G. 1969. Winter territoriality in mountain lions. *J. Wildl. Manage.* 33:457–464.

Howard, H. E. 1920. *Territory in Bird Life*. Dutton, New York, NY.

Howard, W. E., R. L. Fenner, and H. E. Childs, Jr. 1959. Wildlife survival in brush burns. *J. Range Manage.* 12:230–234.

Hunter, M. L., Jr. 1989. What constitutes an old-growth stand? *J. For.* 87:33–35.

Hutchinson, G. E. 1959. Homage to Santa Rosalia, or why are there so many kinds of animals? *Amer. Nat.* 93:145–159.

Jacobson, H. A., D. C. Guynn, and E. J. Hackett. 1979. Impact of the botfly on squirrel hunting in Mississippi. *Wildl. Soc. Bul.* 7:46–48.

James, M. C., A. L. Meehean, and E. J. Douglas. 1944. *Fish Stocking as Related to the Management of Inland Waters*. Cons. Bul. 35, USDI Fish and Wildl. Serv., Wash., DC.

Jefferies, R. 1848. *The Gamekeeper at Home—The Amateur Poacher*. Oxford University Press, London.

Jeffers, J. N. R. 1980. *Modelling, Statistical Checklist*. Inst. Terrest. Ecol., Nat. Envir. Res. Council, Cambridge, Eng.

Jeffers, J. N. R. 1982. *Modelling*. Chapman and Hall, New York, NY.

Jenkins, D. H., and I. H. Bartlett. 1959. *Michigan Whitetails*. Mich. Dept. of Cons., Lansing.

Jenny, H. 1958. Role of the plant factor in the pedogenic functions. *Ecology* 39:6–16.

Johnson, R. 1970. Tree removal along southwestern rivers and effects on associated organisms. Report of Comm. on Res., *Yearbook of the Amer. Phil. Soc.*

Johnston, D. W., and E. P. Odum. 1956. Breeding bird populations in relation to plant succession on the Piedmont of Georgia. *Ecology* 37:50–62.

Karstad, L. 1979. Diseases of wildlife. Pages 123–127 in R. D. Teague and E. Decker, eds. *Wildlife Conservation*. The Wildl. Soc., Wash., DC.

Kellert, S. R. 1980. Americans' attitudes and knowledge of animals. *Trans. N. Amer. Wildl. and Nat. Resour. Conf.* 43:412–423.

Kendeigh, S. C. 1941. Territorial and mating behavior of the house wren. *Ill. Biol. Monogr.* 10:1–20.

Kimmins, J. P. 1987. *Forest Ecology.* Macmillan, New York, NY.

King, R. T. 1938. The essentials of a wildlife range. *J. For.* 36:457–464.

Klein, D. R. 1968. The introduction, increase, and crash of reindeer on St. Matthew Island. *J. Wildl. Manage.* 32:350–367.

Klopfer, P. H. 1969. *Habitats and Territories: A Study of the Use of Space by Animals.* Basic Books, New York, NY.

Knapp, R. 1974. *Cyclic Successions and Ecosystem Approaches in Vegetation Dynamics.* W. Junk Publishers, The Hague.

Knutson, R. M. 1987. *Flattened Fauna: A Field Guide to Common Animals of Roads, Streets, and Highways.* Ten Speed Press, Berkeley, CA.

Komarek, E. V., Sr. 1969. Fire and animal behavior. *Tall Timbers Fire Ecol. Conf.* 9:161–207.

Koya, C. M. 1977. Reproductive biology of rainbow and brown trout in a geothermally heated stream: The Firehole River of Yellowstone Natl. Park. *Trans. Amer. Fisheries Soc.* 106:354–361.

Kozicky, E. L., G. O. Hendrickson, P. G. Homer. 1955. Weather and fall pheasant populations. *J. Wildl. Manage.* 19:136–142.

Kramp, B., W. Brady, and D. R. Patton. 1983. *The Effects of Fire on Wildlife Habitat and Species.* Wildl. Unit Tech. Rep., USDA For. Serv., Southwestern Region, Albuquerque, NM.

Krebs, C. J. 1972. *Ecology: The Experimental Analysis of Distribution and Abundance.* Harper and Row, New York, NY.

Krebs, C. J., M. S. Gaines, B. L. Keller, J. H. Myers, and R. H. Tamarin. 1973. Population cycles in small rodents. *Science* 179:35–41.

Kroenke, D. M., and D. E. Nilson. 1986. *Database Processing for Microcomputers.* Science Research Associates, Chicago, IL.

Krull, J. N. 1970. Small mammal populations in cut and uncut northern hardwood forests. *N. Y. Fish and Game J.* 17:128–130.

Lawrence, G. E. 1966. Ecology of vertebrate animals in relation to chaparral fire in the Sierra Nevada foothills. *Ecology* 47:278–291.

Lee, R. G. 1990. The redesigned forest. *J. For.* 88:32–34.

Lehmkuhl, J., and D. R. Patton. 1984. *User's Manual for the RUN WILD III Data Storage and Retrieval System.* Wildl. Unit Tech. Rep., USDA For. Serv., Southwestern Region, Albuquerque, NM.

Leopold, A. 1930. Environmental controls for game through modified silviculture. *J. For.* 28:321–326.

———. 1933. *Game Management.* Charles Scribner's Sons, New York, NY.

———. 1939. A biotic view of land. *J. For.* 37:727–730.

———. 1943. Wildlife in American culture. *J. Wildl. Manage.* 7:1–6.

———. 1949. *A Sand County Almanac.* Oxford Univ. Press, New York, NY.

Leopold, A., L. K. Sowls, and D. L. Spencer. 1947. A survey of overpopulated deer ranges in the United States. *J. Wildl. Manage.* 11:162–177.

Leopold, A. S. 1950. Deer in relation to plant succession. *Trans. N. Amer. Wildl. Conf.* 15:571–580.

_____. 1977. *The California Quail*. Univ. of Calif. Press, Berkeley.

Leopold, A. S., S. A. Cain, C. M. Cottam, I. N. Gabrielson, and T. I. Kimball. 1964. Predator and rodent control in the United States. *Trans. N. Amer. Wildl. and Nat. Resour. Conf.* 29:27–49.

Lewin, R. 1986. In ecology, change brings stability. Research News. *Science* 234:1071–1073.

Linduska, J. P., ed. 1964. *Waterfowl Tomorrow*. USDI Fish and Wildl. Serv., Wash., DC.

Lotka, A. J. 1925. *Elements of Physical Biology*. Williams and Wilkins, Baltimore, MD.

Lovelock, J. E. 1988. The earth as a living organism. Pages 486–489 in E. O. Wilson, ed. *Biodiversity*. National Academy Press, Wash., DC.

Lund, H. G. 1978. *Type Maps, Stratified Sampling, and PPS*. Res. Inven. Note BLM 15, USDI Bureau of Land Manage., Denver Serv. Center, Denver, CO.

_____. 1986. *A Primer on Integrating Resource Inventories*. Gen. Tech. Rep. WO-49, USDA For. Serv., Wash., DC.

Lund, H. G., V. J. LeBau, P. F. Ffolliott, D. W. Robinson. 1978. *Integrated Inventories of Renewable Natural Resources*. Gen. Tech. Rep. RM-55, USDA For. Serv., Rocky Mountain For. and Range Exp. Sta., Fort Collins, CO.

Lyon, L. J. 1984. Road effects and impacts on wildlife and fisheries. Paper presented at Forest Trans. Symposium, 11–13 December, Casper, WY. USDA For. Serv., Region 2, Denver, CO.

Lyon, L. J., H. S. Crawford, E. Czuhai, R. L. Fredriksen, R. F. Harlow, L. J. Metz, and H. A. Pearson. 1978. *Effects of Fire on Fauna: A State-of-Knowledge Review*. W.O. Gen. Tech. Rep. 6, USDA For. Serv., Wash., DC.

MacArthur, R. H. 1965. Patterns of species diversity. *Biol. Rev.* 40:510–533.

MacArthur, R. H., and E. O. Wilson. 1967. *The Theory of Island Biogeography*. Princeton Univ. Press., Princeton, NJ.

MacClintock, L. R., R. Whitcomb, and B. Whitcomb. 1977. Evidence for the value of corridors and minimization of isolation in preservation of biotic diversity. *Amer. Birds* 31:6–16.

McCormick, J. 1966. *The Life of a Forest*. McGraw-Hill, New York, NY.

McCulloch, C. Y., and R. L. Brown. 1986. *Rates and Causes of Mortality Among Radio Collared Mule Deer of the Kaibab Plateau, 1978–1983*. Federal Aid in Wildlife Restoration Project W-78-R. Ariz. Game and Fish Dept., Phoenix.

McCullough, D. R. 1979. *The George Reserve Deer Herd: Population Ecology of a K-selected Species*. Univ. of Mich. Press, Ann Arbor.

McMillan, I. I. 1964. Annual population changes in California quail. *J. Wildl. Manage.* 28:702–711.

McTague, J. R., and D. R. Patton. 1989. Stand density index and its application in describing wildlife habitat. *Wildl. Soc. Bul.* 17:58–62.

McWhirter, N. 1981. *Guinness Book of World Records*. Sterling Publishing Co., New York, NY.

Malthus, T. R. 1798. *An Essay on the Principle of Population*. Johnson, London.

Mannan, R. W., M. L. Morrison, and E. C. Meslow. 1984. The use of guilds in forest bird management. *Wildl. Soc. Bul.* 12:426–430.

Marcot, B. G. 1986. Concepts of risk analysis as applied to viable population assessment and planning. Pages 89–101 in B. A. Wilcox, P. F. Brussard, and B. G. Marcot, eds. *The Management of Viability: Theory, Qpplications, and Case Studies.* Center for Cons. Biol., Dept. of Biol. Sci., Stanford Univ., Stanford, CA.

Marmelstein, A. 1977. *Classification, Inventory, and Analysis of Fish and Wildlife Habitat.* FWS/OBS-78/76, USDI Fish and Wildl. Serv., Office of Biol. Serv., Wash., DC.

Martin, A. C., H. S. Zim, and A. L. Nelson. 1951. *American Wildlife and Plants: A Guide to Wildlife Food Habits.* Dover Publications, New York, NY.

Martin, W. E., and R. L. Gum. 1978. Economic value of hunting, fishing, and general rural outdoor recreation. *Wildl. Soc. Bul.* 6:3–7.

Maser, C., R. G Anderson, K. C. Cromakc, Jr., J. T. Williams, and R. E. Martin. 1979. Dead and down woody material. Pages 78–95 in J. W. Thomas, ed. *Wildlife Habitats in Managed Forests: The Blue Mountains of Oregon and Washington.* Hdbk. 553, U.S. Dept. Agric., Wash., DC.

Maser, C., and J. M. Trappe. 1984. *The Seen and Unseen World of the Fallen Tree.* Gen. Tech. Rep. PNW-164, USDA For. Serv., Pacific Northwest For. and Range Exp. Sta. in cooperation with USDI Bureau of Land Manage. Portland, OR.

Mason, R. D. 1970. *Statistical Techniques in Business and Economics.* Richard D. Irwin, Homewood, IL.

May, R. M. 1973. *Stability and Complexity in Model Ecosystems.* Princeton Univ. Press, Princeton, NJ.

Mealey, S. P., J. F. Lipscomb, and K. N. Johnson. 1982. Solving the habitat dispersion problem in forest planning. *Trans. N. Amer. Wildl. and Nat. Resour. Conf.* 47:142–153.

Mech, L. D., L. Frenzel, Jr., R. Meam, and J. Winship. 1971. *Ecological Studies of the Timber Wolf in Northeastern Minnesota.* Res. Paper NC-52, USDA For. Serv., North Central For. Exp. Sta., St Paul, MN.

Meehan, W. R., F. J. Swanson, and J. R. Sedell. 1977. Influences of riparian vegetation on aquatic ecosystems with particular reference to salmonid fishes and their food supply. Pages 137–145 in R. Johnson and D. A. Jones, tech. coords. *Importance, Preservation, and Management of Riparian Habitat.* Gen. Tech. Rep. RM-43, USDA For. Serv., Rocky Mountain For. and Range Exp. Sta., Fort Collins, CO.

Mitchell, G. E. 1950. Wildlife-forest relationships in the Pacific Northwest region. *J. For.* 48:26–30.

Mitchell, T. R., and B. M. Kent. 1987. Characterization of the FORPLAN analysis system. Pages 3–14 in T. W. Hoekstra, A. A. Dyer, and D. C. Le Master, tech. eds. *FORPLAN: An Evaluation of a Forest Planning Tool.* Gen. Tech. Rep. RM-140, USDA For. Serv., Rocky Mountain For. and Range Exp. Sta., Fort Collins, CO.

Mollohan, C. 1988. *Merriam's Turkey Expert Opinion Survey.* Ariz. Game and Fish Dept. Rep., Phoenix.

Mosby, H. S., ed. 1963. *Wildlife Investigational Techniques.* The Wildl. Soc., Wash., DC.

Myers, W. L., and R. L. Shelton. 1980. *Survey Methods for Ecosystem Management*. John Wiley and Sons, New York, NY.

National Academy of Science. 1962. Nutrient requirements of laboratory animals. In *Nutrient Requirements of Domestic Animals*. Natl. Res. Council. Pub. 990, Wash., DC.

———. 1963. Nutrient requirements of beef cattle. In *Nutrient Requirements of Domestic Animals*. Natl. Res. Council. Pub. 1137, Wash., DC.

———. 1964. Nutrient requirements of sheep. In *Nutrient Requirements of Domestic Animals*. Natl. Res. Council. Pub. 1193, Wash., DC.

———. 1970. *Land Use and Wildlife Resources*. Natl. Res. Council, Div. of Biol. and Agric., Wash., DC.

———. 1982. *Impacts of Emerging Agricultural Trends on Fish and Wildlife Habitat*. National Academy Press, Wash., DC.

National Geographic Society. 1983. *Birds of North America. Field Guide*. Natl. Geo. Soc., Wash., DC.

National Wildlife Federation. 1988. *Forests Are More Than Trees*. National Wildlife Week Poster. Wildl. Manage. Inst., Wash., DC.

Needham, J. G., and J. T. Lloyd. 1937. *The Life of Inland Waters*. Comstock Publishing Co., Ithaca, NY.

Nelson, R. D., and H. Salwasser. 1982. The Forest Service wildlife and fish habitat relationships program. *Trans. N. Amer. Wildl. and Nat. Resource Conf.* 47:174–182.

Nestler, R. B. 1949. Nutrition of bobwhite quail. *J. Wildl. Manage.* 13:342–358.

New Mexico Game and Fish Dept. 1973. *Big Game Habitat Analysis for New Mexico*. N.M. Game and Fish Dept., Santa Fe. Mimeo.

Nice, M. 1941. The role of territory in bird life. *Amer. Mid. Nat.* 26:441–487.

Nichols, G. F. 1923. A working basis for the ecological classification of plant communities. *Ecology* 4:11–23, 154–172.

Nixon, C. M., S. P. Haver, and L. P. Hansen. 1980. Initial response of squirrels to forest changes associated with selection cutting. *Wildl. Soc. Bul.* 8:298–306.

Nixon, C. M., M. W. McClain, and R. W. Donohoe. 1980. Effects of clearcutting on gray squirrels. *J. Wildl. Manage.* 44:403–412.

Noble, I. R., and R. O. Slatyer. 1977. Post-fire succession of plants in mediterranean ecosystems. Pages 27–63 in H. A. Money and C. E. Conrad, eds. *Environmental Consequences of Fire and Fuel Management in Mediterranean Ecosystems*. Gen. Tech. Rep. WO-3, USDA For. Serv., Wash., DC.

Norman, R. L., L. A. Roper, P. D. Olson, and R. L. Evans. 1976. *Using Wildlife Values in Benefit/Cost Analysis and Mitigation of Wildlife Losses*. Colorado Div. of Wildl., Denver.

Odum, E. P. 1969. The strategy of ecosystem development. *Science* 164:262–270.

———. 1971. *Fundamentals of Ecology*. W. B. Saunders Co., Philadelphia, PA.

Office of Tech. Assessment. 1987. *Technologies to Maintain Biological Diversity*. Congress of the United States. USGPO, Wash., DC.

Oliver, C. D. 1986. Silviculture: The next 30 years: The past 30 years. *J. For.* 84(4):32–42.

O'Neil, L. J. 1985. *Habitat Evaluation Methods Notebook.* Instruction Rep. EL 85-3, Dept. of the Army, Corps of Eng. Waterways Exp. Sta., Vicksburg, MS.

Orwell, G. 1946. *Animal Farm.* Harcourt Brace, New York, NY.

Owen-Smith, R. N. 1982. Management of large mammals in African conservation areas. *Symposium Proceedings, Centre for Resource Ecology,* Univ. of Witwatersrand, Johannesburg, S.A.

Pattee, O. H., and S. K. Hennes. 1983. Bald eagles and waterfowl, the lead shot connection. *Trans. N. Amer. Wildl. and Nat. Resour. Conf.* 48:230–237.

Patton, D. R. 1974. Patch cutting increases deer and elk use of a pine forest in Arizona. *J. For.* 72:764–766.

_____. 1975. A diversity index for quantifying habitat "edge." *Wildl. Soc. Bul.* 3:171–173.

_____. 1978. *Run Wild: A Storage and Retrieval System for Wildlife Habitat Information.* Gen. Tech. Rep. RM-51, USDA For. Serv., Rocky Mountain For. and Range Exp. Sta., Fort Collins, CO.

_____. 1979a. *How to Use RUN WILD Data Files Stored on Microfiche.* Res. Note RM-377, USDA For. Serv., Rocky Mountain For. and Range Exp. Sta., Fort Collins, CO.

_____. 1979b. RUN WILD II: A Storage and Retrieval System for Wildlife Data. *Trans. N. Amer. Wildl. Conf.* 44:425–430.

_____. 1982. Wildlife habitat in land management planning: Some ideas and principles. Pages 33–38 in P. F. Ffolliott, L. K. Sowls, and J. C. Tash, tech. coords. *The Effects of Land Management Practices on Fish and Wildlife in Southwestern Conifer Forests.* Workshop Proceedings, Univ. of Ariz., School of Renewable Nat. Resour., Tucson.

_____. 1984. A model to evaluate Abert squirrel habitat in unevenaged ponderosa pine. *Wildl. Soc. Bul.* 12:408–414.

_____. 1987. Is the use of management indicator species feasible? *West. J. Applied For.* 2:33–34.

_____. 1990. *A Classification Scheme To Group Wildlife With Common Habitat Requirements.* Final Rep., Cooperative Agreement 28-C7-437, USDA For. Serv., Rocky Mountain For. and Range Exp. Sta., Fort Collins, Co.

Patton, D. R., and M. Ertle. 1982. *Wildlife Food Plants of the Southwest.* Wildl. Unit Tech. Rep., USDA For. Serv., Southwestern Region, Albuquerque, NM.

Patton, D. R., and J. M. Hall. 1966. Evaluating key areas by browse age and form class. *J. Wildl. Manage.* 30:476–480.

Patton, D. R., and K. E. Severson. 1989. WILDHARE: A wildlife habitat relationships data model for southwestern ponderosa pine. Pages 268–276 in A. Techle, W. W. Covington, and R.H. Hamre, tech. coords. *Multiresource Management of Ponderosa Pine Forests.* Gen. Tech. Rep. RM-185, USDA For. Serv., Rocky Mountain For. and Range Exp. Sta., Fort Collins, CO.

Payne, N. F., and F. Copes. 1986. *Wildlife and Fisheries Habitat Improvement Handbook.* Wildlife and Fisheries Administrative Rep. (unnumbered), USDA For. Serv., Wash., DC.

Pengelly, W. L. 1972. Clearcutting: Detrimental aspects for wildlife resources. *J. Soil Water Cons.* 27:255–258.

Pennsylvania Game Commission. 1985. Recorded highway mortality, 1984: Deer and bear. *Pa. Game News* (June).

Pennsylvania State University. 1986. *A Trout Stream in Winter.* Movie. 25 minutes. University Park, PA.

Perry, C., and R. Overly. 1976. Impact of roads on big game distribution in portions of the Blue Mountains of Washington. Pages 62–68 in J. M. Peek, Chair. *Elk, Logging, Roads Symposium Proceedings.* Univ. of Idaho, Moscow.

Peterson, G. W., and A. Randall. 1984. *Valuation of Wildland Resource Benefits.* Westview Press in cooperation with USDA For. Serv., Boulder, CO.

Phillips, E. A. 1959. *Methods of Vegetation Study.* Henry Holt and Company, New York, NY.

Pimlot, D. H. 1969. The value of diversity. *Trans. N. Amer. Wildl. and Nat. Resour. Conf.* 34:265–280.

Pinchot, G. 1947. *Breaking New Ground.* Harcount, Brace & Co., New York, NY.

Rasmussen, D. I. 1941. Biotic communities of the Kaibab plateau. *Ecol. Monogr.* 3:229–275.

Ream, C. H., and G. E. Gruell. 1980. Influences of harvesting and residue treatments on small mammals and implications for forest management. Pages 455–467 in *Environmental Consequences for Timber Harvesting in Rocky Mt. Coniferous Forests.* Gen. Tech. Rep. INT-90, USDA For. Serv., Intermountain For. and Range Exp. Sta., Missoula, MT.

Reardon, P. O., L. B. Merrill, and C. A. Taylor, Jr. 1978. White-tailed deer preferences and hunter success under various grazing systems. *J. Range Manage.* 31:40–42.

Reglin, W. L., O. C. Wallmo, J. Nagy, and D. R. Dietz. 1974. Effect of logging on forage values for deer in Colorado. *J. For.* 72:282–285.

Reid, G. K. 1961. *Ecology of Inland Waters and Estuaries.* Reinhold Publishing Co., New York, NY.

Reineke, L. H. 1933. Perfecting a stand density index for even-aged forests. *J. Agric. Res.* 46:627–638.

Reynolds, H. G. 1966. *Use of a Ponderosa Pine Forest in Arizona by Deer, Elk, and Cattle.* Res. Note RM-63, USDA For. Serv., Rocky Mountain For. and Range Exp. Sta., Fort Collins, CO.

———. 1969. Improvement of deer habitat on southwestern forest lands. *J. For.* 67:803–805.

Reynolds, H. G., and R. R. Johnson. 1964. *Habitat Relations of Vertebrates of the Sierra Ancha Experimental Forest.* Res. Paper RM-4, USDA For. Serv., Rocky Mountain For. and Range Exp. Sta., Fort Collins, CO.

Rippe, D. J., and R. L. Rayburn. 1981. *Land Use and Big Game Population Trends in Wyoming.* W/CRAM-81-W22, USDI Fish and Wildl. Serv., Wash., DC.

Rogers, J. L., J. M. Prosser, and L. D. Garrett. 1981. Modeling on-site multiresources effects of silviculture management prescriptions. Pages 107–117 in *Forest Management Planning: Present Practice and*

Future Decisions, IUFRO Symposium Proceedings, Pub. FWS-1-81, School of For. and Wildl. Resour., Virginia Polytechnic Inst. and State Univ., Blacksburg.

Root, R. B. 1967. The niche exploitation pattern of the blue-gray gnatcatcher. *Ecol. Monogr.* 37:317–349.

Rosene, W. 1969. *The Bobwhite Quail: Its Life and Management.* Rutgers Univ. Press, New Brunswick, NJ.

Rudd, R. L., and R. E. Genelly. 1956. *Pesticides: Their Use and Toxicity in Relation to Wildlife.* Game Bul. No. 7, Calif. Dept. of Game and Fish, Sacramento.

Salwasser, H. 1979. *The Ecology and Management of the Devil's Garden Interstate Deer Herd and Its Range.* Ph.D. Dissertation, Univ. of Calif., Berkeley.

Salwasser, H., C. K. Hamilton, W. B. Krohn, J. Lipscomb, and C. H. Thomas. 1983. Monitoring wildlife and fish, mandates and their implications. *Trans. N. Amer. Wildl. and Nat. Resour. Conf.* 48:297–307.

Schamberger, M., A. H. Farmer, and J. W. Terrell. 1982. *Habitat Suitability Index Models.* FWS/OBS-82/10, USDI Fish and Wildl. Serv., Office of Biol. Serv., Div. of Ecol. Serv., Wash., DC.

Schantz, H. L. 1911. *Natural Vegetation as an Indicator of the Capabilities of Land Crop Production in the Great Plains Area.* Bul. 210, Bureau of Plant Industry, U.S. Dept. Agric., Wash., DC.

Schemnitz, S. D. 1980. *Wildlife Management Techniques Manual.* 4th ed. Wildl. Soc., Wash., DC.

Schmidt, R. L., C. P. Hibler, T. R. Spraker, and W. H. Rutherford. 1979. An evaluation of drug treatment for lungworm in bighorn sheep. *J. Wildl. Manage.* 43:461–467.

Schmidt, R. H. 1990. Why do we debate animal rights? *Wildl. Soc. Bul.* 18:459–461.

Schoener, T. W. 1968. Sizes of feeding territories among birds. *Ecology* 49:123–141.

Schonewald-Cox, C. M. 1983. Guidelines to management, a beginning attempt. Pages 414–445 in C. M. Schonewald-Cox, S. M. Chambers, B. MacBryde, and L. Thomas, eds. *Genetics and Conservation, a Reference for Managing Wild Animal and Plant Populations.* Benjamin-Cummings, Menlo Park, CA.

Schwaller, M. R., and B. T. Dealy. 1986. Landsat and GIS, two friends grow up together. *J. For.* 84(9):40–41.

Scott, V. E., and G. L. Crouch. 1987. *Response of Breeding Birds to Commercial Clearcutting of Aspen in Southwestern Colorado.* Res. Note RM-475, USDA For. Serv., Rocky Mountain For. and Range Exp. Sta., Fort Collins, CO.

Scott, V. E., and G. J. Gottfried. 1983. *Bird Response to Timber Harvest in a Mixed Conifer Forest in Arizona.* Res. Paper RM-245, USDA For. Serv., Rocky Mountain For. and Range Exp. Sta., Fort Collins, CO.

Scott, V. E., and D. R. Patton. 1989. *Cavity-nesting Birds of Arizona and New Mexico Forests.* Gen. Tech. Rep. RM-10, USDA For. Serv., Rocky Mountain For. and Range Exp. Sta., Fort Collins, CO.

Seton, E. T. 1929. *Lives of Game Animals.* Doubleday and Co., Garden City, NY.

Severinghaus, C. W. 1972. Weather and deer populations. *The Conservationist* 27:28–31.

Severinghaus, W. D. 1981. Guild theory development as a mechanism for assessing environmental impacts. *Envir. Manage.* 5:187–190.

Severson, K. E., and A. L. Medina. 1983. Deer and elk management in the southwest. *J. Range Manage.*, Monogr. No. 2., Soc. for Range Manage., Denver, CO.

Shafer, E. L. 1963. The twig-count method for measuring hardwood deer browse. *J. Wildl. Manage.* 27:428–437.

Short, H. L., and K. P. Burnham. 1982. *Technique for Structuring Wildlife Guilds to Evaluate Impacts on Wildlife Communities.* Special Sci. Rep. Wildl. 244, USDI Fish and Wildl. Serv., Fort Collins, CO.

Siderits, K., and R. E. Radtke. 1977. Enhancing forest wildlife habitat through diversity. *Trans. N. Amer. Wildl. and Nat. Resour. Conf.* 42:425–433.

Singer, P. 1980. Animals and human beings as equals. *Anim. Regul. Stud.* 2:165–174.

Skinner, T. H., and J. O. Klemmedson. 1978. *Abert Squirrel Influences Nutrient Transfer Through Litterfall in a Ponderosa Pine Forest.* Res. Note RM-353, USDA For. Serv., Rocky Mountain For. and Range Exp. Sta., Fort Collins, CO.

Smith, C. M., E. D. Michael, and H. V. Wiant, Jr. 1975. Size of West Virginia deer as related to soil fertility. *W. Va. Agric. and For.* 6:12–13.

Smith, D. M. 1986. *The Practice of Silviculture.* John Wiley and Sons, New York, NY.

Smith, F. W., and J. N. Long. 1987. Elk hiding and thermal cover guidelines in the context of lodgepole pine stand density. *West. J. Appl. For.* 2:6–10.

Smith, R. L. 1986. *Elements of Ecology.* Harper and Row, New York, NY.

Smith, S. H., and A. H. Rosenthal, eds. 1978. *Concepts and Practices in Fish and Wildlife Administration.* Occ. Paper No. 2, USDI Fish and Wildl. Serv., Wash., DC.

Snedecor, G. W. 1956. *Statistical Methods.* Iowa State Univ. Press, Ames.

Society of American Foresters. 1980. *Choices in Silviculture for American Forests.* Soc. of Amer. Foresters in cooperaton with Wildl. Soc., Wash., DC.

Society for Range Manage. 1974. *A Glossary of Terms Used in Range Management.* Soc. for Range Manage., Denver, CO.

Soule, M. E., ed. 1986. *Conservation Biology: The Science of Scarcity and Diversity.* Sinour and Assoc., Sunderland, MS.

Spight, T. M. 1967. Species diversity: A comment of the role of the predator. *Amer. Nat.* 101:467–474.

Sprugel, D. G. 1975. Dynamic structure of wave-regenerated *Abies balsamia* forests in the northeastern United States. *J. Ecology* 64:889–911.

Starfield, A. M., and A. L. Bleloch. 1986. *Building Models for Conservation and Wildlife Management.* Macmillan Publishing Co., New York, NY.

States, J.S. 1979. *Squirrel-Trees-Truffles: A Model of Interactive Dependency.* Progress Report, Ecological Studies of Hypogeous Fungi, Project No. 4., Dept. Biol. Sci., Northern Arizona Univ., Flagstaff.

Stebbins, R. C. 1985. *A Field Guide to Western Reptiles and Amphibians*. Houghton Mifflin Co., Boston, MA.

Steger, T. D. 1986. *Topographic Maps*. USDI Geo. Sur., Wash., D.C.

Stephenson, R. L. 1974. *Reproduction Biology and Food Habits of Abert's Squirrels in Central Arizona*. M.S. Thesis, Ariz. State Univ., Tempe.

Stickney, P. F. 1966. Browse utilization based on percentage of twig numbers browsed. *J. Wildl. Manage*. 30:204–206.

Stoddard, H. L. 1931. *The Bobwhite Quail: Its Habits, Preservation and Increase*. Charles Scribner's Sons, New York, NY.

Stoddart, L. A., A. D. Smith, and T. W. Box. 1975. *Range Management*. McGraw-Hill, New York, NY.

Stokes, A. W. 1974. *Territory*. Benchmark papers in animal behavior, Vol. 2. Dowden, Hutchington and Ross, Stroudsburg, PA.

Sutton, A., and M. Sutton. 1987. *Eastern Forests*. Audubon Soc. Nature Guides. Alfred A. Knopf, New York, NY.

Swanston, D. N. 1980. *Influence of Forest and Rangeland Management on Anadromous Fish Habitat in Western North America: Impacts of Natural Events*. Gen. Tech. Rep. PNW-104, USDA For. Serv., Pacific Northwest For. and Range Exp. Sta., Portland, OR.

Szaro, R. C. 1985. Guild Theory: Fact or Fiction. *Envir. Manage*. 10:681–688.

Taber, W. D., and R. P. Dasmann. 1958. *Black-tailed Deer of the Chaparral*. Game Bul. No. 8, Calif. Dept. of Game and Fish, Sacramento.

Talbott, M. W. 1937. *Indicators of Southwestern Range Conditions*. Farmers Bul. 1782, U.S. Dept. Agric., Wash., DC.

Tansley, A. G. 1935. The use and abuse of vegetational concepts and terms. *Ecology* 16:284–307.

Taylor, W. P. 1940. Ecological classification of the mammals and birds of Walker County, Texas, and some adjoining areas. *Trans. N. Amer. Wildl. Conf.* 5:170–176.

Teer, J. G., J. W. Thomas, and E. A. Walker. 1965. *Ecology and Management of White-tailed Deer in the Llano Basin of Texas*. Wildl. Monogr. No. 15, The Wildl. Soc., Wash., DC.

Thomas, J. W., 1979a. Introduction. Pages 10–21 in J. W. Thomas, ed. *Wildlife Habitats in Managed Forests: The Blue Mountains of Oregon and Washington*. Hdbk. 553, U.S. Dept. Agric., Wash., DC.

Thomas, J. W., ed. 1979b. *Wildlife Habitats in Managed Forests: The Blue Mountains of Oregon and Washington*. Hdbk. 553, U.S. Dept. Agric., Wash., DC.

Thomas, J. W., R. G. Anderson, C. Maser, and E. L. Bull. 1979. Snags. Pages 60–77 in J. W. Thomas, ed. *Wildlife Habitats in Managed Forests: The Blue Mountains of Oregon and Washington*. Hdbk. 553, U.S. Dept. Agric., Wash., DC.

Thomas, J. W., H. Black, Jr., R. J. Scherzinger, and R. J. Pederson. 1979. Deer and elk. Pages 104–127 in J. W. Thomas, ed. *Wildlife Habitats in Managed Forests: The Blue Mountains of Oregon and Washington*. Hdbk. 553, U.S. Dept. Agric., Wash., DC.

Thomas, J. W., C. Maser, and J. E. Rodiek. 1979a. Riparian zones. Pages 40–47 in J. W. Thomas, ed. *Wildlife Habitats in Managed Forests: The Blue Mountains of Oregon and Washington*. Hdbk. 553, U.S. Dept. Agric., Wash., DC.

_____. 1979b. Edges. Pages 48–59 in J. W. Thomas, ed. *Wildlife Habitats in Managed Forests: The Blue Mountains of Oregon and Washington.* Hdbk. 553, U.S. Dept. Agric., Wash., DC.

Thomas, J. W., R. J. Miller, C. Maser, R. G. Anderson, and B. E. Carter. 1979. Plant communities and successional stages. Pages 22–39 in J. W. Thomas, ed. *Wildlife Habitats in Managed Forests: The Blue Mountains of Oregon and Washington.* Hdbk. 553, U.S. Dept. Agric., Wash., DC.

Thomas, J. W., L. F. Ruggiero, R. W. Mannan, J. W. Schoen, and R. A. Lancia. 1988. Management and conservation of old-growth forest in the United States. *Wildl. Soc. Bul.* 16:252–262.

Thomas, J. W., and D. E. Toweill, eds. 1982. *Elk of North America: Ecology and Management.* Wildl. Manage. Inst. Stackpole Books, Harrisburg, PA.

Titterington, R. W., H. S. Crawford, and B. N. Burgason. 1979. Songbird responses to commercial clear-cutting in Maine spruce-fir forests. *J. Wildl. Manage.* 43:602–609.

Trappe, J. M., and C. Maser. 1978. Ectomycorrhizal fungi: interaction of mushrooms and truffles with beasts and trees. Pages 163–179 in T. Walters, ed. *Mushrooms and Man, an Interdisciplinary Approach to Mycology.* Linn-Benton Community College, Albany, OR.

Trippensee, R. E. 1948. *Wildlife Management: Upland Game and General Principles.* Vol. 1. McGraw-Hill, New York, NY.

Tuchman, B. W. 1978. *A Distant Mirror: The Calamitous 14th Century.* Alfred A. Knopf, New York, NY.

Turner, J. C., Jr. 1973. *Water, Energy and Electrolyte Balance in the Desert Bighorn Sheep, Ovis canadensis.* Ph.D. Dissertation, Univ. of Calif., Riverside.

Udall, S. L. 1963. *The Quiet Crisis.* Holt, Rinehart and Winston, New York, NY.

Udvardy, M. D. F. 1975. *A Classification of the Biogeographical Provinces of the World.* Occ. Paper No. 18, International Union for Conservation of Nature, Morges, Switzerland.

United Nations. 1973. *International Classification and Mapping of Vegetation.* Ecol. and Cons. Series 6, UNESCO, Paris, France.

U.S. Congress. 1973. *Endangered Species Act.* Public Law 93–205 as amended; 16 U.S.C. 1531–1536, 1538–1540, Wash., DC.

U.S. Dept. of Agriculture. 1958. *Forest Regions of the United States.* National Atlas Map, For. Serv., Wash., DC.

_____. 1967. *National Handbook for Range and Related Grazing Lands.* Soil Cons. Serv., Wash., DC.

_____. 1968. *Forest Regions of the United States.* For. Serv. Map, Wash., DC.

_____. 1973. *Silvicultural Systems for the Major Forest Types of the United States.* Hdkb. No. 445, For. Serv., Div. of Timber Manage., Wash., DC.

_____. 1974. *Wildlife Habitat Evaluation Program (WHEP).* For. Serv., Southeastern Area, State and Private Forestry, Region 8, Atlanta, GA.

_____. 1976. *Fire Management.* Title 5100. For. Serv., Region 3, Supp. 119, Albuquerque, NM.

_____. 1977. *Vegetation and Environmental Features of Forest and Range Ecosystems.* Hdbk. No. 475, For. Serv., Wash., DC.

_____. 1979. *Habitat: A Special Place*. Movie. 30 minutes. National Audio Visual Center, Capital Heights, MD.

_____. 1980. *Land and Resource Management Planning*. Chap. 1920, For. Serv. Manual, Wash., DC.

_____. 1981. *An Assessment of the Forest and Rangeland Situation in the United States*. For. Serv., Wash., DC.

_____. 1982a. *An Analysis of the Timber Situation in the United States: 1952–2030*. Resour. Rep. No. 23, For. Serv., Wash., DC.

_____. 1982b. *Policy on Fish and Wildlife*. Secretary's Memo. 9500–3. Office of the Sec. of Agric., Wash., DC.

_____. 1984. *Wildlife, Fish, and Sensitive Plant Habitat Management*. For. Serv. Hdbk. 2600, Amend. 48., Wash., DC.

_____. 1988. *Integrated Resource Management*. 2nd ed. For. Serv., Southwestern Region, Albuquerque, NM.

_____. 1990. *Conserving Our Heritage, America's Biodiversity*. For. Serv., Wash., DC.

U.S. Dept. of Commerce. 1977. *Current Population Reports: 1980–2025*. Bureau of the Census, Wash., DC.

U.S. Dept. of Interior. 1966. *Fish, Wildlife and Pesticides*. Fish and Wildl. Serv., Wash., DC.

_____. 1980. *Habitat as a Basis for Environmental Assessment*. ESM 101, Release 4–80, Fish and Wildl. Serv., Div. of Ecol. Serv., Wash., DC.

_____. 1985. *National Survey of Hunting, Fishing and Wildlife Associated Recreation*. Fish and Wildl. Serv., Wash., DC.

_____. 1987. *Restoring America's Wildlife, 1937–1987*. Fish and Wildl. Serv., Wash., DC.

_____. 1990. *Endangered Species*. Tech. Bul. 15(9). Fish and Wildl. Serv., Wash., DC.

Vandermeer, J. 1981. *Elementary Mathematical Ecology*. John Wiley and Sons, New York, NY.

Van Lear, D. H. 1987. *Silviculture Effects on Wildlife Habitat in the South: An Annotated Bibliography, 1980–1985*. Tech. Paper 17, Dept. of For., Clemson Univ., Clemson, SC.

Vannote, R. L., G. W. Minshall, K. W. Cummins, J. R. Sedell, and C. E. Cushing. 1980. The river continuum concept. *Can. J. Fish. Aquat. Sci.* 37:130–137.

Verme, L. J. 1965. Swamp conifer deer yards in Northern Michigan. *J. For.* 63:522–529.

Verme, L. J., and J. J Ozoga. 1981. Changes in small mammal populations following clear-cutting in upper Michigan conifer swamps. *Can. Field Nat.* 65:253–256.

Verner, J. 1983. An integrated system for monitoring wildlife on the Sierra National Forest. *Trans. N. Amer. Wildl. and Nat. Resour. Conf.* 48:355–366.

_____. 1984. The guild concept applied to management of bird populations. *Envir. Manage.* 8:1–14.

Verner, J., and A. Boss. 1980. *California Wildlife and Their Habitats: Western Sierra Nevada*. Gen. Tech. Rep. PSW-37, USDA For. Serv., Pacific Southwest For. and Range Exp. Sta., Berkeley, CA.

Verner, J., M. L. Morrison, and C. J. Ralph. 1984. *Wildlife 2000: Modeling Habitat Relationships of Terrestrial Vertebrates.* Univ. of Wisc. Press, Madison.

Vilkitis, J. R. 1968. *Characteristics of Big Game Violators and Extent of Their Activity in Idaho.* M.S. Thesis, Univ. of Idaho, Moscow.

Vogl, R. J. 1977. Fire: A destructive menace or natural process. Pages 261–289 in Cairns et al., eds. *Recovery and Restoration of Damaged Ecosystems.* Univ. of Virginia Press, Charlottesville.

Volterra, V. 1931. Variations and fluctuations of the number of individuals in animal species living together. In R. N Chapman, *Animal Ecology.* McGraw-Hill, New York, NY.

von Liebig, J. 1840. *Chemistry in Its Application to Agriculture and Physiology.* Taylor and Walton, London.

Wagner, F. H., and L. C. Stoddart. 1972. Influence of coyote predation on black-tailed jackrabbit populations in Utah. *J. Wildl. Manage.* 36:329–343.

Wallace, M. C., and P. R. Krausman. 1987. Elk, mule deer, and cattle habitats in Central Arizona. *J. Range Manage.* 40:80–83.

Wallmo, O. C. 1969. *Response of Deer to Alternate-strip Clearcutting of Lodgepole Pine and Spruce-fir Timber in Colorado.* Res. Note RM-141, USDA For. Serv., Rocky Mountain For. and Range Exp. Sta., Fort Collins, CO.

———, ed. 1981. *Mule and Black-tailed Deer of North America.* Wildl. Manage. Inst. and Univ. of Nebraska Press, Lincoln.

Wallmo, O. C., L. H. Carpenter, W. L. Regelin, R. B. Gill, and D. L. Baker. 1977. Evaluation of deer habitat on a nutritional basis. *J. Range Manage.* 30:122–127.

Ward, A. L. 1976. Elk behavior in relation to timber harvest operations and traffic on the Medicine Bow Range in South-Central Wyoming. Pages 32–43 in J. M. Peek, chair. *Elk, Logging, Roads Symposium Proceedings.* Univ. of Idaho, Moscow.

Watt, A. S. 1947. Pattern and process in the plant community. *J. Ecol.* 35:1–22.

———. 1955. Bracken versus heather, a study in plant sociology. *J. Ecol.* 43:490–506.

Watt, K. E. 1972. Man's efficient rush toward deadly dullness. *Nat. Hist. Magazine.* Amer. Museum of Nat. Hist., Wash., DC.

Weaver, J. E. 1954. *North American Prairie.* Johnson Publishing Co., Lincoln, NE.

Weaver, W. 1975. The religion of a scientist. Pages 296–305 in L. Rosten, ed. *Religions of America.* Simon and Schuster, New York, NY.

Webb, W. L., D. F Behrand, and B. Saisorn. 1977. Effect of *Logging on Songbird Populations in a Northern Hardwood Forest.* Wildl. Monogr. No. 55, The Wildl. Soc., Wash., DC.

Webster's Seventh New Collegiate Dictionary. 1966. G. and C. Merriam Co., Springfield, MS.

Welch, P. S. 1948. *Limnological Methods.* McGraw-Hill, New York, NY.

Wenger, K. F., ed. 1984. *Forestry Handbook.* John Wiley and Sons, New York, NY.

Wesley, D. E., K. L. Knox, and J. G. Nagy. 1970. Energy flux and water kinetics in young pronghorn antelope. *J. Wildl. Manage.* 34:908–912.

Wheeler, R. J., Jr. 1948. *The Wild Turkey in Alabama.* P-R Project, Ala. Dept. Cons., Montgomery.

Whitaker, J. O., Jr. 1988. *The Audubon Society Field Guide to North American Mammals.* Alfred A. Knopf, New York, NY.

White, G. C., D. R. Anderson, K. P. Burnham, and D. L. Otis. 1982. *Capture-Recapture and Removal Methods for Sampling Closed Populations.* LA-8787-NERP, Los Alamos Natl. Lab., Los Alamos, NM.

Whiting, S. 1985. *Western Forests.* Audubon Soc. Nature Guide. Alfred A. Knopf, New York, NY.

Whittaker, R. H. 1953. A consideration of climax theory: The climax as a population and pattern. *Ecol. Monogr.* 23:41–78.

Wildlife Management Institute. 1975. *Placing American Wildlife Management in Perspective.* Wildl. Manage. Inst., Wash., DC.

_____ . 1987. *Organization, Authority and Programs of State Fish and Wildlife Agencies.* Wildl. Manage. Inst., Wash., DC.

_____ . 1990. Endangered species: A dilemma. *Outdoor News Bulletin* (Nov. 23). Wildl. Manage. Inst., Wash., DC.

Williams, G. L., K. R. Russell, and W. K. Seitz. 1977. Pattern recognition as a tool in the ecological analysis of habitat. Pages 521–531 in A. Marmelson, chair. *Classification, Inventory, and Analysis of Fish and Wildlife Habitat.* Symposium Proceedings, UDSI Fish and Wildl. Serv. OBS, Wash., DC.

Wilson, E. O. 1975. *Sociobiology: The New Synthesis.* Belknap Press, Cambridge, MA.

_____ , ed. 1986. *Biodiversity.* National Academy of Science, National Academy Press, Wash., DC.

Witter, D. J. 1980. Wildlife values: Applications and information needs in state wildlife management agencies. Pages 83–98 in W. W. Shaw and E. H. Zube, eds. *Wildlife Values.* Inst. Rep. No. 1, Center for Assessment of Noncommodity Nat. Resour. Values, Univ. of Arizona, Tucson.

Wolff, J. O. 1975. *Red Squirrel Response to Clearcut and Shelterwood Systems in Interior Alaska.* USDA For. Serv. Res. Note PNW-255, Pacific Northwest For. and Range Exp. Sta., Portland, OR.

Wood, G. W. 1990. The art and science of wildlife management. *J. For.* 88:8–12.

Wright, H. A., and A. W. Bailey. 1982. *Fire Ecology, United States and Southern Canada.* John Wiley and Sons, New York, NY.

Wynne-Edwards, V. C. 1962. *Animal Dispersion in Relation to Social Behavior.* Hafner, New York, NY.

Yeager, L. E. 1961. Classification of North American mammals and birds according to forest habitat preference. *J. For.* 59:671–674.

Yeatter, R. E., and D. H. Thompson. 1952. Tularemia, weather, and rabbit populations. *Ill. Nat. Hist. Sur. Bul.* 25:351–382.

INDEX

CONVERSION FACTORS

Multiply	By	To Convert Into
Length		
chains (ch)	20.1168	meters
feet (ft)	0.3048	meters
inches (in)	2.54	centimeters (cm)
kilometers (km)	0.62137	miles
meters (m)	3.28083	feet
meters	0.04971	chains
millimeters (mm)	0.03937	inches
miles (mi)	1.60934	kilometers
yards (yd)	0.9144	meters
Area		
acres (ac)	0.40468	hectares
acres	43,560.0	square feet
hectares (ha)	2.47104	acres
hectares	10,000.0	square meters
square feet (ft^2)	0.0929	square meters
square kilometers (km^2)	0.3861	square miles
square kilometers	100.0	hectares
square kilometers	247.104	acres
square meters (m^2)	10.76387	square feet
square miles (mi^2)	2.58999	square kilometers
square miles	640.0	acres
Basal Area		
square feet/ac (ft^2/ac)	0.2296	square meters/ha
square meters/ha (m^2/ha)	4.356	square feet/ac
Weight		
kilograms (kg)	2.20462	pounds
kilograms/hectare (kg/ha)	0.8922	pounds/ac
pounds (lb)	453.592	grams (g)
pounds/acre (lb/ac)	1.1208	kilograms/ha
Volume		
board ft	0.00566	cubic meters
board ft/acre (bd ft/ac)	0.01399	cubic meters/ha
cubic feet (ft^3)	0.02831	cubic meters
cubic feet/acre (ft^3/ac)	0.06997	cubic meters/ha
cubic meters (m^3)	35.3145	cubic feet
cubic meters	176.57	board feet
cubic meters/hectare (m^3/ha)	71.4571	board feet/ac
cubic meters/hectare	14.2913	cubic feet/ac
gallons (gal)	3.785	liters
liters (l)	0.26417	gallons
Temperature		
°Centigrade (C)	0.5556 (F − 32)	Fahrenheit
°Fahrenheit (F)	32 + (1.80 C)	Centigrade